Advances in Novel Flame Retardant Technologies for Fire-Safe Polymeric Materials

Advances in Novel Flame Retardant Technologies for Fire-Safe Polymeric Materials

Editors

Xin Wang
Weiyi Xing
Gang Tang

Basel • Beijing • Wuhan • Barcelona • Belgrade • Novi Sad • Cluj • Manchester

Editors

Xin Wang
State Key Laboratory
of Fire Science
University of Science and
Technology of China
Hefei
China

Weiyi Xing
State Key Laboratory
of Fire Science
University of Science and
Technology of China
Hefei
China

Gang Tang
School of Architecture and
Civil Engineering
Anhui University
of Technology
Ma'anshan
China

Editorial Office
MDPI
St. Alban-Anlage 66
4052 Basel, Switzerland

This is a reprint of articles from the Special Issue published online in the open access journal *Molecules* (ISSN 1420-3049) (available at: www.mdpi.com/journal/molecules/special_issues/X4G6V4Z10M).

For citation purposes, cite each article independently as indicated on the article page online and as indicated below:

Lastname, A.A.; Lastname, B.B. Article Title. *Journal Name* **Year**, *Volume Number*, Page Range.

ISBN 978-3-7258-1206-6 (Hbk)
ISBN 978-3-7258-1205-9 (PDF)
doi.org/10.3390/books978-3-7258-1205-9

© 2024 by the authors. Articles in this book are Open Access and distributed under the Creative Commons Attribution (CC BY) license. The book as a whole is distributed by MDPI under the terms and conditions of the Creative Commons Attribution-NonCommercial-NoDerivs (CC BY-NC-ND) license.

Contents

Preface vii

Xin Wang, Weiyi Xing and Gang Tang
Advances in Novel Flame-Retardant Technologies for Fire-Safe Polymeric Materials
Reprinted from: *Molecules* **2024**, *29*, 573, doi:10.3390/molecules29030573 1

Xulong Ma, Ni Kang, Yonghang Zhang, Yang Min, Jianhua Yang and Daming Ban et al.
Enhancing Flame Retardancy and Smoke Suppression in Epoxy Resin Composites with Sulfur–Phosphorous Reactive Flame Retardant
Reprinted from: *Molecules* **2023**, *29*, 227, doi:10.3390/molecules29010227 5

Yao Yuan, Weiliang Lin, Yi Xiao, Bin Yu and Wei Wang
Advancements in Flame-Retardant Systems for Rigid Polyurethane Foam
Reprinted from: *Molecules* **2023**, *28*, 7549, doi:10.3390/molecules28227549 16

Liping Jin, Chenpeng Ji, Shun Chen, Zhicong Song, Juntong Zhou and Kun Qian et al.
Multifunctional Textiles with Flame Retardant and Antibacterial Properties: A Review
Reprinted from: *Molecules* **2023**, *28*, 6628, doi:10.3390/molecules28186628 35

Simeng Xiang, Jiao Feng, Hongyu Yang and Xiaming Feng
Synthesis and Applications of Supramolecular Flame Retardants: A Review
Reprinted from: *Molecules* **2023**, *28*, 5518, doi:10.3390/molecules28145518 57

Victor F. Kablov, Oksana M. Novopoltseva, Daria A. Kryukova, Natalia A. Keibal, Vladimir Burmistrov and Vladimir G. Kochetkov
Functionally Active Microheterogeneous Systems for Elastomer Fire- and Heat-Protective Materials
Reprinted from: *Molecules* **2023**, *28*, 5267, doi:10.3390/molecules28135267 85

Oleg P. Korobeinichev, Egor A. Sosnin, Artem A. Shaklein, Alexander I. Karpov, Albert R. Sagitov and Stanislav A. Trubachev et al.
The Effect of Flame-Retardant Additives DDM-DOPO and Graphene on Flame Propagation over Glass-Fiber-Reinforced Epoxy Resin under the Influence of External Thermal Radiation
Reprinted from: *Molecules* **2023**, *28*, 5162, doi:10.3390/molecules28135162 95

Xiaming Feng, Xiang Lin, Kaiwen Deng, Hongyu Yang and Cheng Yan
Facile Ball Milling Preparation of Flame-Retardant Polymer Materials: An Overview
Reprinted from: *Molecules* **2023**, *28*, 5090, doi:10.3390/molecules28135090 113

Po Hu, Weixi Li, Shuai Huang, Zongmian Zhang, Hong Liu and Wang Zhan et al.
Effect of Layered Aminovanadic Oxalate Phosphate on Flame Retardancy of Epoxy Resin
Reprinted from: *Molecules* **2023**, *28*, 3322, doi:10.3390/molecules28083322 132

Huiyu Chai, Weixi Li, Shengbing Wan, Zheng Liu, Yafen Zhang and Yunlong Zhang et al.
Amino Phenyl Copper Phosphate-Bridged Reactive Phosphaphenanthrene to Intensify Fire Safety of Epoxy Resins
Reprinted from: *Molecules* **2023**, *28*, 623, doi:10.3390/molecules28020623 145

Mingxin Zhu, Sujie Yang, Zhiying Liu, Shunlong Pan and Xiuyu Liu
Flame-Retarded Rigid Polyurethane Foam Composites with the Incorporation of Steel Slag/Dimelamine Pyrophosphate System: A New Strategy for Utilizing Metallurgical Solid Waste
Reprinted from: *Molecules* **2022**, *27*, 8892, doi:10.3390/molecules27248892 157

Zihui Xu, Jing Zhan, Zhirong Xu, Liangchen Mao, Xiaowei Mu and Ran Tao
A Bridge-Linked Phosphorus-Containing Flame Retardant for Endowing Vinyl Ester Resin with Low Fire Hazard
Reprinted from: *Molecules* **2022**, *27*, 8783, doi:10.3390/molecules27248783 . . . . **171**

Preface

Fires are one of the most common disasters threatening public safety and social development. Preventing the occurrence and spread of fire is of great significance for protecting public property and life safety. Fire-safe polymeric materials are the key components of fire prevention and control, mainly including flame-retardant materials, fire-resistant and thermal insulation materials, low-smoke-emission materials, etc. With the development of science and technology, fire-safe polymeric materials are increasingly widely used in transportation, construction, aerospace, electronics and electrical devices, and many other areas of people's lives. Therefore, the design and preparation of novel flame retardants for these fire-safe polymeric materials is urgent.

This Special Issue entitled "Advances in Novel Flame Retardants for Fire-Safe Polymeric Materials" welcomes original research and reviews on experimental or theoretical/computational studies from all subjects, including but not limited to the following:

- Synthesis of novel flame retardants;
- Preparation of flame-retardant polymer nanocomposites;
- Low heat release and/or low smoke emission;
- Bio-based flame retardants;
- Novel flame-retardant coatings;
- Eco-friendly flame-retardant textiles.

Xin Wang, Weiyi Xing, and Gang Tang
Editors

Editorial

Advances in Novel Flame-Retardant Technologies for Fire-Safe Polymeric Materials

Xin Wang [1,*], Weiyi Xing [1] and Gang Tang [2]

1 State Key Laboratory of Fire Science, University of Science and Technology of China, Hefei 230026, China
2 School of Architecture and Civil Engineering, Anhui University of Technology, Ma'anshan 243002, China
* Correspondence: wxcmx@ustc.edu.cn

1. Introduction

This Special Issue, titled "Advances in Novel Flame-Retardant Technologies for Fire-Safe Polymeric Materials", aims to detail the recent advances in the design and preparation of novel flame retardants for use in fire-safe polymeric materials. Due to developments in science and technology, fire-safe polymeric materials are increasingly widely used in transportation, construction, aerospace, electronics, and electrical devices [1–3].

The developments that have occurred in creating flame-retardant polymeric materials thus far can be roughly divided into four stages (Figure 1): (1) The early flame-retardant polymeric materials used halogenated flame retardants such as tetrabromobisphenol A. Despite their high efficiency as flame retardants, tetrabromobisphenol A generates toxic and corrosive gases such as hydrogen halide, which are harmful to human health and the environment [4]. Therefore, halogen-free flame-retardant polymeric materials have become the principal trend in recent years. (2) Halogen-free (mainly silicon-, nitrogen-, phosphorus-, and boron-containing) flame-retardant polymeric materials have emerged owing to their improved flame retardancy and environmental friendliness, but their mechanical strength and/or thermal stability usually deteriorate due to high loading. (3) Flame-retardant polymer nanocomposites exhibit the combined advantages of a low heat release rate, smoke toxicity suppression, and good mechanical and thermal properties. However, their level of flame retardancy struggles to meet industrial requirements such as a UL-94 V-0 rating. (4) The latest generation of flame-retardant polymers, also known as fire-safe polymer technology, not only achieves a UL-94 V-0 rating but also possesses a low heat release rate and low toxic smoke emissions.

Fire-safe polymers are difficult to ignite and have a low heat release rate and low smoke and toxic gas emissions. The fact that they are difficult to ignite reduces the probability of materials catching fire; a low heat release rate can suppress the flame spread rate and increase the amount of time personnel have to escape in the event of fire; and lower smoke and toxic gas emissions can reduce the number of casualties in fire accidents. Therefore, the application of fire-safe technology is a pertinent future development trend in polymeric materials.

This Special Issue contains seven original research contributions and four review articles related to fire-safe polymeric materials, including epoxy resins, rigid polyurethane foams, vinyl ester resins, and textiles.

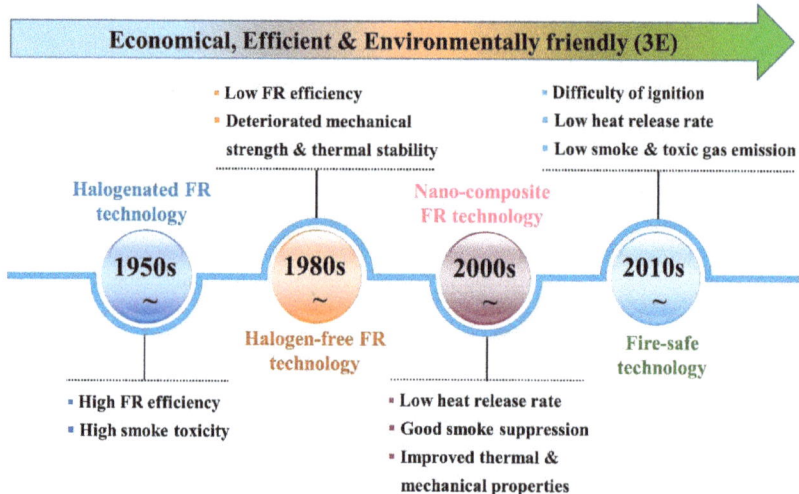

Figure 1. The development stages of flame-retardant polymeric materials.

2. An Overview of the Published Articles

In the Special Issue's first contribution, Ma et al. synthesized a novel sulfur–phosphorus reactive flame retardant (SPMS) for use in epoxy resins (EPs). The co-addition of the SPMS and ammonium polyphosphate (APP) effectively promoted flame retardancy and smoke suppression in the EPs. Specifically, the peak heat release rate (pHRR), peak smoke production rate (pSPR), and total heat release (THR) of the EP containing 6.67% of the SPMS and 13.33% APP were 82.4%, 93.5%, and 61.4% lower than those of the pure EP, respectively. The results demonstrated that this sulfur–phosphorus flame retardant was highly effective at alleviating the fire risk in EPs.

Next, in the research paper by Korobeinichev et al. (contribution 2), the effect of a combination of 9,10-dihydro-9-oxa-10-phosphaphenanthrene-10-oxide-4,4′-diamino-diphenyl methane (DDM-DOPO) and graphene on the flammability of glass-fiber-reinforced epoxy composites (GFREPs) was investigated. The GFREP samples with flame-retardant additives showed a lower downward shift in the flame propagation rate than those without additives. Additionally, the measured thermal structure of the actual flames was compared with the results of numerical simulations of the flame propagation in the GFREPs.

Furthermore, in the study by Hu et al. (contribution 3), a layered ammonium vanadium oxalate-phosphate (AVOPh) was synthesized using a hydrothermal method. With the incorporation of 8 wt% of the AVOPh into the EP, the pHRR, total smoke production (TSP), and peak of CO production (PCOP) were 32.7%, 20.4%, and 37.1% lower than those of the pure EP, respectively. As a result, AVOPh was determined to be a promising high-efficiency flame retardant for use in EPs.

Chai et al. (contribution 4) synthesized a hybrid flame retardant (CuPPA-DOPO) by surface-grafting of amino phenyl copper phosphate with 9, 10-dihydro-9-oxygen-10-phospha-phenanthrene-10-oxide. With the addition of 6 wt% of the CuPPA-DOPO to the EP, the EP managed to reach the UL-94 V-1 classification and a limiting oxygen index (LOI) of 32.6%. In addition, compared with those of the pure EP, the pHRR and pSPR of the EP/6 wt% CuPPA-DOPO combination decreased by 52.5% and 26.1%, respectively. This favorable inhibition of the fire risk in the EP due to the CuPPA-DOPO was ascribed to the synergistic effects of the release of phosphorus free radicals during the gaseous phase and the catalytic charring ability of metal oxides during the condensed phase.

Meanwhile, in the research paper by Zhu et al. (contribution 5), a combination of steel slag (SS) and dimelamine pyrophosphate (DMPY) was used as a flame retardant for rigid polyurethane (RPUF). The pHRR and THR values of RPUF-3 when using the DMPY/SS system were reduced by 54.5% and 42.7% compared to when using unmodified RPUF, respectively. This study offers a novel strategy for the preparation of fire-safe RPUFs by utilizing metallurgical solid waste.

In the research paper by Xu et al. (contribution 6), a new phosphorus-based flame retardant, 6, 6′-(1-phenylethane-1,2 diyl) bis(dibenzo[c,e][1,2]oxaphosphinine 6-oxide) (PBDOO), was synthesized for use in vinyl ester resin (VE). The incorporation of 15 wt% of the PBDOO into the VE composites supported a UL-94 V-0 rating and a high LOI value of 31.5%. Additionally, the pHRR and THR of the VE containing 15 wt% of PBDOO were 76.71% and 40.63% lower than those of the unmodified VE, respectively. This study's findings can be applied as an efficient approach to improving the fire safety of VE.

Kablov et al. (contribution 7) prepared elastomeric composites based on aluminosilicate microspheres, carbon microfibers, and a phosphor–nitrogen–organic modifier as part of their research, studying the composites' fire- and heat-protective properties. The results showed that their linear burning speed was reduced by 6–17% compared to that of their known counterparts, which was attributed to the improved coke strength and catalyzed carbonization processes.

As regards literature reviews rather than novel studies on the topic, Yuan et al. (contribution 8) covered the recent progress in the field of flame retardancy and smoke suppression of RPUFs. Both conventional methods and innovative trends with respect to manufacturing fire-safe RPUFs, including reactive flame retardants, additive flame retardants, inorganic nanoparticles, and protective coatings, were analyzed in detail. Additionally, this review paper also proposed several challenges and future trends in this field.

Furthermore, Jin et al. (contribution 9) summarized the latest advances in multifunctional textiles, with a special focus on their flame-retardant and antibacterial properties. They describe how various treatment strategies, including the spray method, dip-coating, and pad-dry-cure methods; layer-by-layer (LBL) deposition; the sol-gel process; and chemical grafting modification, have been applied to endow textiles with multiple functions. Finally, the merits and drawbacks of these treatment strategies were compared, helping guiding future research and promoting the translation of the research into industry practice.

A review of the synthesis of supramolecular flame retardants (SFRs) based on non-covalent interactions was conducted by Xiang et al. (contribution 10). In this paper, first, different categories of SFRs of various dimensions were defined. Then, the influence of these SFRs on the fire-safe characteristics of typical polymers was emphasized. Additionally, the effects of the SFRs on the properties of polymeric materials, including their mechanical properties, were also evaluated.

In the final review paper, Feng et al. (contribution 11) explained the recent progress made in utilizing simple ball milling to fabricate flame retardants and flame-retardant polymer composites. First, they described how high-performance flame retardants were crushed, exfoliated, modified, and reacted using a ball mill. Then, they emphasized the incorporation of flame retardants into polymer composites using ball milling, with a special focus on the formation of multifunctional segregated structures. Their paper paves the way for simply and feasible developing fire-safe polymer materials.

3. Conclusions

Due to the increasing fire safety demands in the majority of their application contexts, polymeric materials need to be made flame-retardant. Consequently, a vast body of literature has become available on fire-safe polymeric materials over the past few decades. The emerging trend is undoubtedly toward eco-friendly, sustainable, and multi-functional technologies related to forming fire-safe polymeric materials [5,6]. Although many efficient flame retardants have been reported in the literature, transferring these newly arising fire-safe technologies from the research to an industry context will require further progress.

Acknowledgments: The editors would like to express their great appreciation to all the authors, reviewers, and technical assistants that contributed to the Special Issue.

Conflicts of Interest: The authors declare no conflicts of interest.

List of Contributions:

1. Ma, X.; Kang, N.; Zhang, Y.; Min, Y.; Yang, J.; Ban, D.; Zhao, W. Enhancing Flame Retardancy and Smoke Suppression in Epoxy Resin Composites with Sulfur–Phosphorous Reactive Flame Retardant. *Molecules* **2024**, *29*, 227.
2. Korobeinichev, O.P.; Sosnin, E.A.; Shaklein, A.A.; Karpov, A.I.; Sagitov, A.R.; Trubachev, S.A.; Shmakov, A.G.; Paletsky, A.A.; Kulikov, I.V. The Effect of Flame-Retardant Additives DDM-DOPO and Graphene on Flame Propagation over Glass-Fiber-Reinforced Epoxy Resin under the Influence of External Thermal Radiation. *Molecules* **2023**, *28*, 5162.
3. Hu, P.; Li, W.X.; Huang, S.; Zhang, Z.M.; Liu, H.; Zhan, W.; Chen, M.Y.; Kong, Q.H. Effect of Layered Aminovanadic Oxalate Phosphate on Flame Retardancy of Epoxy Resin. *Molecules* **2023**, *28*, 3322.
4. Chai, H.Y.; Li, W.X.; Wan, S.B.; Liu, Z.; Zhang, Y.F.; Zhang, Y.L.; Zhang, J.H.; Kong, Q.H. Amino Phenyl Copper Phosphate-Bridged Reactive Phosphaphenanthrene to Intensify Fire Safety of Epoxy Resins. *Molecules* **2023**, *28*, 623.
5. Zhu, M.X.; Yang, S.J.; Liu, Z.Y.; Pan, S.L.; Liu, X.Y. Flame-Retarded Rigid Polyurethane Foam Composites with the Incorporation of Steel Slag/Dimelamine Pyrophosphate System: A New Strategy for Utilizing Metallurgical Solid Waste. *Molecules* **2022**, *27*, 8892.
6. Xu, Z.H.; Zhan, J.; Xu, Z.R.; Mao, L.C.; Mu, X.W.; Tao, R. A Bridge-Linked Phosphorus-Containing Flame Retardant for Endowing Vinyl Ester Resin with Low Fire Hazard. *Molecules* **2022**, *27*, 8783.
7. Kablov, V.F.; Novopoltseva, O.M.; Kryukova, D.A.; Keibal, N.A.; Burmistrov, V.; Kochetkov, V.G. Functionally Active Microheterogeneous Systems for Elastomer Fire- and Heat-Protective Materials. *Molecules* **2023**, *28*, 5267.
8. Yuan, Y.; Lin, W.L.; Xiao, Y.; Yu, B.; Wang, W.; Sonnier, R. Advancements in Flame-Retardant Systems for Rigid Polyurethane Foam. *Molecules* **2023**, *28*, 7549.
9. Jin, L.P.; Ji, C.P.; Chen, S.; Song, Z.C.; Zhou, J.T.; Qian, K.; Guo, W.W. Multifunctional Textiles with Flame Retardant and Antibacterial Properties: A Review. *Molecules* **2023**, *28*, 6628.
10. Xiang, S.M.; Feng, J.; Yang, H.Y.; Feng, X.M. Synthesis and Applications of Supramolecular Flame Retardants: A Review. *Molecules* **2023**, *28*, 5518.
11. Feng, X.M.; Lin, X.; Deng, K.W.; Yang, H.Y.; Yan, C. Facile Ball Milling Preparation of Flame-Retardant Polymer Materials: An Overview. *Molecules* **2023**, *28*, 5090.

References

1. Laoutid, F.; Bonnaud, L.; Alexandre, M.; Lopez-Cuesta, J.M.; Dubois, P. New prospects in flame retardant polymer materials: From fundamentals to nanocomposites. *Mater. Sci. Eng. R* **2009**, *63*, 100–125. [CrossRef]
2. Dasari, A.; Yu, Z.Z.; Cai, G.P.; Mai, Y.W. Recent developments in the fire retardancy of polymeric materials. *Prog. Polym. Sci.* **2013**, *38*, 1357–1387. [CrossRef]
3. Wang, X.; Kalali, E.N.; Wan, J.T.; Wang, D.-Y. Carbon-family materials for flame retardant polymeric materials. *Prog. Polym. Sci.* **2017**, *69*, 22–46. [CrossRef]
4. Lu, S.Y.; Hamerton, I. Recent developments in the chemistry of halogen-free flame retardant polymers. *Prog. Polym. Sci.* **2002**, *27*, 1661–1712. [CrossRef]
5. Costes, L.; Laoutid, F.; Brohez, S.; Dubois, P. Bio-based flame retardants: When nature meets fire protection. *Mater. Sci. Eng. R* **2017**, *117*, 1–25. [CrossRef]
6. Malucelli, G.; Carosio, F.; Alongi, J.; Fina, A.; Frache, A.; Camino, G. Materials engineering for surface-confined flame retardancy. *Mater. Sci. Eng. R* **2014**, *84*, 1–20. [CrossRef]

Disclaimer/Publisher's Note: The statements, opinions and data contained in all publications are solely those of the individual author(s) and contributor(s) and not of MDPI and/or the editor(s). MDPI and/or the editor(s) disclaim responsibility for any injury to people or property resulting from any ideas, methods, instructions or products referred to in the content.

Article

Enhancing Flame Retardancy and Smoke Suppression in Epoxy Resin Composites with Sulfur–Phosphorous Reactive Flame Retardant

Xulong Ma [1], Ni Kang [1], Yonghang Zhang [1], Yang Min [1], Jianhua Yang [1], Daming Ban [1,*] and Wei Zhao [2,*]

[1] School of Chemistry and Materials Science, Guizhou Normal University, Guiyang 550001, China; 21010080287@gznu.edu.cn (X.M.); kmy087@126.com (J.Y.)
[2] Technology and Engineering Center for Space Utilization, Chinese Academy of Sciences, Beijing 100094, China
* Correspondence: bdaming@gznu.edu.cn (D.B.); zhaowei@csu.ac.cn (W.Z.)

Abstract: The presence of massive amounts of toxic volatiles and smoke during combustion is a very serious problem facing epoxy resin (EP) composites. Therefore, flame retardants (FRs) can simultaneously enhance flame retardancy and reduce the release of smoke and fatal gases. Herein, a novel sulfur–phosphorous reactive flame retardant (SPMS) was synthesized for epoxy resin. The high efficiency of smoke suppression and flame retardancy of the EP/SPMS-APP hybrid was investigated using a cone calorimeter, a vertical burning test, and limited oxygen index measurements. Compared with those of pure EP, the composite with 20 wt% SPMS-APP reduced the peak heat release rate (pHRR), the peak smoke production rate (SPR), and total smoke production rate (TSR) by 82%, 94%, and 84%, respectively. The results showed a remarkable suppressed effect of alleviating the fire hazard of EP using a sulfur–phosphorus flame retardant.

Keywords: smoke suppression; phosphaphenanthrene; suppressed effect; flame retardant

Citation: Ma, X.; Kang, N.; Zhang, Y.; Min, Y.; Yang, J.; Ban, D.; Zhao, W. Enhancing Flame Retardancy and Smoke Suppression in Epoxy Resin Composites with Sulfur–Phosphorous Reactive Flame Retardant. *Molecules* **2024**, *29*, 227. https://doi.org/10.3390/molecules29010227

Academic Editors: Giulio Malucelli and Rodolphe Sonnier

Received: 25 November 2023
Revised: 28 December 2023
Accepted: 29 December 2023
Published: 31 December 2023

Copyright: © 2023 by the authors. Licensee MDPI, Basel, Switzerland. This article is an open access article distributed under the terms and conditions of the Creative Commons Attribution (CC BY) license (https://creativecommons.org/licenses/by/4.0/).

1. Introduction

Epoxy resin has been widely used in coatings, civil engineering, construction, and other fields due to its outstanding advantages such as adhesion, good mechanical properties, and chemical stability [1–3]. However, high flammability and massive smoke are among the main disadvantages of epoxy resins, which severely restrict their application. Therefore, fire safety EP with low smoke production urgently needs to be developed [4–6].

In recent decades, halogen-containing compounds have been widely utilized to optimize the inflammability of epoxy resin, but halogen-containing compounds often release toxic hydrogen halide gases, organic halides, and dioxins during combustion. Recently, in consideration of environmental aspects, several halogen-containing flame retardants have been gradually prohibited by many countries [7,8].

Phosphorus-based flame retardants are alternatives to halogen compounds. Several small-molecule organic-phosphorus-containing flame retardants have been widely used for EP. However, owing to their gaseous nature, these materials exhibit high flame retardancy accompanied by decreased smoke suppression, increased smoke density, or the release of corrosive gases [9,10]. Therefore, improving flame retardancy and smoke suppression performance attracted increasing attention simultaneously. There are many smoke suppressants, such as molybdenum trioxide, zinc borate, and copper oxide [11]. However, these smoke suppressants exhibit decreased efficacy because they emit water vapor mixed with smoke when used in an EP matrix [12].

With an increase in fire retardant requirements and an increase in environmental protection awareness, the hotspots of flame retardants are trending toward being halogen-free, environmentally friendly, smoke-suppressive, and having low toxicity [13]. Phosphaphenanthrene heterocyclic compounds are characterized by noncoplanarity, interactions with intramolecular or intermolecular groups, large-volume structures, and molecular polarity [14,15]. They easily bind to specific groups and retain flame retardant properties, increasing the organic solubility of new molecules [16]. The active hydrogen of phosphaphenanthrene can react with a variety of electron-deficient groups and then act as a reactive flame retardant involved in the polymerization of embedded polymer molecules [17–21]. Moreover, because flame retardants have a biphenyl rigid structure, the heat resistance and mechanical properties of epoxy resin may improve. Ammonium polypophosphate (APP) can play a vital role in the condensed phase, helping the matrix to form massive char residue under fire conditions and obtain a higher LOI value. It is worth mentioning that such compounds are halogen-free flame retardants, unlike halogen-based flame retardants, which can reduce environmental pollution [22].

In this study, a phosphorous flame retardant 10-(2,5-dicarbonylpropyl) 9,10 dihydro-9-oxa-10-phosphaphenanthrene-10-sulfide (SPMS) was synthesized via a reaction between 6H-dibenzo[c,e][1,2]oxaphosphinine-6-sulfide (DOPS) and itaconic acid (ITA), as shown in Scheme 1. By incorporating SPMS-APP into EP, heat and smoke release decreased dramatically. The high efficiency of SPMS-APP in terms of smoke suppression was revealed for the first time. The smoke suppression and flame retardant mechanism of EP/SPMS-APP are discussed in detail.

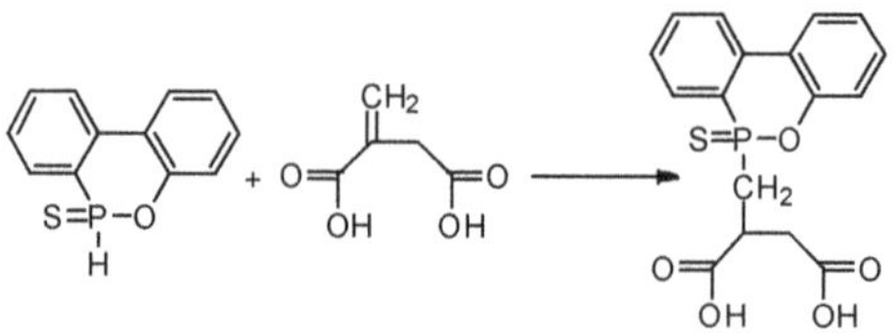

Scheme 1. Schematic presentation of SPMS.

2. Results and Discussion

2.1. Characterization of the SPMS

The FT-IR spectra of DOPS and SPMS are shown in Figure 1. The corresponding characteristic bands were as follows: absorption at approximately 3427 cm^{-1} corresponded to --OH; -CH_2- stretching vibration absorption appeared at approximately 2970 cm^{-1}; C=O absorption at approximately 1711 cm^{-1}; and absorption at approximately 1194, 1147, and 934 cm^{-1} corresponding to P-O-C(aromatic) stretching vibration. The P-C stretching vibration absorption band appeared at approximately 1472 and 1427 cm^{-1}. Compared with the FT-IR spectrum of DOPS, the characteristic absorption band of the P-H bond at 2367 cm^{-1} disappeared, proving that a reaction occurred between DOPS and ITA. The ^{1}H-NMR spectrum of SPMS is shown in Figure 2. ^{1}H NMR (TMS, 400 MHz) δ: 9.13 (s, 2 H), 8.02 (dd, J = 14.2 Hz, J = 7.7 Hz, 1 H), 7.90 (d, J = 6.4 Hz, 1 H), 7.77 (s, 1 H), 7.74 (t, J = 7.8 Hz, 1 H), 7.59 (s, 1 H), 7.44~7.31 (m, 1 H), 7.29 (s, 1 H), 7.26 (s, 1 H), 3.74 (t, J = 7.2 Hz, 2 H), 2.25 (s, 1 H), 1.25 (s, 2 H). These results confirmed the successful synthesis of SPMS.

2.2. Combustion Properties of the Cured Epoxy Resins

The formulations and flame retardant properties of the EP/SPMS-APP composites are listed in Table 1.

The neat EP was highly combustible with an LOI value of 19.8% and no UL-94 grade. Moreover, the specimens were burned with flammable dripping during the test. Table 1 shows that the flame retardancy of the cured EP gradually increased with flame retardant concentration. All the LOI values of the modified EP were greater than those of the EP. Considering the same loading, EP2-5 had higher LOI values than EP-5 and EP1-5. This difference may be ascribed to the concurrent action of SPMS and APP, which acted both in the condensed and gaseous phases. Moreover, EP2-5 and EP3-10 reached UL-94 V0 ratings at 5 wt% and 10 wt% loading, respectively. Most importantly, although APP alone raised the UL-94 rating more slowly than SPMS-APP did, it helped EP achieve a higher LOI value at the same loading. Unlike EP7-20 and EP8-20, EP9-20 reached an LOI value of 34.5.

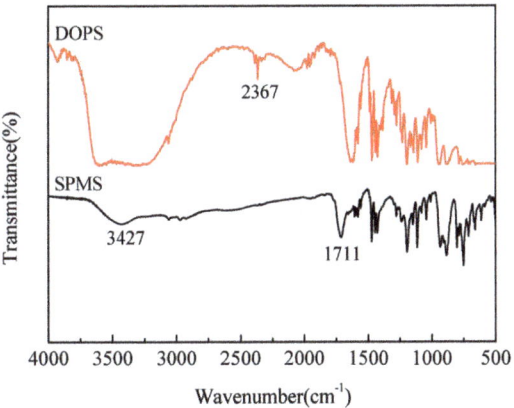

Figure 1. FT−IR spectra of DOPS and SPMS.

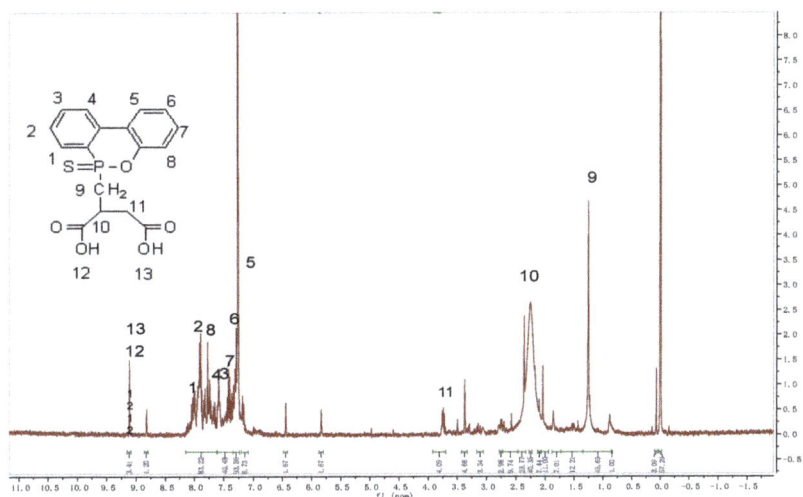

Figure 2. ^{1}H−NMR spectrum of SPMS.

Table 1. Formulations and flame retardancy of epoxy resin thermosets.

Sample	Mass Fraction (%)			LOI (%)	UL-94
	EP + PDA	SPMS	APP		
EP0	100	0.00	0.00	19.8	N.R.
EP-5	95	0.00	5.00	21.4	N.R.
EP1-5	95	5.00	0.00	21.1	V-1
EP2-5	95	1.67	3.33	20.8	V-0
EP3-10	90	10.0	0.00	21.5	V-0
EP4-10	90	3.33	6.67	21.7	V-0
EP5-15	85	15.0	0.00	24.7	V-0
EP6-15	85	5.00	10.0	26.4	V-0
EP7-20	80	20.0	0.00	25.0	V-0
EP8-20	80	6.67	13.33	27.3	V-0
EP9-10	90	0.00	10.0	26.8	V-0
EP10-20	80	0.00	20.0	34.5	V-0

2.3. Thermal Properties of Cured Epoxy Resins

The thermal stability of the hybrids is shown in Figure 3. Several important parameters, including the onset degradation temperature (T_d) of cured epoxy resins, the temperature of midpoint degradation ($T_{50\%}$), the temperature at the maximum weight loss rate (T_{max}), and char residue at 700 °C, are summarized in Table 2.

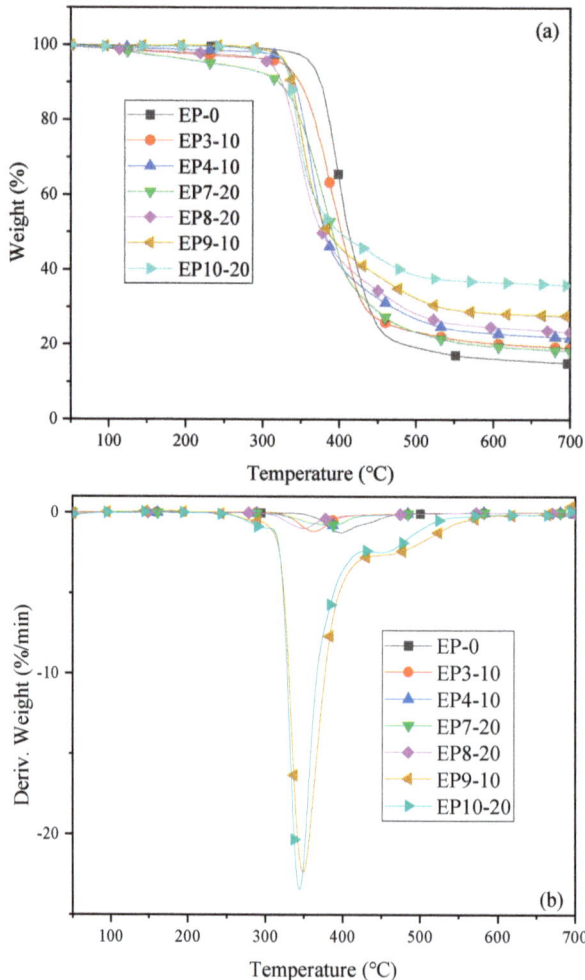

Figure 3. TGA (**a**) and DTG (**b**) curves of EP composites in an N_2 atmosphere.

The TGA curves in Figure 3 show a one-step degradation trend. With increasing FR, the Td and Tmax of EP tended to decrease under a nitrogen atmosphere, especially EP7-20, which decreased above 100 °C, which was caused by the earlier degradation of S=P-C and P-O covalent bonds. Moreover, the steric hindrance effect induced by the bulky and rigid phosphaphenanthrene group in SPMS decreased the cross-linking density of the thermoset [23]. Moreover, the flame retardant epoxy resin reached T10% and Tmax earlier than EP. Figure 3 shows that the maximum weight loss rates (R_{max}) of EP0, EP3-10, EP4-10, EP7-20, EP8-20, EP9-10, and EP10-20 are 1.26, 0.97, 1.20, 0.80, 1.00, 22.37, and 23.44%/min, respectively. Considering the R_{max} values, the SPMS behaved more efficiently than the SPMS-APP, and APP alone lost weight more sharply than other samples. Therefore, the incorporation of flame retardants successfully reduced the thermal degradation rate in the low-temperature region. The epoxy resin only retained 15% of the char residues, as listed in Table 2. The char residues of EP3-10, EP4-10, EP7-20, EP8-20, EP9-10, and EP10-20 were 19, 21, 18, 23, 27 and 36%, respectively. The presence of SPMS-APP might aid in the formation of char residues, resulting in the suppression of flame spread, a reduction in flammable volatiles, and the inhibition of drip melting. APP showed high efficiency in char formation. A thick char residue layer can prevent further thermal degradation of the matrix and produce fewer inflammable volatiles.

Table 2. Related TGA data for EP composites.

Sample	T_d (°C)	$T_{50\%}$ (°C)	T_{max} (°C)	R_{max} (%/min)	R_{700} (%)
EP0	357	408	398	1.26	15
EP3-10	324	401	382	0.97	19
EP4-10	329	379	363	1.20	21
EP7-20	233	389	387	0.80	18
EP8-20	306	375	348	1.00	23
EP9-10	330	385	349	22.37	27
EP10-20	328	399	344	23.44	36

2.4. Fire Hazard Analysis

Cone calorimeter (CC) tests were also conducted to investigate the combustion behaviors of SPMS and SPMS/APP on epoxy resin thermosets. The time to ignition (TTI), peak heat release rate (pHRR), peak smoke production rate (pSPR), total heat release (THR), total smoke rate (TSR), and fire performance index (FPI) are shown in Table 3 and Figure 4.

Table 3. Cone calorimeter data for the flame-retarded EP composites.

Sample	TTI (s)	pHRR (kW·m^{-2})	pSPR (m^2·s^{-1})	THR at 400 s (MJ·m^{-2})	TSR at 400 s (m^2·m^{-2})	FPI (s·m^2·kW^{-1})
EP0	70	933	0.77	88	11,640	0.075
EP3-10	59	767	0.58	71	7880	0.076
EP4-10	65	311	0.14	70	2695	0.20
EP7-20	48	567	0.44	61	6529	0.084
EP8-20	60	164	0.05	34	1891	0.36
EP9-20	59	354	0.16	47	2341	0.167

The TTI was used to determine the influence of flame retardants on ignitability. As revealed in Table 3, the TTI of the pure epoxy resin was 70 s, whereas that of the flame-retarded EP composites decreased to some extent. This difference may be attributed to the early decomposition of flame retardants, which promoted degradation of the epoxy resin matrix at lower temperatures. The FPI is a parameter derived from the TTI and pHRR. The higher the values are, the better the composite behavior during fire hazard. EP8-20 had an FPI almost five times greater than that of neat EP, so it showed excellent fire performance in all tests. As shown in Figure 4a, the pHRR of the modified EP decreased over time, especially for the EP8-20 sample.

As shown in Figure 4a,b, and Table 3, pHRR (933 kW/m^2) and THR at 400 s (88 MJ/m^2) were obtained for pure EP. With the addition of 20 wt% SPMS, the pHRR and THR decreased to 567 kW/m^2 and 61 MJ/m^2, respectively. APP alone showed in-between data between SPMS and SPMS/APP. When flame retardant SPMS was replaced by APP in sample EP8-20 at the same loading of 20 wt% SPMS/APP, the pHRR and THR decreased sharply to 164 kW/m^2 and 34 MJ/m^2, respectively. It can be demonstrated that SPMS/APP decomposes at low temperatures to form a protective char residue layer on the surface of the sample. These results are consistent with the results of TGA.

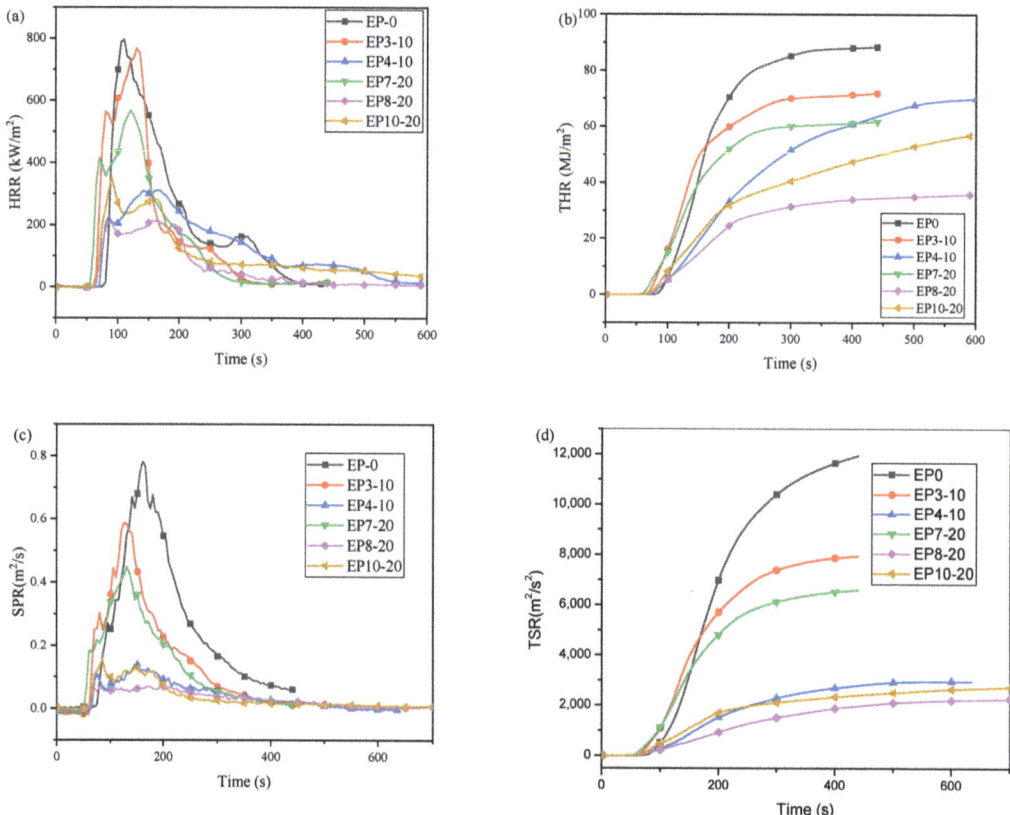

Figure 4. Cone calorimeter curves of the EP and SPMS/APP/EP composites: (**a**) heat release rate, (**b**) total heat release, (**c**) smoke production release, and (**d**) total smoke release.

As shown in Figure 4c,d and Table 3, the SPR and TSR at 400 s for the EP composites decreased quickly after the incorporation of SPMS and APP. The peak SPR energy of EP7-20 was 0.44 m^2/s, which was 43% lower than that of EP0 (0.77 m^2/s). Moreover, the TSR at 400 s was reduced by 45%. Moreover, the samples containing both SPMS and APP had lower peak SPR values and TSR values at 400 s than those of SPMS and APP alone. The TSR at 400 s for EP4-10 was 2695 m^2/m^2, which was reduced by 77%. In sample EP8-20, the peak SPR and TSR at 400 s decreased dramatically, by 94% and 84%, to 0.05 m^2/s and 1891 m^2/m^2, respectively. And APP alone showed a medium datum of 2341 m^2/m^2 between that of EP7-20 and EP8-20. The main reason may be that the suppressed effect of SPMS/APP helps greatly in promoting charring formation, and the formation of a thick carbon layer causes the number of combustible volatiles and smoke-forming materials to decrease rapidly in the gas phase during combustion. This finding implies that the application of SPMS/APP/EP composites could increase the likelihood of safe escape in cases of fire hazard due to the high FPI and reduced TSR.

2.5. Residual Char Analysis

Digital photographs of char residues for the EP composites after CC are shown in Figure 5. There is a minimum quantity of char residue left behind in pure EP, which corresponds to the highest SPR, HRR, and mass loss among all the samples. Compared with EP0, EP3-10 containing only SPMS had relatively high and continuous char residue. However, there were some collapses in the char residue layer generated by the volatile compounds, and the inner surface was covered by small homogeneous bricks. As APP was incorporated into EP4-10, the amount of char residue increased significantly, and the outer surface of the char residue became increasingly compact and tough while the homogeneous bricks aggregated to form larger bricks. This revealed that APP can help char residue become more stable. It can be seen from the front view that EP4-10 and EP8-20 generated a more compact carbonaceous layer with less cracking and greater intumescence. Moreover, after combustion, the amount of residual carbon in the EP composites cured with both SPMS and APP was significantly greater than that in the EP composites cured with SPMS. The reason was that there was a suppressed effect between SPMS and APP, and the degradation of flame retardants produced a phosphorus-based acid, such as phosphoric acid, hypophosphite, and noncombustible gases, which reacted with the decomposing matrix by esterification and dehydration to promote the formation of a char layer and the noncombustible gases filled the char layer. The char layer in the molten state was expanded and foamed, and when the reaction was close to completion, the system was solidified to form a porous and foam char layer. The char layer could play a role in heat insulation and oxygen separation, so the char layer could effectively protect the polymer beneath the burn so that the amount of carbon residue was obviously increased.

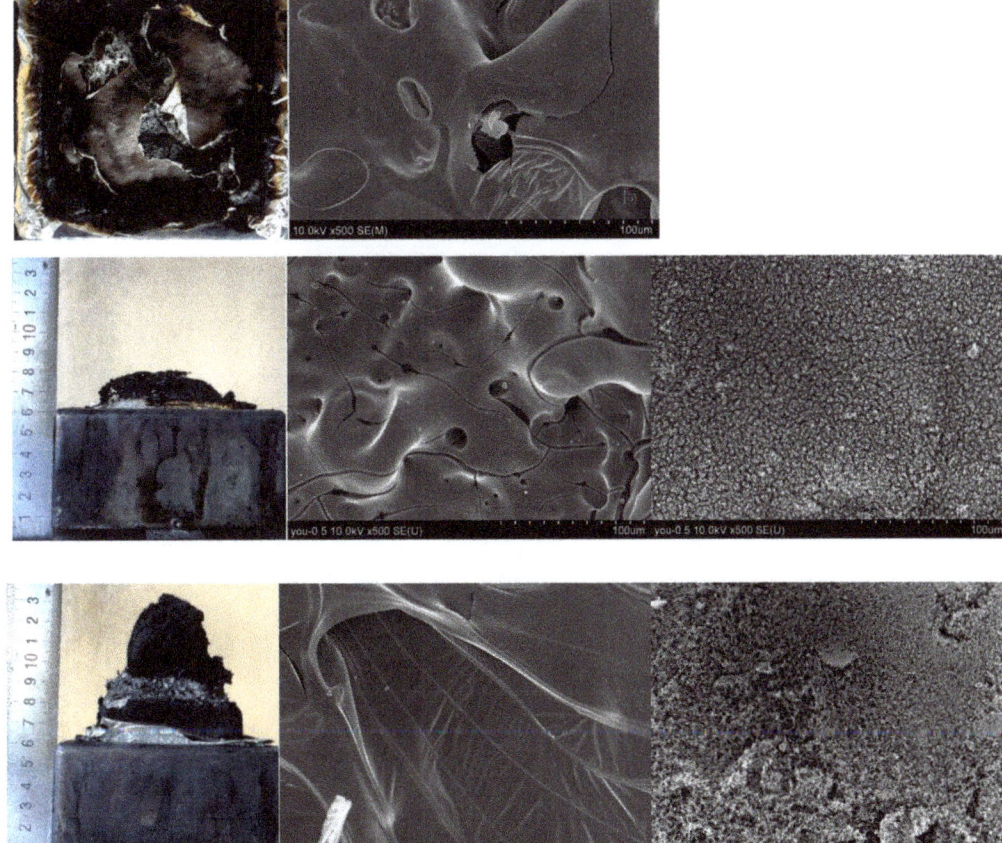

Figure 5. *Cont.*

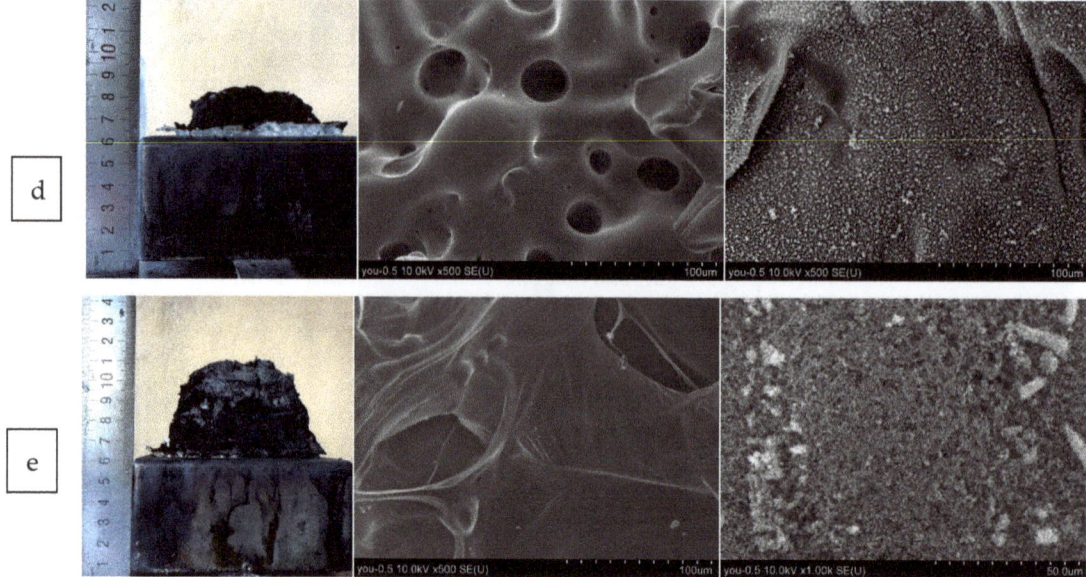

Figure 5. Digital, inner, and outer surface SEM images of SPMS/APP-EP after the cone test: (**a**) EP0, (**b**) EP3-10, (**c**) EP4-10, (**d**) EP7-20, and (**e**) EP8-20.

2.6. Mechanical Properties

The influence of flame retardants on the mechanical properties of the cured EP composites was characterized via a vertical drawing test. A portion of the cured epoxy resin exhibited lower tensile strengths than did pure EP, and the tensile strength of pure EP was 39.16 MPa; however, the corresponding strength values of EP3-10 and EP4-10 decreased to 22.52 MPa and 30.94 MPa, respectively. The reason may be that EP3-10 and EP4-10 have lower phosphorous contents and rigid structures. This kind of rigid structure causes steric hindrance, and the material does not easily deform during tensile testing; moreover, a small amount of flame retardants causes stress concentration. These factors deteriorate the mechanical properties of cured EP composites.

Another part of the cured epoxy resins exhibited higher tensile strengths than pure EP. With the addition of 20 wt% SPMS/APP and SPMS to the composites, the tensile strength increased from 39.16 MPa to 49.78 MPa and 53.69 MPa, respectively. The rigid structure served as a physical cross-linking point in the cured epoxy resins, which could transfer the stress from different polymer chains to maintain the structure of the materials when a small amount of molecular chain breaks. Therefore, the tensile strength of the cured epoxy resins increased, and the flame retardants may also play a role in strengthening and toughening.

3. Materials and Methods

3.1. Materials

6H-dibenzo[c,e][1,2]oxaphosphinine-6-sulfide (DOPS) was synthesized in our laboratory [24]. Epoxy resin (DGEBA E-44), a commercial product, was purchased from Blue Star Chemical New Materials Co., Ltd. (Beijing, China). Itaconic acid (ITA) was supplied by Sinopharm Group Chemical reagents, Shanghai, China. Tetrahydrofuran (THF) and acetone were purchased from Tianjin Fuyu Fine Chemical Co., Ltd. (Tianjin, China). APP was obtained from Shifang Taifeng New Flame Retardant Co., Ltd. (Chengdu, China). M-phenylenediamine (PDA) was purchased from Tianjin Guangfu Fine Chemical Research Institute (Tianjin, China). All chemicals were used as received and without further purification.

3.2. Synthesis of SPMS

The schematic route of SPMS is illustrated in Scheme 1, and the details are as follows: 3.1 g of DOPS, 1.83 g of ITA, and 60 mL of THF were introduced into a four-neck 250 mL flask equipped with a nitrogen inlet and a mechanical stirrer. The resulting mixture was heated to reflux with agitation under a nitrogen gas atmosphere for 5.5 h. After cooling to room temperature, the precipitate was removed by filtration; the resulting material was subsequently dried and recrystallized. A white solid of SPMS was obtained with a yield of 75.7% and a m.p. of 178~181 °C.

3.3. Preparation of the EP/SPMS-APP Hybrid

The cured EP was prepared via a thermal curing process. Briefly, the epoxy resin was heated to melt first. Subsequently, flame retardants were added separately according to the mass ratio in Table 1, and the mixture was sufficiently stirred. The mixture was subsequently placed in a preheated vacuum oven to degas at 80 °C. After that, PDA was added accurately under vigorous stirring until it was completely dissolved before being poured into the mold. The mixture was subsequently cured in an air-drying oven at 80 °C for 2 h, followed by 100 °C for 2 h, and then postcuring at 120 °C for 2 h. Thereafter, the curd EP was permitted to cool slowly to room temperature. Then, the splines and trimmed parts were removed so that they could meet the requirements of the corresponding performance test.

3.4. Characterization

The limiting oxygen index (LOI) was measured on a JF-5 oxygen index instrument (Jiangning Analytical Instrument Factory, Nanjing, China) according to the ASTM D-2863 testing procedure, with sample dimensions of $80 \times 6.5 \times 3.2$ mm^3. The vertical burning test (UL-94) was performed on a CZF-6 instrument (the same factory used for JF-5) according to the ASTM D-3801, with sample dimensions of $125 \times 13 \times 3.2$ mm^3. A mechanical performance test was carried out using a universal testing machine (Jinan East Special Equipment Co., Ltd. Jinan, China) according to the GB T13022-91 testing procedure, with a stretch rate of 1 mm/min. Thermal gravimetric analysis (TGA) was performed with a thermal analyzer (NETZSCH STA 409 PC/PG, Bavaria, Germany) at a heating rate of 20 °C/min from 20 °C to 700 °C under a nitrogen flow of 60 mL/min. Cone calorimeter measurements were performed according to the ISO5660-1 protocol at an incident flux of 35 kW/m^2, with sample dimensions of $100 \times 100 \times 1.2$ mm^3. Scanning electron microscopy (SEM) images of the inner and outer surfaces of the char residue after CC were obtained by using a Hitachi S-4800. Char layers were sputter coated with a thin layer of gold, and SEM was performed at an accelerating voltage of 10 kV.

4. Conclusions

In this article, a novel phosphorous-containing flame retardant, SPMS, was successfully synthesized and characterized. The flame retardants in the epoxy resins exhibited excellent flame retardancy and intumescent effects. The flame retardancy was obviously improved, and the cured epoxy resins had higher char yields, which showed that there was a concurrent action of the SPMS and APP. The CC results showed that the pHRR, av-HRR, THR, and TSR values of the epoxy resins decreased significantly. This kind of reactive flame retardant is promising for other polymer matrices that interact with diacid functional groups or react with other functional group flame retardants.

Author Contributions: Conceptualization, D.B.; methodology, D.B., X.M. and W.Z.; software, N.K. and Y.Z.; validation, D.B. and W.Z.; formal analysis, Y.M.; resources, W.Z.; data curation, J.Y.; writing—original draft preparation, D.B. and X.M.; writing—review and editing, D.B.; visualization, X.M.; project administration, D.B.; funding acquisition, D.B. All authors have read and agreed to the published version of the manuscript.

Funding: This research was funded by the National Natural Science Foundation of China, grant number 51763005, and the Guiyang Science Project, grant number Zhu Ke Hetong[2021]43-4.

Institutional Review Board Statement: Not applicable.

Informed Consent Statement: Not applicable.

Data Availability Statement: Data are contained within the article.

Conflicts of Interest: The authors declare no conflicts of interest.

References

1. Huo, S.Q.; Song, P.A.; Yu, B.; Ran, S.Y.; Chevali, V.S.; Liu, L.; Fang, Z.P.; Wang, H. Phosphorus-containing flame retardant epoxy thermosets: Recent advances and future perspectives. *Prog. Polym. Sci.* **2021**, *114*, 101366. [CrossRef]
2. Jamsaz, A.; Goharshadi, E.K. Graphene-based flame-retardant polyurethane: A critical review. *Polym. Bull.* **2023**, *80*, 11633–11669. [CrossRef]
3. Gao, T.Y.; Wang, F.D.; Xu, Y.; Wei, C.X.; Zhu, S.E.; Yang, W.; Lu, H.D. Luteolin-based epoxy resin with exceptional heat resistance, mechanical and flame retardant properties. *Chem. Eng. J.* **2022**, *428*, 131173. [CrossRef]
4. Liu, B.-W.; Zhao, H.-B.; Wang, Y.-Z. Advanced Flame-Retardant Methods for Polymeric Materials. *Adv. Mater.* **2022**, *34*, 2107905. [CrossRef]
5. Zhang, J.H.; Mi, X.Q.; Chen, S.Y.; Xu, Z.J.; Zhang, D.H.; Miao, M.H.; Wang, J.S. A bio-based hyperbranched flame retardant for epoxy resins. *Chem. Eng. J.* **2020**, *381*, 122719. [CrossRef]
6. Liu, X.D.; Zheng, X.T.; Dong, Y.Q.; He, L.X.; Chen, F.; Bai, W.B.; Lin, Y.C.; Jian, R.K. A novel nitrogen-rich phosphinic amide towards flame-retardant, smoke suppression and mechanically strengthened epoxy resins. *Polym. Degrad. Stab.* **2022**, *196*, 109840. [CrossRef]
7. Wen, Y.; Cheng, Z.; Li, W.; Li, Z.; Liao, D.; Hu, X.; Pan, N.; Wang, D.; Hull, T.R. A novel oligomer containing DOPO and ferrocene groups: Synthesis, characterization, and its application in fire retardant epoxy resin. *Polym. Degrad. Stab.* **2018**, *156*, 111–124. [CrossRef]
8. Deng, P.; Liu, Y.; Liu, Y.; Xu, C.; Wang, Q. Preparation of phosphorus-containing phenolic resin and its application in epoxy resin as a curing agent and flame retardant. *Polym. Adv. Technol.* **2018**, *29*, 1294–1302. [CrossRef]
9. Luo, Y.; Xie, Y.H.; Jiang, H.; Chen, Y.; Zhang, L.; Sheng, X.X.; Xie, D.L.; Wu, H.; Mei, Y. Flame-retardant and form-stable phase change composites based on MXene with high thermostability and thermal conductivity for thermal energy storage. *Chem. Eng. J.* **2021**, *420*, 130466. [CrossRef]
10. Vlad-Bubulac, T.; Hamciuc, C.; Serbezeanu, D.; Macsim, A.-M.; Lisa, G.; Anghel, I.; Preda, D.-M.; Kalvachev, Y.; Rîmbu, C.M. Simultaneous Enhancement of Flame Resistance and Antimicrobial Activity in Epoxy Nanocomposites Containing Phosphorus and Silver-Based Additives. *Molecules* **2023**, *28*, 5650. [CrossRef]
11. Xia, W.J.; Wang, S.W.; Xu, T.; Jin, G.L. Flame retarding and smoke suppressing mechanisms of nano composite flame retardants on bitumen and bituminous mixture. *Constr. Build. Mater.* **2021**, *266*, 121203. [CrossRef]
12. Teng, F.; Li, W.X.; Zhong, W.Y. Flame-responsive aryl ether nitrile structure towards multiple fire hazards suppression of thermoplastic polyester. *J. Hazard. Mater.* **2021**, *403*, 123714. [CrossRef]
13. Li, L.; Li, S.; Wang, H.; Zhu, Z.; Yin, X.; Mao, J. Low flammability and smoke epoxy resins with a novel DOPO based imidazolone derivative. *Polym. Adv. Technol.* **2021**, *32*, 294–303. [CrossRef]
14. Wang, Z.Z.; Gao, X.; Li, W.F. Epoxy resin/cyanate ester composites containing DOPO and wollastonite with simultaneously improved flame retardancy and thermal resistance. *High. Perform. Polym.* **2020**, *32*, 710–718. [CrossRef]
15. Jian, R.K.; Ai, Y.F.; Xia, L.; Zhao, L.J.; Zhao, H.B. Single component phosphamide-based intumescent flame retardant with potential reactivity towards low flammability and smoke epoxy resins. *J. Hazard. Mater.* **2019**, *371*, 529–539. [CrossRef]
16. Qi, Z.; Zhang, W.C.; He, X.D.; Yang, R.J. High-efficiency flame retardancy of epoxy resin composites with perfect T-8 caged phosphorus containing polyhedral oligomeric silsesquioxanes (P-POSSs). *Compos. Sci. Technol.* **2016**, *127*, 8–19. [CrossRef]
17. Wang, P.; Yang, F.; Li, L.; Cai, Z. Flame retardancy and mechanical properties of epoxy thermosets modified with a novel DOPO-based oligomer. *Polym. Degrad. Stab.* **2016**, *129*, 156–167. [CrossRef]
18. Liu, C.; Yang, D.; Sun, M.N.; Deng, G.J.; Jing, B.H.; Wang, K.; Shi, Y.Q.; Fu, L.B.; Feng, Y.Z.; Lv, Y.C.; et al. Phosphorous-Nitrogen flame retardants engineering MXene towards highly fire safe thermoplastic polyurethane. *Compos. Commun.* **2022**, *29*, 101055. [CrossRef]
19. Xu, M.J.; Xu, G.R.; Leng, Y.; Li, B. Synthesis of a novel flame retardant based on cyclotriphosphazene and DOPO groups and its application in epoxy resins. *Polym. Degrad. Stab.* **2016**, *123*, 105–114. [CrossRef]
20. Wang, H.; Li, S.; Zhu, Z.M.; Yin, X.Z.; Wang, L.X.; Weng, Y.X.; Wang, X.Y. A novel DOPO-based flame retardant containing benzimidazolone structure with high charring ability towards low flammability and smoke epoxy resins. *Polym. Degrad. Stab.* **2021**, *183*, 109426. [CrossRef]
21. Chi, Z.Y.; Guo, Z.W.; Xu, Z.C.; Zhang, M.J.; Li, M.; Shang, L.; Ao, Y.H. A DOPO-based phosphorus-nitrogen flame retardant bio-based epoxy resin from diphenolic acid: Synthesis, flame-retardant behavior and mechanism. *Polym. Degrad. Stab.* **2020**, *176*, 109151. [CrossRef]

22. Xiao, L.; Sun, D.; Niu, T.; Yao, Y. Syntheses and characterization of two novel 9,10-dihydro-9-oxa-10-phosphaphenanthrene-10-oxide-based flame retardants for epoxy resin. *High. Perform. Polym.* **2014**, *26*, 52–59. [CrossRef]
23. Nabipour, H.; Wang, X.; Song, L.; Hu, Y. Intrinsically Anti-Flammable Apigenin-derived Epoxy Thermosets with High Glass Transition Temperature and Mechanical Strength. *Chin. J. Polym. Sci.* **2022**, *40*, 1259–1268. [CrossRef]
24. Chen, S.M.; Lai, F.; Pei, L.; Wei, G.; Hai, F.; Yin, X.G.; Ban, D.M. Synthesis of Flame Retardant Based on Phosphaphenanthrene and Flame Retardancy Study of Epoxy Resin Modified by Intumescent Flame Retardant System Composed of Ammonium Polyphosphate. *Acta Polym. Sin.* **2017**, *8*, 1358–1365.

Disclaimer/Publisher's Note: The statements, opinions and data contained in all publications are solely those of the individual author(s) and contributor(s) and not of MDPI and/or the editor(s). MDPI and/or the editor(s) disclaim responsibility for any injury to people or property resulting from any ideas, methods, instructions or products referred to in the content.

Review

Advancements in Flame-Retardant Systems for Rigid Polyurethane Foam

Yao Yuan [1,*], Weiliang Lin [1], Yi Xiao [1], Bin Yu [2] and Wei Wang [3,*]

1. Fujian Provincial Key Laboratory of Functional Materials and Applications, School of Materials Science and Engineering, Xiamen University of Technology, Xiamen 361024, China; weilianglin318315@163.com (W.L.); ahxiaoyi@163.com (Y.X.)
2. State Key Laboratory of Fire Science, University of Science and Technology of China, Hefei 230026, China; yubin@ustc.edu.cn
3. School of Mechanical and Manufacturing Engineering, University of New South Wales, Sydney, NSW 2052, Australia
* Correspondence: yuanyao@xmut.edu.cn (Y.Y.); wei.wang15@unsw.edu.au (W.W.)

Abstract: The amplified employment of rigid polyurethane foam (RPUF) has accentuated the importance of its flame-retardant properties in stimulating demand. Thus, a compelling research report is essential to scrutinize the recent progression in the field of the flame retardancy and smoke toxicity reduction of RPUF. This comprehensive analysis delves into the conventional and innovative trends in flame-retardant (FR) systems, comprising reactive-type FRs, additive-type FRs, inorganic nanoparticles, and protective coatings for flame resistance, and summarizes their impacts on the thermal stability, mechanical properties, and smoke toxicity suppression of the resultant foams. Nevertheless, there are still several challenges that require attention, such as the migration of additives, the insufficient interfacial compatibility between flame-retardant polyols or flame retardants and the RPUF matrix, and the complexity of achieving both flame retardancy and mechanical properties simultaneously. Moreover, future research should focus on utilizing functionalized precursors and developing biodegradable RPUF to promote sustainability and to expand the applications of polyurethane foam.

Keywords: rigid polyurethane foam; flame retardancy; smoke toxicity suppression; flame retardants

Citation: Yuan, Y.; Lin, W.; Xiao, Y.; Yu, B.; Wang, W. Advancements in Flame-Retardant Systems for Rigid Polyurethane Foam. *Molecules* **2023**, *28*, 7549. https://doi.org/10.3390/molecules28227549

Academic Editor: Rodolphe Sonnier

Received: 28 September 2023
Revised: 27 October 2023
Accepted: 8 November 2023
Published: 11 November 2023

Copyright: © 2023 by the authors. Licensee MDPI, Basel, Switzerland. This article is an open access article distributed under the terms and conditions of the Creative Commons Attribution (CC BY) license (https://creativecommons.org/licenses/by/4.0/).

1. Introduction

Rigid polyurethane foam (RPUF) is a highly adaptable type of polymer foam that finds utility across a diverse array of industries and everyday household applications, such as building insulation and the structural portions of roofs, walls, floors, and furniture, primarily due to its excellent insulation properties, durability, lightweight nature, and flexibility in customization [1–7]. According to a recent report, the worldwide polyurethane (PU) market is predicted to attain a value of USD 88 million by the year 2026, exhibiting a compound annual growth rate (CAGR) of 6.0% [8]. It is evident from the data that the number of scientific articles related to PU and PU foam has more than doubled during the period of 2010–2020 compared to 1996–2009, serving as a clear indication. The primary usage of PU in the market is in PU foam, which accounts for up to 67% of the world's total foam consumption [9].

Rigid polyurethane foam is a type of foam insulation made by combining two liquid components, a polyol (a type of alcohol) and an isocyanate (a type of chemical compound), which react and expand to form a solid foam material [10–13]. In regard to flammability, the cellular structure and organic composition of RPUF make it combustible and susceptible to burning upon exposure to high temperatures or flames [14–17]. Moreover, the combustion of RPUF may result in the emission of toxic smoke and gases, such as carbon monoxide (CO), nitrous oxides (NO_x), and hydrogen cyanide (HCN), which can be hazardous to human health [18]. For this reason, it is crucial to explore flame-retardant RPUF and to expand its practical applications [5,19–23].

There are three primary approaches being utilized to address the limitations of RPUF, as depicted in Table 1. The first method is copolymerization, which entails the chemical modification of the surface and core of the polymer [24,25]. The final two methods involve adding flame-retardant (FR) additives to the polymer matrix through mixing or coating [5,21,26–29]. Meanwhile, the first option has been demonstrated to be the durable and feasible approach for pristine RPUF. These flame retardants are primarily composed of organic compounds which possess a flame-retardant segment that can create covalent bonds with RPUF [30]. As a result, the integration of flame retardants into RPUF composites can lead to a considerable improvement in the compatibility between the polymer and the flame retardants as well as provide additional benefits such as enhanced mechanical properties, improved thermal stability, and better compatibility with other additives [31,32].

Numerous reviews have been published regarding the utilization of additive-type flame retardants to enhance the flame resistance of RPUF, and this has been established as the most practical and economical approach for untreated RPUF [28,33,34]. Phosphorus-based and nitrogen-based flame retardants are among the most frequently utilized additives in the fabrication of RPUF composites [35]. Phosphorus-based and nitrogen-based FRs are the most used FRs in RPUF composites. Phosphorus-based FRs, such as ammonium polyphosphate (APP) and melamine polyphosphate (MPP), have excellent flame-retardant properties and generate char upon exposure to heat, which further protects the polymer matrix from combustion [33,36]. Nitrogen-based FRs, such as melamine cyanurate (MC) and guanidine phosphate (GP), generate nonflammable gases and effectively suppress the flames during combustion. Furthermore, the use of nanocomposites that contain nanofillers, like layered double hydroxides (LDHs) and graphene oxide (GO), has demonstrated potential as effective flame retardants and smoke suppressants for RPUF composites [37,38]. These nanofillers have a large surface-area-to-volume ratio, improving their ability to interact with the polymer matrix and impede the spread of flames.

The growing utilization of RPUF has intensified the significance of its flame-retardant properties in driving demand. Consequently, an imperative research report is warranted to investigate recent advancements in the field of the flame retardancy and smoke toxicity suppression of RPUF, given the limited extant literature on this topic. This comprehensive review centers on conventional and emerging trends in flame-retardant (FR) systems, including reactive-type and additive-type FRs, inorganic nanoparticles, and protective coatings, aimed at enhancing the flame retardancy and smoke toxicity suppression of RPUF. This review also covers an overview of the preparation and properties of RPUF and the current trends in flame-retardant strategies for RPUF and a discussion on the future outlook of flame-retardant RPUF.

Table 1. Examples of RPUF composites with various flame retardants.

Flame-Retardant Type	Composite	Remarks	Ref.
Reactive flame retardants	RPUF/AMPO (polyol-bis(hydroxymethyl)-N, N-bis(2-hydroxyethyl) aminomethylphosphine oxide)	- The incorporation of 10.5 wt.% AMPO resulted in an increase in the LOI value from 20.0% in the pure RPUF to 23.4%. - However, the inclusion of AMPO in RPUF resulted in a 21.2% decrease in compressive strength.	[39] (1982)

Table 1. Cont.

Flame-Retardant Type	Composite	Remarks	Ref.
Reactive flame retardants	RPUF/GEP (glycerol/ethanol phosphate)	- The incorporation of 8.0 wt.% glycerol/ethanol phosphate (GEP) resulted in an increase in the LOI of RPUF to 23.5% - However, the total heat release (THR) and peak smoke production rate (PSPR) were increased by 52.0% and 170.0%.	[40] (2015)
	RPUF/PPGE (phenylphosphoryl glycol ether oligomer)	- By incorporating 10.0 wt.% of the additive, RPUF demonstrated an LOI of 24.5%, obtained a UL-94 V-1 rating, and experienced a notable 1.5% enhancement in compressive strength.	[41] (2019)
Additive flame retardants	RPUF/pEG-P(MA) (pulverized expandable graphite (pEG)-poly(methyl methacrylate-acrylic acid) copolymer)	- The RPUF containing 10 wt.% of flame-retardant particles exhibited excellent flame retardancy, as evidenced by a high LOI of 26 vol.%. - The RPUF/pEG-P(MA) composites demonstrated a compressive modulus of 48.4 MPa and a compressive strength of 2.8 MPa.	[42] (2011)
	RPUF/MATMP (melamine amino trimethylene phosphate)	- The incorporation of 15.0 wt.% of MATMP in RPUF resulted in a UL-94 V-0 rating accompanied by a 34.0% decrease in PHRR and an LOI of 25.5%. - At lower loading levels (<10.0 wt.%), MATMP simultaneously improved the compressive strength and thermal insulating properties of RPUF.	[43] (2017)
	RPUF/MFAPP (melamine–formaldehyde resin-microencapsulated ammonium polyphosphate)	- RPUF/MFAPP30 attained a V-1 rating in the UL-94 test, exhibiting an LOI of 21.3 vol.% compared to RPUF/APP30 with an equal load of APP. - The RPUF/MFAPP30 demonstrated a compressive strength of 0.295 MPa, exhibiting a 13.5% increase compared to RPUF/APP30.	[44] (2020)
Flame-retardant coating	RPUF/alginate/clay aerogel	- The RPUF coated with alginate/clay aerogel exhibited an impressive LOI of 60.0%, with 32.0% and 37.0% reductions in PHRR and TSR as compared to the untreated foam. - However, the presence of the aerogel within the porous foam noticeably compromised its thermal conductivity.	[45] (2016)

Table 1. Cont.

Flame-Retardant Type	Composite	Remarks	Ref.
Flame-retardant coating	RPUF/poly(VS-co-HEA) (copolymerization of hydroxyethyl acrylate (HEA) and sodium vinylsulfonate (VS))	- The resulting RPUF obtained a UL-94 V-0 rating; an LOI of 35.5%; and significant reductions of 87.0% and 71.0% in PHRR and TSR, respectively, compared to the untreated foam. - The introduction of HGM to the coating enabled the coated PU foam to maintain a low thermal conductivity.	[2] (2021)

2. Reactive-Type Flame Retardants

The process of forming RPUF involves various reactions, including urethane formation, crosslinking reactions, and foaming reactions facilitated using a chemical blowing agent. The formation of the urethane linkage is illustrated in Figure 1a, while Figure 1b shows how the urethane group reacts with an isocyanate group to create allophanate, which results in chemical crosslinking. Reactive flame retardants demonstrate favorable interfacial compatibility with the matrix due to their chemical bonding interactions, resulting in minimal impact on the mechanical properties of RPUF. Meanwhile, reactive flame retardants containing multiple hydroxyl, amino, or epoxy groups can serve as polyols in the curing process of RPUF, providing flame retardant properties through the presence of phosphorus, nitrogen, or sulfur elements in their structure [8,41,46–48]. Additionally, reactive flame retardants with chemical bonding interactions are highly durable in industrial applications, as they prevent migration from the RPUF matrix. This section focuses on recent advancements in reactive flame retardants.

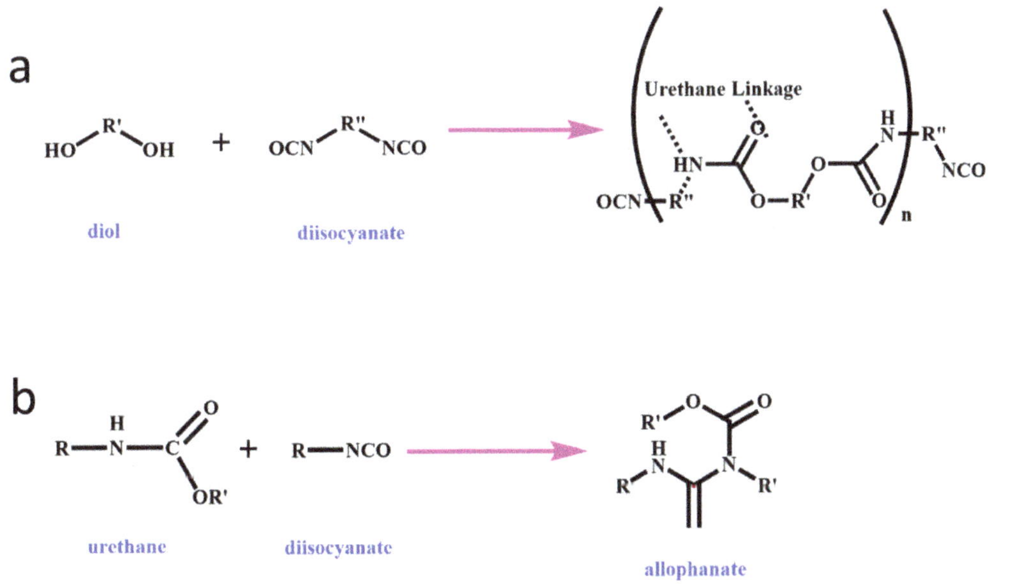

Figure 1. Basic reaction scheme for (**a**) urethane and (**b**) allophanate formation.

2.1. Incorporation of Phosphorus-Containing Groups

Phosphorous-containing flame-retardant polyols are a type of reactive flame retardants, which can be used as a substitute for conventional polyols in the preparation of RPUF to enhance its flame-retardant properties [41,49]. The effectiveness of phosphorus-based flame retardants in minimizing the flammability of polyurethane foam has been well established, leading to its widespread adoption in industries ranging from construction and transportation to electronics [32,50]. Polyols containing phosphorus-based flame retardants possess multiple hydroxyl groups that can actively engage in the curing reaction of polyurethane foam. The incorporation of phosphorus-containing flame retardants in polyols enables the formation of chars and reduces the release of flammable gases during combustion [51].

One of the studies in this area focused on the application of biobased flame-retardant polyols. In their study, Bhoyate et al. [52] explored the fire safety of a polyol sourced from limonene, which was chemically modified with phenyl phosphonic acid. According to their findings, adding 1.5 wt.% of phosphorus through chemical modification could decrease the self-extinguishing time from 81 s to 11.2 s. Zhang et al. [53] employed the approach illustrated in Figure 2a to produce castor oil phosphate flame-retardant polyol (COFPL), a flame-retardant polyol derived by incorporating a phosphate group into the polyol. The process involves the preparation of glycerolized castor oil (GCO) and the epoxidation of GCO, which is ultimately converted to COFPL through a reaction with diethyl phosphate. Wang et al. [54] developed two biobased flame-retardant polyols (CODEOA and CPPA) from a modified vegetable oil. By incorporating epoxidized polyols (BIO_2) and derived carbon materials into the RPUF matrix, the heat release rate (HRR), total heat release (THR), and total smoke production (TSP) can be significantly reduced.

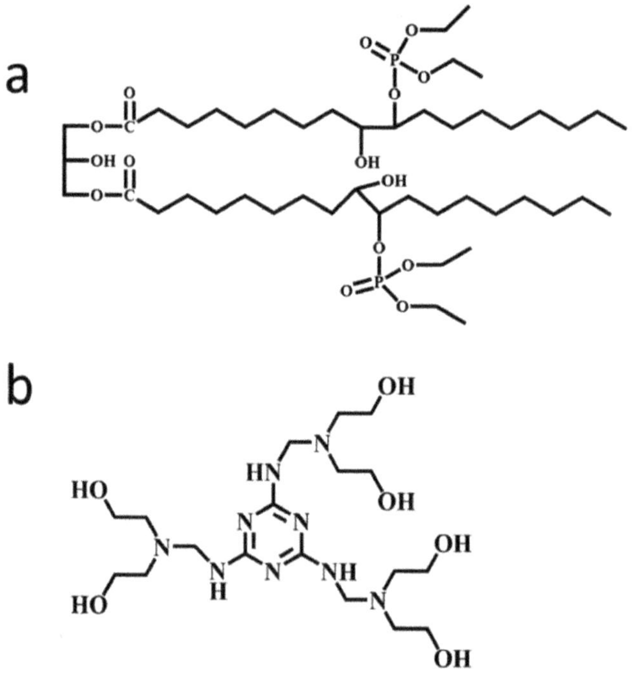

Figure 2. Structure of different flame retardants: (**a**) COFPL and (**b**) MADP.

Currently, green materials such as plant oil, tung oil, and lignin have been chosen as the primary synthetic resources. However, the flame-retardant performance of biobased polyols is hindered by their long-chain structure, resulting in the low content of flame-retardant elements (phosphorus, nitrogen, silicon, etc.). Zhou et al. [55] synthesized tung-oil-based polyols through the ring-opening reaction of epoxidized tung oil and silane-coupling agents. Although the limiting oxygen index of the biobased RPUF prepared from these polyols increased from 19.0 vol.% to 22.6 vol.%, the improvement in the flame retardancy was not significant. This can be attributed to the low phosphorus content in long-chain biobased polyols.

In another approach, flame-retardant polyols with short chains are utilized for the synthesis and application of RPUF due to their elevated phosphorus content. Zou et al. [56] developed a hard-segment flame retardant (HSFR) to enhance the fire resistance of RPUF. By incorporating 13.8 wt.% of THPO, the resulting system was able to achieve a UL-94 V-0 rating, with an LOI ranging from 17.0 to 25.5%. During combustion, the HSFR can produce PO• and PO$_2$• radicals in the vapor phase, which then react with flammable free radicals and impede segment decomposition. In the study conducted by Hu et al. [40], a flame-retardant polyol with a substantial phosphorus content was synthesized through a dehydrochlorination reaction. Hu et al. [57] synthesized a flame-retardant polyol with a high phosphorus content by performing a dehydrochlorination reaction. The addition of expandable graphite (EG) to the RPUF/BHPP system significantly enhanced the flame-retardant characteristics of the RPUF composites, achieving a high LOI value of 30.0%.

To achieve highly flame-retardant RPUF, Wang et al. [58] successfully synthesized a short-chain flame-retardant polyol. They placed particular emphasis on analyzing its compatibility with the conventional polyols 4110 and PEG400. The research findings demonstrated that, when these two mixed polyols were thoroughly ultrasonically blended to achieve full miscibility, their compatibility was significantly enhanced. This enhancement can impart RPUF with superior flame-retardant and mechanical properties.

2.2. Incorporation of Nitrogen-Containing Groups

The incorporation of nitrogen-containing groups is another effective approach for flame retardancy in polyurethane foam [59–61]. These compounds are usually incorporated into the polyol component during the production process, which then react with isocyanates to form polyurethane foam with improved flame retardancy [62]. During combustion, these compounds release nonflammable gases, such as nitrogen or ammonia, when exposed to high temperatures, diluting the flammable gases and reducing the combustibility of the foam [63–65]. Nevertheless, the flame-retardant effectiveness of nitrogen-based compounds is typically inferior to that of phosphorus-containing ones due to their single flame-retardant function.

Melamine-based polyols are a type of reactive flame retardant used in the production of RPUF due to their high nitrogen content [66,67]. These polyols are synthesized by reacting melamine with an excess of formaldehyde and an alcohol or polyol, resulting in a highly crosslinked and thermally stable polymer. Hu et al. [62] developed a melamine-based polyol (MADP, Figure 2b) and incorporated it into the RPUF matrix, and the interactions between the PU NCO groups and DOPO are shown in Figure 3. The findings from their study demonstrated a notable improvement in the LOI values (increasing from 19.0 to 28.5%) with the implementation of the flame-retardant system. This enhancement promotes the development of protective char layers and reduces the concentration of flammable gases in the gaseous phase. The synthesis of a Mannich base polyol derived from cardanol (MCMP, Figure 4) was conducted by Zhang et al. [68]. The incorporation of melamine into the molecular structure of MCMP led to improvements in the mechanical properties, thermal stability, and flame resistance of the resulting RPUF. Li et al. [69] developed an ecofriendly melamine-based polyether polyol referred to as GPP. They found that incorporating GPP in RPUF synthesis greatly enhances the flame retardancy of the resulting foam. The compressive strength of the RPUF samples increased by 106.0%, and an LOI value of 30.4% was achieved by fully incorporating GPP during the preparation process.

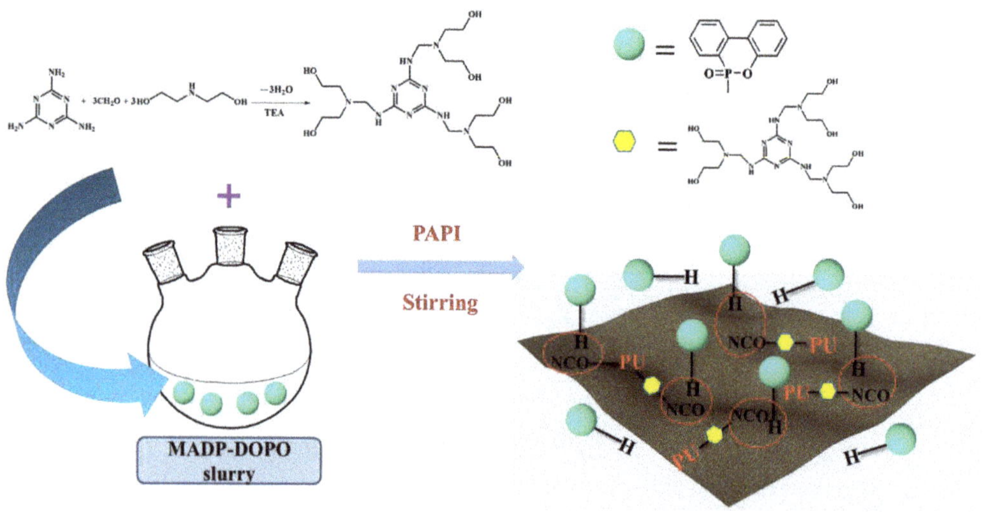

Figure 3. Schematic presentation of the interactions between PU—-NCO groups and DOPO [62] (Copyright 2018). Reproduced with permission from Elsevier Science Ltd.

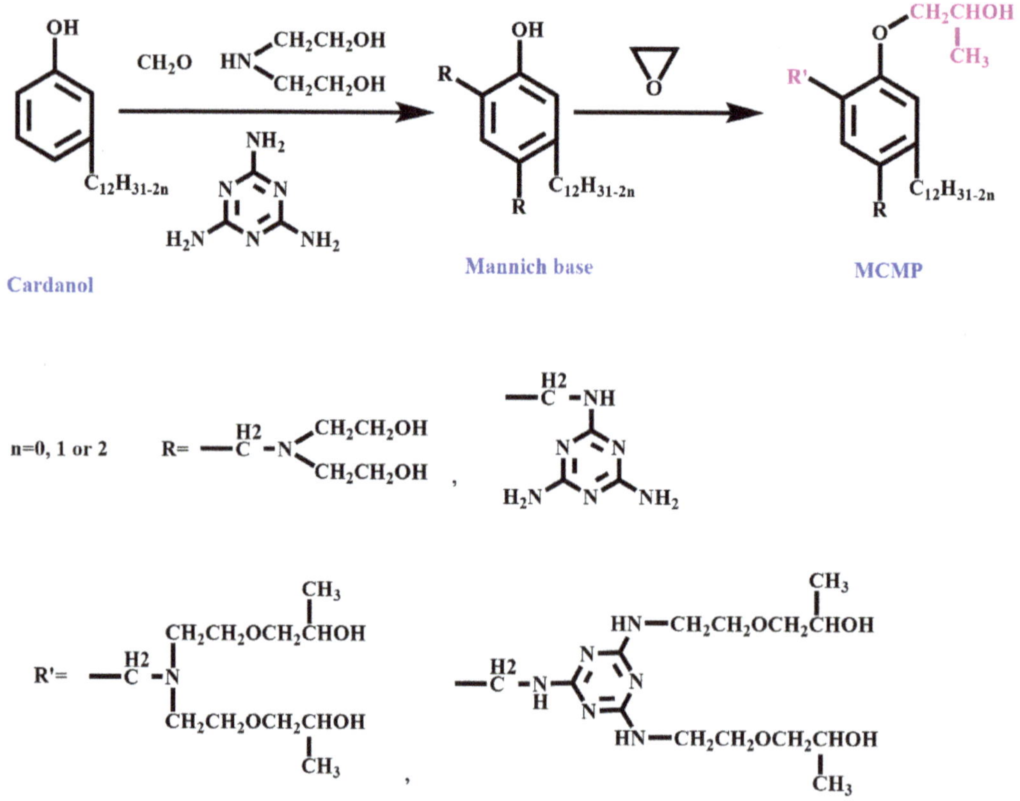

Figure 4. The synthesis of MCMP.

2.3. Incorporation of Sulfur-Containing Groups

Sulfur-based flame retardants represent an alternative category of reactive flame retardants employed in the manufacture of polyurethane foam. These compounds have the advantages of having a low cost, being effective, and having a low impact on the mechanical properties of the foam. A sulfur-containing polyol was synthesized by Bhoyate et al. [49] to create flame-retardant polyurethane foams with improved compressive strength without affecting the foam morphology or closed cell content. The modified RPUF exhibited a self-extinguishing time that was decreased from 94.0 s to 1.7 s when the phosphorus content reached 1.5 wt.% in contrast to the pristine RPUF.

3. Additive-Type Flame Retardants

Additive-type flame retardants for rigid polyurethane foam are typically nonreactive compounds that are added directly to the foam formulation to enhance its thermal stability and flame retardancy [27,28,70–72]. They can be either halogenated or nonhalogenated and can be divided into various subcategories based on their chemical composition, such as phosphorus-based, nitrogen-based, or metal-based flame retardants [73]. Additive flame retardants have various benefits compared to reactive flame retardants, such as their ease of integration into foam formulations and lower costs. Nevertheless, their constrained interfacial compatibility with the matrix results in a decline in the mechanical strength and thermal conductivity of the foam [74]. Furthermore, they may present certain disadvantages, such as potential migration from the matrix, which can influence the mechanical and thermal characteristics of RPUF, in addition to environmental concerns linked to their utilization.

3.1. Addition of Phosphorous-Containing Flame Retardants

The addition of phosphorous-containing flame retardants (P-FRs) in rigid polyurethane foam is a common approach to enhance its fire resistance. Phosphorous-containing compounds are considered effective flame retardants due to their ability to promote char formation in the condensed phase and to decrease the release of flammable gases in the condensed phase during combustion [75]. Examples of commonly used P-FRs in rigid polyurethane foam include ammonium polyphosphate (APP), dimethyl methylphosphonate (DMMP), red phosphorous, and resorcinol bis(diphenyl phosphate) [63,76]. These flame retardants can provide varying degrees of fire resistance depending on their chemical structure, loading level, and processing conditions. It has been observed that the effectiveness of flame retardancy can be influenced by the valence state of phosphorus.

DOPO and its derivatives have recently emerged as P-FRs for the RPUF matrix, which release phosphorus species (PO•) and scavenge H• and OH• radicals in the flame to prevent the thermal degradation of polymers [77]. In a study by Zhang et al. [78], a new flame retardant called DOPO-BA was synthesized and added to a rosin-based RPUF matrix. When 20 wt.% DOPO-BA was incorporated, the LOI value increased from 20.1% to 28.1%. However, there was a significant reduction in the total smoke release.

Ranaweera et al. [79] addressed the migration issue of liquid flame-retardant dimethyl methylphosphonate (DMMP) in RPUF by synthesizing a biobased polyol from limonene and incorporating it with DMMP. They found that the addition of 2 wt.% DMMP can significantly reduce the burning time of RPUF by 83%. In addition, RPUF with TSPB exhibited improved water resistance. Wu et al. [80] observed that adding 10.6 wt.% toluidine spirocyclic pentaerythritol bisphosphonate (TSPB) to RPUF resulted in an improved LOI value and a UL-94 V-0 rating, which was attributed to the char-forming effect.

3.2. Addition of Phosphorus–Nitrogen-Based Flame Retardants

Nitrogen-containing flame retardants (N-FRs) are another class of commonly used flame retardants in the RPUF matrix that work by releasing nitrogen species in the gas phase during combustion, which can act as diluents fuels, oxygen, and free radicals in the flame [59]. The most used N-FRs in RPUF include melamine, melamine cyanurate, melamine phosphate, and guanidine derivatives. These N-FRs can provide excellent an flame retardancy performance in RPUF, especially when used in combination with phosphorus-containing flame retardants. The addition of N-FRs can also improve other properties of RPUF, such as its mechanical properties and thermal stability, while minimizing the smoke and toxic gas released during combustion. Xu et al. [81] investigated the smoke suppression mechanism of melamine in rigid polyurethane foam using a smoke density chamber, cone calorimetry, and a Py/GC-MS analysis.

A new phosphorus–nitrogen intumescent flame retardant (DPPM) was synthesized by Guo et al. [32]. The addition of only 9% DPPM was sufficient to enable RPUF to achieve a UL-94 V-0 rating and a limit oxygen index of 29%. Hexa(phosphitehydroxylmethylphenoxyl) cyclotriphosphazene (HPHPCP, depicted in Figure 5) is another flame retardant that has been successfully synthesized and incorporated into rigid polyurethane foam [32]. HPHPCP contains multifunctional groups that introduce crosslinking into the foam structure, thereby improving its thermal stability and compressive strength. The addition of HPHPCP to DPPM-RPUF resulted in an LOI of 29.5% and a UL-94 V-0 rating, which can be attributed to the surface pyrolysis of RPUF.

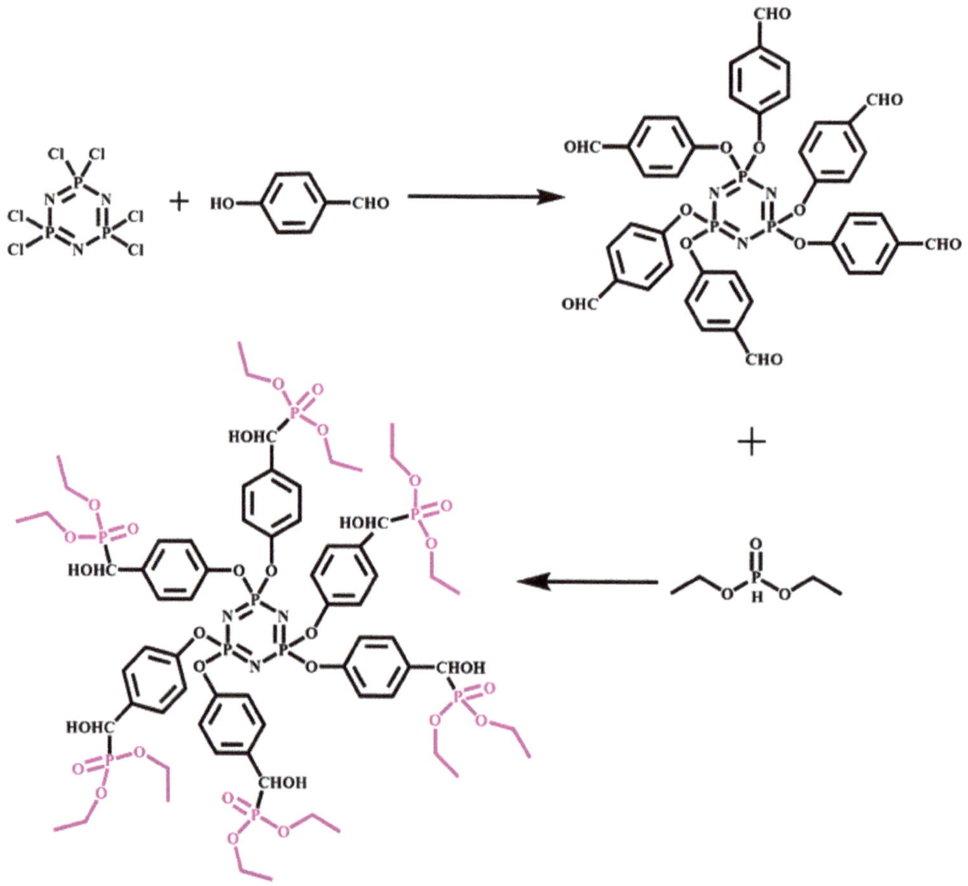

Figure 5. The synthesis of reactive flame-retardant HPHPCP.

3.3. Addition of Expandable Graphite and Derivatives

Expandable graphite (EG) and derivatives have emerged as a promising flame retardant for rigid polyurethane foam (RPUF) due to their exceptional fire-resistant properties [82–84].

Expandable graphite (EG) is produced by introducing sulfuric acid, acetic acid, or nitric acid into the crystalline structure of graphite. This process results in a unique material with exceptional thermal expansion properties when exposed to heat (Figure 6) [85]. As a result, EG effectively decreases the flammability and heat release of RPUF. Wang et al. [86] employed a heterocoagulating method to encapsulate EG using magnesium hydroxide (MH). The incorporation of 11.5 wt.% core-shell EG@MH increased the limiting oxygen index (LOI) of the rigid polyurethane foam (RPUF) to 32.6% and improved the storage modulus by approximately 55.0%.

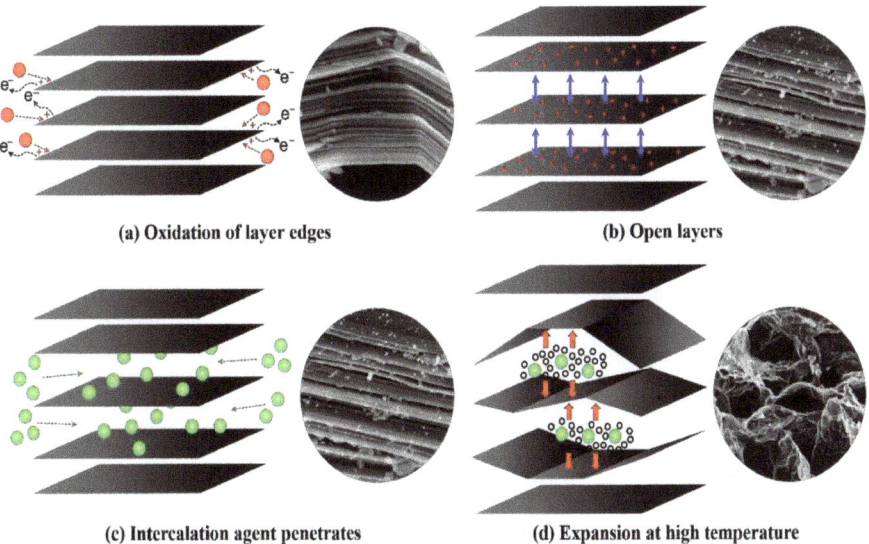

Figure 6. Preparation and expansion processes of expandable graphite [85] (Copyright 2019). Reproduced with permission from Elsevier Science Ltd.

When considering the utilization of RPUF in refrigerators, it is essential that the material exhibits a thermal conductivity falling within the range of 19 to 22 mW/(m K) and a compressive strength exceeding 110 kPa. Akdogan et al. [87] developed a flame-retardant RPUF using 15 wt.% EG and 5 wt.% ammonium pentaborate (APB). The outcomes revealed that this RPUF exhibited a remarkable 42.8% reduction in the THR and a 77.0% decrease in the TSR compared to the pristine foam. Furthermore, it demonstrated an LOI of 27.9% coupled with a low thermal conductivity of 20.41 mW/(m K) and a high compressive strength of approximately 125 kPa, thus displaying promising characteristics for its potential utilization in refrigeration applications.

3.4. Addition of Nanoclay and Other Nanoparticles

Adding nanoclay and other nanoparticles to both flexible and rigid polyurethane foam is a promising strategy for decreasing the production of smoke particles and toxic gases during combustion [88–90]. Clay nanosheets are frequently used as nanoparticle fillers in polymer nanocomposites because of their low cost, widespread availability, and flexibility. Incorporating nanoclay (a two-dimensional nanoparticle) into rigid polyurethane foam can form a physical barrier that obstructs gas diffusion and heat transfer, resulting in reduced smoke production and increased thermal stability [34,35,91]. Furthermore, the platelet structure of nanoclay is attributed to the significant enhancement in the mechanical properties of RPUF, such as its strength, stiffness, and toughness [92]. Adilah Alis et al. [93] conducted a study on the flame retardancy and thermal stability of clay nanosheets, where halloysite nanotubes (HNTs) were synthesized and added to biobased RPUF. The study showed that increasing the HNT load from 1 wt.% to 5 wt.% led to a corresponding increase in the thermal stability of the RPUF. Moreover, RPUF blended with 5 wt.% HNTs exhibited the highest residual char of 18.1% at 600 °C compared to pristine RPUF with only 7.6%, indicating that the addition of HNTs significantly improved the char-forming ability of RPUF.

Similarly, nanoparticles, such as cuprous oxide (Cu_2O), titanium dioxide (TiO_2), nickel oxide (NiO), and silica (SiO_2), have also been investigated for their potential in reducing smoke production and toxic gas emissions during rigid polyurethane foam's combustion [94–96]. These nanoparticles act as flame retardants by catalyzing the formation of integral and compact chars, reducing the production of volatile compounds, and increasing the thermal stability of RPUF. A comparative study was conducted by Hu et al. [97] to investigate highly efficient catalysts for reducing toxic gas generation in RPUF nanocomposites at various temperatures. The primary objective of the study was to determine the most effective catalyst for minimizing the release of harmful gases during the combustion of RPUF. The findings indicated that both NiO and $NiMoO_4$ were efficient catalysts in reducing toxic gas emissions. Furthermore, a comprehensive quantitative analysis of the gaseous degradation products, such as HCN, NO_x, and CO, was performed for various polyurethane composite materials, including PU/Cu_2O, PU/NiO, PU/MoO_3, $PU/CuMoO_4$, and $PU/NiMoO_4$. This analysis was conducted using a tubular furnace method at both 650 °C and 850 °C.

Yuan et al. [98] utilized a straightforward wet chemical method to synthesize cuprous oxide (Cu_2O) crystals of varying sizes and investigated their impact on the combustion performance of RPUF. According to the study, the addition of 2 wt.% Cu_2O to the RPUF matrix resulted in a notable decrease in the peak rate of carbon monoxide (CO) production and total smoke production. Furthermore, as illustrated in Figure 7, the reduction of Cu^{2+}-Cu^+-Cu^0 by degraded gases and the oxidation of Cu^0-Cu^+-Cu^{2+} by oxygen were involved in the conversion of CO to CO_2 and the complete combustion of RPUF.

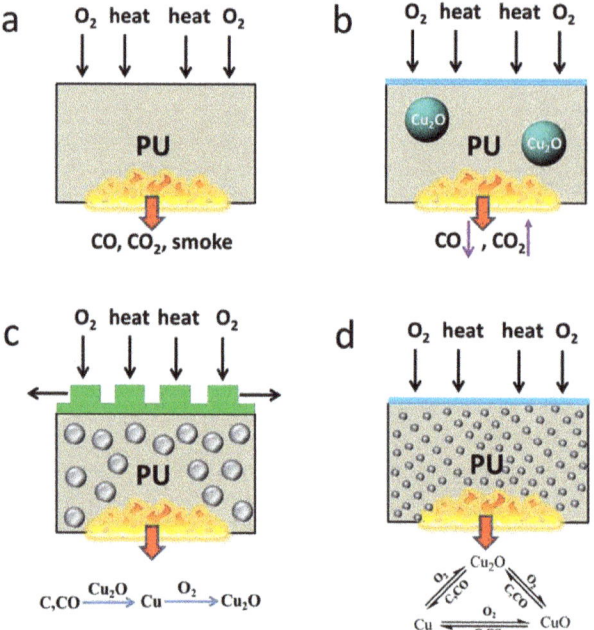

Figure 7. Schematic illustration for the mechanism of catalytic oxidation and catalytic carbonization of pristine RPUF (**a**), $RPUF/Cu_2O$-1µm (**b**), $RPUF/Cu_2O$-100nm (**c**) and $RPUF/Cu_2O$-100nm (**d**) [98] (Copyright 2021). Reproduced with permission from Elsevier Science Ltd.

3.5. Addition of Phase-Change Materials

The addition of phase-change materials (PCMs) to flame-retardant RPUF has been studied as a potential approach to enhance the thermal energy storage and fire retardancy properties of the material [99]. PCMs can absorb and release thermal energy during phase transition, thereby reducing temperature fluctuations and enhancing the thermal stability [100]. Several studies have investigated the effects of different types and concentrations of PCMs on the thermal and fire performance of RPUF, including the reduction in the peak heat release rate, total heat release, and smoke toxicity. The incorporation of PCMs in RPUF has the potential to be used in various applications, such as thermal insulation for buildings, refrigeration and thermal energy storage systems, and transportation industries. Niu et al. [101] focused on embedding a flame-retardant carbon nanotube (d-CNT) modified by DOPO into MPCMs to enhance their thermal stability and flame retardancy. The small room model and infrared thermal imager results also demonstrated that RPUF/d-c-MPCM could make indoor temperature fluctuation gentler, with a maximum indoor temperature of only 26.6 °C. Overall, the addition of PCMs to RPUF shows promise as a way to enhance both the thermal energy storage and fire-retardancy properties of the material, potentially leading to improved energy efficiency and safety in various applications.

4. Flame-Retardant Coatings

Flame-retardant coatings are a popular method to enhance the fire resistance of PUF. These coatings can be applied to the surface of PUF to create a protective layer that slows down the spread of fire and reduces the amount of heat and smoke released during combustion [102,103]. The coatings are typically made up of flame-retardant additives, such as aerogels, alumina trihydrate, metal hydroxides, and other inorganic materials that have a low flammability and good thermal stability. These additives work by releasing water vapor when exposed to heat, which helps cool down the surface of the foam and prevent it from igniting. In addition to enhancing the fire resistance of RPUF, flame-retardant coatings can also improve other properties, such as mechanical strength, chemical resistance, and UV stability.

One example of a flame-retardant coating is the intumescent coating, which can form a char layer when exposed to fire, leading to reduced heat transfer and flame spread. Wang et al. [104] conducted a study in which they developed an intumescent coating by combining a silicone resin (poly-DDPM) and expandable graphite (EG). The resulting poly-DDPM/EG coating tightly adhered to the surface of RPUF and exhibited excellent flame retardancy. Additionally, the compressive strength of the coated RPUF increased significantly, up to 10%. As shown in Figure 8, Huang et al. [105] utilized UV-curable intumescent coatings to treat RPUF. By employing a spray-coating method, they applied a conformal IFR/MXene coating to the foam, aiming to enhance its fire safety.

In addition to intumescent coatings, other types of flame-retardant coatings have also been investigated for their effectiveness in enhancing the flame retardancy of RPUF. For example, a study by Chen et al. [45] reported on the development of a flame-retardant coating using alginate/clay aerogel. The coating was shown to reduce the flame spread rate, the HRR, and the THR of the RPUF during combustion as well as inhibit smoke production. The study showed that the facile and inexpensive posttreatment is a promising approach to improve the thermal stability and flame retardancy of RPUF. As depicted in Figure 9, motivated by the interfacial mechanical interlocking and hydrogen-bonding mechanisms observed in tree frogs and snails, Song et al. [2] successfully produced bioinspired flame-retardant poly(VS-co-HEA) coatings through a free-radical copolymerization process (refer to Figure 9). The thoughtfully engineered microphase-separated micro/nanostructure granted these coatings robust interfacial adhesion to the matrix [106]. The resulting RPUF exhibited notable fire-retardant properties, including a UL-94 V-0 rating and an LOI of 35.5%. Additionally, the treated foam showcased significant reductions in the peak heat release rate (PHRR) by 87.0% and total smoke release (TSR) by 71.0% compared to the untreated foam.

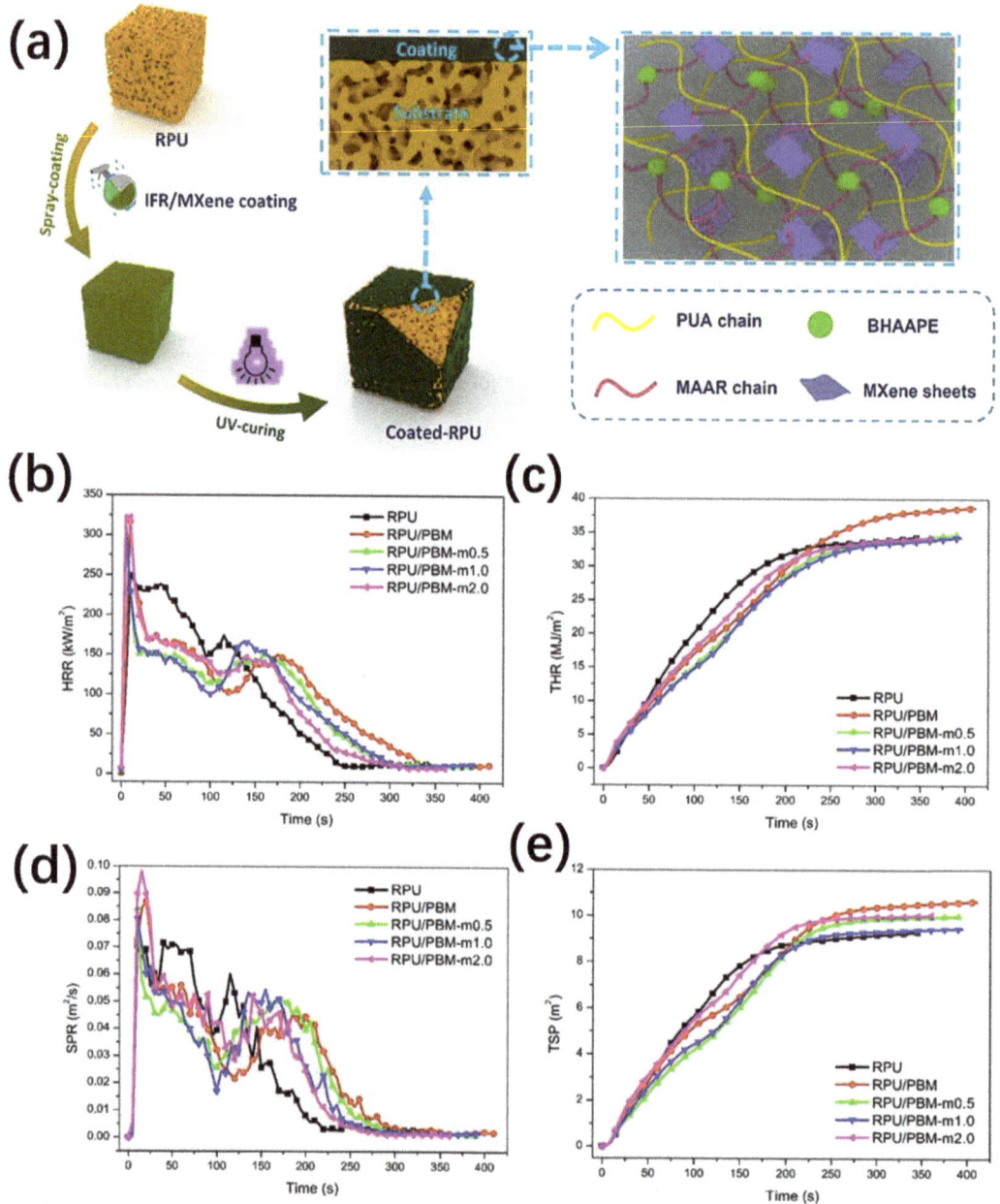

Figure 8. (**a**) Schematic illustration of the fabrication procedure of IFR/MXene-coated RPUF and (**b**) HRR, (**c**) THR, (**d**) SPR, and (**e**) TSP for RPUF and coated RPUF [105] (Copyright 2019). Reproduced with permission from Elsevier Science Ltd.

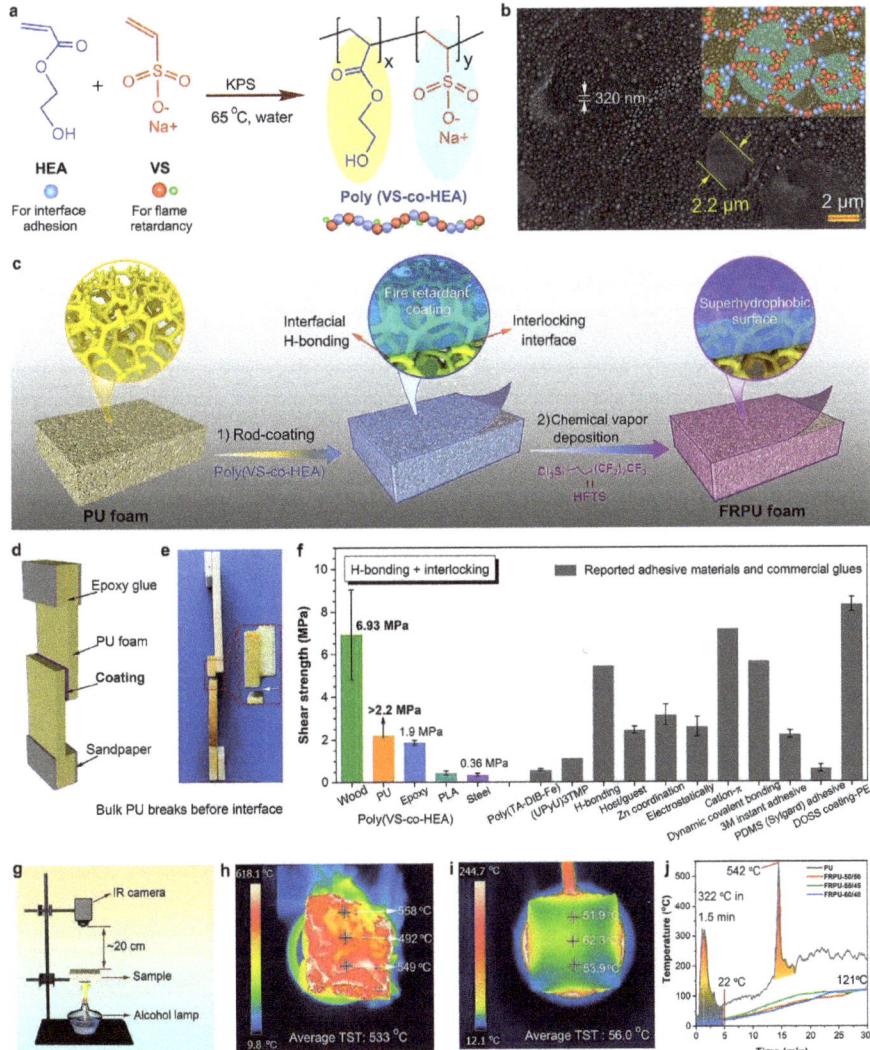

Figure 9. Preparation of RPUF with poly(VS-co-HEA) coatings [2] (Copyright 2021). Reproduced with permission from American Chemical Society. (**a**) Synthesis of bioinspired flame retardant poly(VS-co-HEA) coatings and (**b**) a typical phase-separated micro/nanostructure of poly(VS50-co-HEA50). (**c**) Schematic illustration for the preparation process of flame-retardant rigid PU foam (FRPU). (**d**) The chart illustrating the adhesion or shear strength tests for PU foam. (**e**) Digital image of poly(VS-co-HEA) coatings against PU foam after shear tests, during which bulk PU foam broke before interfaces. (**f**) Shear strength of poly(VS-co-HEA) coatings against different substrates in comparison to some previous and commercial adhesives. (**g**) The homemade setup for determining the flammability of PU foam. The top-surface temperature (TST) determined by the IR camera for (**h**) untreated PU and (**i**) FRPU-60/40–600 μm after ignited for 15 min above an alcohol lamp. The sample thickness is ~3.0 cm. (**j**) TSTs of PU and FRPU as a function of burning time.

5. Concluding Remarks and Future Aspects

The primary objective of this review is to investigate the impact of reactive-type and additive-type flame retardants as well as flame-retardant coatings on the flame retardancy, mechanical properties, and smoke toxicity reduction of rigid polyurethane foam (RPUF). While the techniques discussed in this review have demonstrated favorable outcomes in enhancing the flame retardancy and thermal stability of RPUF, there remain several obstacles to overcome. These include issues such as additive migration, the inadequate interfacial compatibility between additive flame retardants and the RPUF matrix, the harmfulness of some flame-retardant agents, and the challenge of simultaneously achieving flame retardancy and mechanical properties. Furthermore, the inadequate interfacial compatibility between additive flame retardants and the matrix results in the degradation of the mechanical properties and thermal conductivity [107]. One potential opportunity for the future is the development of novel flame-retardant coatings for RPUF. These coatings could be made using sustainable and ecofriendly materials, such as cellulose-based materials, and could offer a promising alternative to conventional coatings. In addition, they could potentially enhance other desirable properties of RPUF, including its mechanical strength and insulation. Hence, one of the future directions in advancing flame-retardant RPUF involves utilizing functionalized precursors capable of yielding inherently flame-retardant foam while preserving the foaming process and physicochemical characteristics. Additionally, efforts should be made towards the development of biodegradable RPUF with flame-retardant properties to promote sustainability and wider applications.

Author Contributions: Conceptualization, Y.Y. and W.W.; methodology, Y.Y.; software, W.L.; validation, Y.Y., W.L., and Y.X.; formal analysis, B.Y.; investigation, W.W.; resources, Y.Y.; writing—original draft preparation, Y.Y.; writing—review and editing, Y.Y. and W.W.; visualization, W.L.; supervision, Y.Y. and W.W.; project administration, Y.Y.; funding acquisition, Y.Y. and W.W. All authors have read and agreed to the published version of the manuscript.

Funding: This research was funded by the Natural Science Foundation of Fujian Province, China (No. 2021J05266); the Talents Introduction Program of the Xiamen University of Technology, China (YKJ19019R); and the Young and Middle-Aged Teachers Education Scientific Research Project of Fujian Province (JAT190657); This research was supported under Australian Research Council/Discovery Early Career Researcher Award (DECRA) funding scheme (project number DE230100180).

Data Availability Statement: Not applicable.

Conflicts of Interest: The authors declare no conflict of interest. The authors declare that they do not have any competing financial interests or personal relationships that might influence their work.

References

1. Chattopadhyay, D.K.; Webster, D.C. Thermal stability and flame retardancy of polyurethanes. *Prog. Polym. Sci.* **2009**, *34*, 1068–1133. [CrossRef]
2. Ma, Z.; Liu, X.; Xu, X.; Liu, L.; Yu, B.; Maluk, C.; Huang, G.; Wang, H.; Song, P. Bioinspired, highly adhesive, nanostructured polymeric coatings for superhydrophobic fire-extinguishing thermal insulation foam. *ACS Nano* **2021**, *15*, 11667–11680. [CrossRef]
3. Tao, J.; Yang, F.; Wu, T.; Shi, J.; Zhao, H.-B.; Rao, W. Thermal insulation, flame retardancy, smoke suppression, and reinforcement of rigid polyurethane foam enabled by incorporating a P/Cu-hybrid silica aerogel. *Chem. Eng. J.* **2023**, *461*, 142061. [CrossRef]
4. Zhang, X.; Sun, S.; Liu, B.; Wang, Z.; Xie, H. Synergistic effect of combining amino trimethylphosphonate calcium and expandable graphite on flame retardant and thermal stability of rigid polyurethane foam. *Int. J. Polym. Anal. Charact.* **2022**, *27*, 302–315. [CrossRef]
5. Singh, H.; Jain, A.; Sharma, T. Effect of phosphorus-nitrogen additives on fire retardancy of rigid polyurethane foams. *J. Appl. Polym. Sci.* **2008**, *109*, 2718–2728. [CrossRef]
6. Septevani, A.A.; Evans, D.A.; Annamalai, P.K.; Martin, D.J. The use of cellulose nanocrystals to enhance the thermal insulation properties and sustainability of rigid polyurethane foam. *Ind. Crops Prod.* **2017**, *107*, 114–121. [CrossRef]
7. Acuña, P.; Lin, X.; Calvo, M.S.; Shao, Z.; Pérez, N.; Villafañe, F.; Rodríguez-Pérez, M.Á.; Wang, D.-Y. Synergistic effect of expandable graphite and phenylphosphonic-aniline salt on flame retardancy of rigid polyurethane foam. *Polym. Degrad. Stab.* **2020**, *179*, 109274. [CrossRef]
8. Zhu, M.; Ma, Z.; Liu, L.; Zhang, J.; Huo, S.; Song, P. Recent advances in fire-retardant rigid polyurethane foam. *J. Mater. Sci. Technol.* **2022**, *112*, 315–328. [CrossRef]

9. Muhammed Raji, A.; Hambali, H.U.; Khan, Z.I.; Binti Mohamad, Z.; Azman, H.; Ogabi, R. Emerging trends in flame retardancy of rigid polyurethane foam and its composites: A review. *J. Cell. Plast.* **2023**, *59*, 65–122. [CrossRef]
10. Henry, C.; Gondaliya, A.; Thies, M.; Nejad, M. Studying the suitability of nineteen lignins as partial polyol replacement in rigid polyurethane/polyisocyanurate foam. *Molecules* **2022**, *27*, 2535. [CrossRef]
11. Yang, R.; Wang, B.; Li, M.; Zhang, X.; Li, J. Preparation, characterization and thermal degradation behavior of rigid polyurethane foam using a malic acid based polyols. *Ind. Crops Prod.* **2019**, *136*, 121–128. [CrossRef]
12. Srihanum, A.; Tuan Noor, M.T.; Devi, K.P.; Hoong, S.S.; Ain, N.H.; Mohd, N.S.; Nek Mat Din, N.S.M.; Kian, Y.S. Low density rigid polyurethane foam incorporated with renewable polyol as sustainable thermal insulation material. *J. Cell. Plast.* **2022**, *58*, 485–503. [CrossRef]
13. Bayer, O.; Siefken, W.; Rinke, H.; Orthner, L.; Schild, H. A process for the production of polyurethanes and polyureas. *Ger. Pat. DRP* **1937**, *728981*, 13.
14. Ma, C.; Qiu, S.; Xiao, Y.; Zhang, K.; Zheng, Y.; Xing, W.; Hu, Y. Fabrication of fire safe rigid polyurethane foam with reduced release of CO and NO_x and excellent physical properties by combining phosphine oxide-containing hyperbranched polyol and expandable graphite. *Chem. Eng. J.* **2022**, *431*, 133347. [CrossRef]
15. Xu, B.; Zhao, S.; Shan, H.; Qian, L.; Wang, J.; Xin, F. Effect of two boron compounds on smoke-suppression and flame-retardant properties for rigid polyurethane foams. *Polym. Int.* **2022**, *71*, 1210–1219. [CrossRef]
16. Günther, M.; Lorenzetti, A.; Schartel, B. From Cells to Residues: Flame-retarded rigid polyurethane foams. *Combust. Sci. Technol.* **2020**, *192*, 2209–2237. [CrossRef]
17. Günther, M.; Lorenzetti, A.; Schartel, B. Fire phenomena of rigid polyurethane foams. *Polymers* **2018**, *10*, 1166. [CrossRef]
18. Chan, Y.Y.; Schartel, B. It Takes Two to Tango: Synergistic Expandable Graphite-Phosphorus Flame Retardant Combinations in Polyurethane Foams. *Polymers* **2022**, *14*, 2562. [CrossRef]
19. Yang, Y.; Shen, H.; Luo, Y.; Zhang, R.; Sun, J.; Liu, X.; Zong, Z.; Tang, G. Rigid polyurethane foam composites based on iron tailing: Thermal stability, flame retardancy and fire toxicity. *Cell. Polym.* **2022**, *41*, 189–207. [CrossRef]
20. Jia, P.; Ma, C.; Lu, J.; Yang, W.; Jiang, X.; Jiang, G.; Yin, Z.; Qiu, Y.; Qian, L.; Yu, X. Design of copper salt@ graphene nanohybrids to accomplish excellent resilience and superior fire safety for flexible polyurethane foam. *J. Colloid Interface Sci.* **2022**, *606*, 1205–1218. [CrossRef] [PubMed]
21. Akar, A.; Değirmenci, B.; Köken, N. Fire-retardant and smoke-suppressant rigid polyurethane foam composites. *Pigm. Resin Technol.* **2023**, *52*, 237–245. [CrossRef]
22. Agrawal, A.; Kaur, R. Effect of nano filler on the flammability of bio-based RPUF. *Integr. Ferroelectr.* **2019**, *202*, 20–28. [CrossRef]
23. Son, M.-H.; Kim, Y.; Jo, Y.-H.; Kwon, M. Assessment of chemical asphyxia caused by toxic gases generated from rigid polyurethane foam (RPUF) fires. *Forensic Sci. Int.* **2021**, *328*, 111011. [CrossRef] [PubMed]
24. Wang, L.; Tawiah, B.; Shi, Y.; Cai, S.; Rao, X.; Liu, C.; Yang, Y.; Yang, F.; Yu, B.; Liang, Y. Highly effective flame-retardant rigid polyurethane foams: Fabrication and applications in inhibition of coal combustion. *Polymers* **2019**, *11*, 1776. [CrossRef] [PubMed]
25. Ching, Y.; Chuah, C.; Ching, K.; Abdullah, L.; Rahman, A. Applications of thermoplastic-based blends. In *Recent Developments in Polymer Macro, Micro and Nano Blends*; Woodhead Publishing: Cambridge, UK, 2017; pp. 111–129.
26. Lubczak, J.; Lubczak, R. Thermally resistant polyurethane foams with reduced flammability. *J. Cell. Plast.* **2018**, *54*, 561–576. [CrossRef]
27. Akdogan, E.; Erdem, M.; Ureyen, M.E.; Kaya, M. Rigid polyurethane foams with halogen-free flame retardants: Thermal insulation, mechanical, and flame retardant properties. *J. Appl. Polym. Sci.* **2020**, *137*, 47611. [CrossRef]
28. Çalışkan, E.; Çanak, T.Ç.; Karahasanoğlu, M.; Serhatlı, I.E. Synthesis and characterization of phosphorus-based flame retardant containing rigid polyurethane foam. *J. Therm. Anal. Calorim.* **2022**, *147*, 4119–4129. [CrossRef]
29. Tsuyumoto, I. Flame-retardant coatings for rigid polyurethane foam based on mixtures of polysaccharides and polyborate. *J. Coat. Technol. Res.* **2021**, *18*, 155–162. [CrossRef]
30. Chen, Y.; Li, L.; Qi, X.; Qian, L. The pyrolysis behaviors of phosphorus-containing organosilicon compound modified APP with different polyether segments and their flame retardant mechanism in polyurethane foam. *Compos. Part B Eng.* **2019**, *173*, 106784. [CrossRef]
31. Acosta, A.P.; Otoni, C.G.; Missio, A.L.; Amico, S.C.; Delucis, R.d.A. Rigid Polyurethane Biofoams Filled with Chemically Compatible Fruit Peels. *Polymers* **2022**, *14*, 4526. [CrossRef]
32. Wang, C.; Wu, Y.; Li, Y.; Shao, Q.; Yan, X.; Han, C.; Wang, Z.; Liu, Z.; Guo, Z. Flame-retardant rigid polyurethane foam with a phosphorus-nitrogen single intumescent flame retardant. *Polym. Adv. Technol.* **2018**, *29*, 668–676. [CrossRef]
33. Wang, J.; Xu, B.; Wang, X.; Liu, Y. A phosphorous-based bi-functional flame retardant for rigid polyurethane foam. *Polym. Degrad. Stab.* **2021**, *186*, 109516. [CrossRef]
34. Han, S.; Zhu, K.; Chen, F.; Chen, S.; Liu, H. Flame-retardant system for rigid polyurethane foams based on diethyl bis (2-hydroxyethyl) aminomethylphosphonate and in-situ exfoliated clay. *Polym. Degrad. Stab.* **2020**, *177*, 109178. [CrossRef]
35. Agrawal, A.; Kaur, R.; Singh Walia, R. Flame retardancy of ceramic-based rigid polyurethane foam composites. *J. Appl. Polym. Sci.* **2019**, *136*, 48250. [CrossRef]
36. Zhu, H.; Xu, S. Preparation of flame-retardant rigid polyurethane foams by combining modified melamine–formaldehyde resin and phosphorus flame retardants. *ACS Omega* **2020**, *5*, 9658–9667. [CrossRef]

37. Li, Y.; Tian, H.; Zhang, J.; Zou, W.; Wang, H.; Du, Z.; Zhang, C. Fabrication and properties of rigid polyurethane nanocomposite foams with functional isocyanate modified graphene oxide. *Polym. Compos.* **2020**, *41*, 5126–5134. [CrossRef]
38. Wang, X.; Kalali, E.N.; Xing, W.; Wang, D.-Y. CO_2 induced synthesis of Zn-Al layered double hydroxide nanostructures towards efficiently reducing fire hazards of polymeric materials. *Nano Adv.* **2018**, *3*, 12–17. [CrossRef]
39. Sivriev, H.; Borissov, G.; Zabski, L.; Walczyk, W.; Jedlinski, Z. Synthesis and studies of phosphorus-containing polyurethane foams based on tetrakis (hydroxymethyl) phosphonium chloride derivatives. *J. Appl. Polym. Sci.* **1982**, *27*, 4137–4147. [CrossRef]
40. Xu, W.; Wang, G. Synthesis of polyhydric alcohol/ethanol phosphate flame retardant and its application in PU rigid foams. *J. Appl. Polym. Sci.* **2015**, *132*, 42298. [CrossRef]
41. Wu, N.; Niu, F.; Lang, W.; Yu, J.; Fu, G. Synthesis of reactive phenylphosphoryl glycol ether oligomer and improved flame retardancy and mechanical property of modified rigid polyurethane foams. *Mater. Des.* **2019**, *181*, 107929. [CrossRef]
42. Zhang, X.G.; Ge, L.L.; Zhang, W.Q.; Tang, J.H.; Ye, L.; Li, Z.M. Expandable graphite-methyl methacrylate-acrylic acid copolymer composite particles as a flame retardant of rigid polyurethane foam. *J. Appl. Polym. Sci.* **2011**, *122*, 932–941. [CrossRef]
43. Liu, L.; Wang, Z.; Xu, X. Melamine amino trimethylene phosphate as a novel flame retardant for rigid polyurethane foams with improved flame retardant, mechanical and thermal properties. *J. Appl. Polym. Sci.* **2017**, *134*, 45234. [CrossRef]
44. Tang, G.; Jiang, H.; Yang, Y.; Chen, D.; Liu, C.; Zhang, P.; Zhou, L.; Huang, X.; Zhang, H.; Liu, X. Preparation of melamine-formaldehyde resin-microencapsulated ammonium polyphosphate and its application in flame retardant rigid polyurethane foam composites. *J. Polym. Res.* **2020**, *27*, 375. [CrossRef]
45. Chen, H.-B.; Shen, P.; Chen, M.-J.; Zhao, H.-B.; Schiraldi, D.A. Highly efficient flame retardant polyurethane foam with alginate/clay aerogel coating. *ACS Appl. Mater. Interfaces* **2016**, *8*, 32557–32564. [CrossRef]
46. Yang, R.; Wang, B.; Han, X.; Ma, B.; Li, J. Synthesis and characterization of flame retardant rigid polyurethane foam based on a reactive flame retardant containing phosphazene and cyclophosphonate. *Polym. Degrad. Stab.* **2017**, *144*, 62–69. [CrossRef]
47. Yang, R.; Hu, W.; Xu, L.; Song, Y.; Li, J. Synthesis, mechanical properties and fire behaviors of rigid polyurethane foam with a reactive flame retardant containing phosphazene and phosphate. *Polym. Degrad. Stab.* **2015**, *122*, 102–109. [CrossRef]
48. Qian, L.; Li, L.; Chen, Y.; Xu, B.; Qiu, Y. Quickly self-extinguishing flame retardant behavior of rigid polyurethane foams linked with phosphaphenanthrene groups. *Compos. Part B Eng.* **2019**, *175*, 107186. [CrossRef]
49. Bhoyate, S.; Ionescu, M.; Kahol, P.; Gupta, R.K. Sustainable flame-retardant polyurethanes using renewable resources. *Ind. Crops Prod.* **2018**, *123*, 480–488. [CrossRef]
50. Huang, X.; Wang, C.; Gao, J.; Zhou, Z.; Tang, G.; Wang, C. Research on two sides horizontal flame spread over rigid polyurethane with different flame retardants. *J. Therm. Anal. Calorim.* **2021**, *146*, 2141–2150. [CrossRef]
51. Zhang, Q.; Chen, F.; Ma, L.; Zhou, X. Preparation and application of phosphorous-containing bio-polyols in polyurethane foams. *J. Appl. Polym. Sci.* **2014**, *131*, 40422. [CrossRef]
52. Bhoyate, S.; Ionescu, M.; Kahol, P.; Chen, J.; Mishra, S.; Gupta, R.K. Highly flame-retardant polyurethane foam based on reactive phosphorus polyol and limonene-based polyol. *J. Appl. Polym. Sci.* **2018**, *135*, 46224. [CrossRef]
53. Zhang, L.; Zhang, M.; Zhou, Y.; Hu, L. The study of mechanical behavior and flame retardancy of castor oil phosphate-based rigid polyurethane foam composites containing expanded graphite and triethyl phosphate. *Polym. Degrad. Stab.* **2013**, *98*, 2784–2794. [CrossRef]
54. Acuña, P.; Zhang, J.; Yin, G.-Z.; Liu, X.-Q.; Wang, D.-Y. Bio-based rigid polyurethane foam from castor oil with excellent flame retardancy and high insulation capacity via cooperation with carbon-based materials. *J. Mater. Sci.* **2021**, *56*, 2684–2701. [CrossRef]
55. Zhou, W.; Hao, S.-J.; Feng, G.-D.; Jia, P.-Y.; Ren, X.-L.; Zhang, M.; Zhou, Y.-H. Properties of rigid polyurethane foam modified by tung oil-based polyol and flame-retardant particles. *Polymers* **2020**, *12*, 119. [CrossRef]
56. Luo, Y.; Miao, Z.; Sun, T.; Zou, H.; Liang, M.; Zhou, S.; Chen, Y. Preparation and mechanism study of intrinsic hard segment flame-retardant polyurethane foam. *J. Appl. Polym. Sci.* **2021**, *138*, 49920. [CrossRef]
57. Yuan, Y.; Yang, H.; Yu, B.; Shi, Y.; Wang, W.; Song, L.; Hu, Y.; Zhang, Y. Phosphorus and nitrogen-containing polyols: Synergistic effect on the thermal property and flame retardancy of rigid polyurethane foam composites. *Ind. Eng. Chem. Res.* **2016**, *55*, 10813–10822. [CrossRef]
58. Wang, S.-X.; Zhao, H.-B.; Rao, W.-H.; Huang, S.-C.; Wang, T.; Liao, W.; Wang, Y.-Z. Inherently flame-retardant rigid polyurethane foams with excellent thermal insulation and mechanical properties. *Polymer* **2018**, *153*, 616–625. [CrossRef]
59. Zhang, Z.; Li, D.; Xu, M.; Li, B. Synthesis of a novel phosphorus and nitrogen-containing flame retardant and its application in rigid polyurethane foam with expandable graphite. *Polym. Degrad. Stab.* **2020**, *173*, 109077. [CrossRef]
60. Vakili, M.; Nikje, M.M.A.; Hajibeygi, M. The effects of a phosphorus/nitrogen-containing diphenol on the flammability, thermal stability, and mechanical properties of rigid polyurethane foam. *Colloid Polym. Sci.* **2023**, 1–12. [CrossRef]
61. Zhou, W.; Jia, P.; Zhang, M.; Zhou, Y. Preparation and characterization of nitrogen-containing heterocyclic tung oil-based rigid polyurethane foam. *Chem. Ind. For. Prod.* **2019**, *39*, 53–58.
62. Yuan, Y.; Ma, C.; Shi, Y.; Song, L.; Hu, Y.; Hu, W. Highly-efficient reinforcement and flame retardancy of rigid polyurethane foam with phosphorus-containing additive and nitrogen-containing compound. *Mater. Chem. Phys.* **2018**, *211*, 42–53. [CrossRef]
63. Zhu, H.; Xu, S.-a. Synthesis and properties of rigid polyurethane foams synthesized from modified urea-formaldehyde resin. *Constr. Build. Mater.* **2019**, *202*, 718–726. [CrossRef]
64. Lewin, M.; Weil, E.D. Mechanisms and modes of action in flame retardancy of polymers. *Fire Retard. Mater.* **2001**, *1*, 31–68.

65. Levchik, S.V.; Weil, E.D. Overview of recent developments in the flame retardancy of polycarbonates. *Polym. Int.* **2005**, *54*, 981–998. [CrossRef]
66. Jia, D.; Yang, J.; He, J.; Li, X.; Yang, R. Melamine-based polyol containing phosphonate and alkynyl groups and its application in rigid polyurethane foam. *J. Mater. Sci.* **2021**, *56*, 870–885. [CrossRef]
67. Liu, Y.; He, J.; Yang, R. The synthesis of melamine-based polyether polyol and its effects on the flame retardancy and physical–mechanical property of rigid polyurethane foam. *J. Mater. Sci.* **2017**, *52*, 4700–4712. [CrossRef]
68. Zhang, M.; Zhang, J.; Chen, S.; Zhou, Y. Synthesis and fire properties of rigid polyurethane foams made from a polyol derived from melamine and cardanol. *Polym. Degrad. Stab.* **2014**, *110*, 27–34. [CrossRef]
69. Li, X.; Yu, Z.; Zhang, L. Synthesis of a green reactive flame-retardant polyether polyol and its application. *J. Appl. Polym. Sci.* **2021**, *138*, 50154. [CrossRef]
70. Gong, Q.; Qin, L.; Yang, L.; Liang, K.; Wang, N. Effect of flame retardants on mechanical and thermal properties of bio-based polyurethane rigid foams. *RSC Adv.* **2021**, *11*, 30860–30872. [CrossRef] [PubMed]
71. Tang, G.; Liu, M.; Deng, D.; Zhao, R.; Liu, X.; Yang, Y.; Yang, S.; Liu, X. Phosphorus-containing soybean oil-derived polyols for flame-retardant and smoke-suppressant rigid polyurethane foams. *Polym. Degrad. Stab.* **2021**, *191*, 109701. [CrossRef]
72. Yang, H.; Wang, X.; Song, L.; Yu, B.; Yuan, Y.; Hu, Y.; Yuen, R.K. Aluminum hypophosphite in combination with expandable graphite as a novel flame retardant system for rigid polyurethane foams. *Polym. Adv. Technol.* **2014**, *25*, 1034–1043. [CrossRef]
73. Chaudhary, B.; Barry, R.; Cheung, Y.; Ho, T.; Guest, M.; Stobby, W. Halogenated Fire-Retardant Compositions and Foams and Fabricated Articles Therefrom. U.S. Patent Application No. 09/728, 8 August 2002.
74. Chen, Y.; Bai, Z.; Xu, X.; Guo, J.; Chen, X.; Hsu, S.L.; Lu, Z.; Wu, H. Phosphonitrile decorating expandable graphite as a high-efficient flame retardant for rigid polyurethane foams. *Polymer* **2023**, *283*, 126268. [CrossRef]
75. Chen, Y.; Li, L.; Wu, X. Construction of an efficient ternary flame retardant system for rigid polyurethane foam based on bi-phase flame retardant effect. *Polym. Adv. Technol.* **2020**, *31*, 3202–3210. [CrossRef]
76. Yang, H.; Song, L.; Hu, Y.; Yuen, R.K. Diphase flame-retardant effect of ammonium polyphosphate and dimethyl methyl phosphonate on polyisocyanurate-polyurethane foam. *Polym. Adv. Technol.* **2018**, *29*, 2917–2925. [CrossRef]
77. Xu, J.; Wu, Y.; Zhang, B.; Zhang, G. Synthesis and synergistic flame-retardant effects of rigid polyurethane foams used reactive DOPO-based polyols combination with expandable graphite. *J. Appl. Polym. Sci.* **2021**, *138*, 50223. [CrossRef]
78. Zhang, M.; Luo, Z.; Zhang, J.; Chen, S.; Zhou, Y. Effects of a novel phosphorus–nitrogen flame retardant on rosin-based rigid polyurethane foams. *Polym. Degrad. Stab.* **2015**, *120*, 427–434. [CrossRef]
79. Ranaweera, C.; Ionescu, M.; Bilic, N.; Wan, X.; Kahol, P.; Gupta, R.K. Biobased polyols using thiol-ene chemistry for rigid polyurethane foams with enhanced flame-retardant properties. *J. Renew. Mater.* **2017**, *5*, 1. [CrossRef]
80. Wu, D.H.; Zhao, P.H.; Liu, Y.Q.; Liu, X.Y.; Wang, X.F. Halogen Free flame retardant rigid polyurethane foam with a novel phosphorus− nitrogen intumescent flame retardant. *J. Appl. Polym. Sci.* **2014**, *131*, 39581. [CrossRef]
81. Xu, Q.; Zhai, H.; Wang, G. Mechanism of smoke suppression by melamine in rigid polyurethane foam. *Fire Mater.* **2015**, *39*, 271–282. [CrossRef]
82. Chen, Y.; Luo, Y.; Guo, X.; Chen, L.; Xu, T.; Jia, D. Structure and flame-retardant actions of rigid polyurethane foams with expandable graphite. *Polymers* **2019**, *11*, 686. [CrossRef]
83. Chen, Y.; Bai, Z.; Xu, X.; Guo, X.; Guo, J.; Chen, X.; Lu, Z.; Wu, H. Fabrication of a novel P-N flame retardant and its synergistic flame retardancy with expandable graphite on rigid polyurethane foam. *J. Appl. Polym. Sci.* **2023**, *140*, e54013. [CrossRef]
84. Liu, M.; Shen, H.; Luo, Y.; Zhang, R.; Tao, Y.; Liu, X.; Zong, Z.; Tang, G. Rigid polyurethane foam compounds with excellent fire performance modified by a piperazine pyrophosphate/expandable graphite synergistic system. *Fire Mater.* **2023**, *47*, 925–937. [CrossRef]
85. Dai, C.; Gu, C.; Liu, B.; Lyu, Y.; Yao, X.; He, H.; Fang, J.; Zhao, G. Preparation of low-temperature expandable graphite as a novel steam plugging agent in heavy oil reservoirs. *J. Mol. Liq.* **2019**, *293*, 111535. [CrossRef]
86. Wang, W.; Wang, F.; Dong, Q.; Yuan, W.; Liu, P.; Ding, Y.; Zhang, S.; Yang, M.; Zheng, G. Expandable graphite encapsulated by magnesium hydroxide nanosheets as an intumescent flame retardant for rigid polyurethane foams. *J. Appl. Polym. Sci.* **2018**, *135*, 46749. [CrossRef]
87. Akdogan, E.; Erdem, M.; Ureyen, M.E.; Kaya, M. Synergistic effects of expandable graphite and ammonium pentaborate octahydrate on the flame-retardant, thermal insulation, and mechanical properties of rigid polyurethane foam. *Polym. Compos.* **2020**, *41*, 1749–1762. [CrossRef]
88. Yuan, Y.; Wang, W.; Shi, Y.; Song, L.; Ma, C.; Hu, Y. The influence of highly dispersed Cu_2O-anchored MoS_2 hybrids on reducing smoke toxicity and fire hazards for rigid polyurethane foam. *J. Hazard. Mater.* **2020**, *382*, 121028. [CrossRef] [PubMed]
89. Xu, W.; Wang, G.; Xu, J.; Liu, Y.; Chen, R.; Yan, H. Modification of diatomite with melamine coated zeolitic imidazolate framework-8 as an effective flame retardant to enhance flame retardancy and smoke suppression of rigid polyurethane foam. *J. Hazard. Mater.* **2019**, *379*, 120819. [CrossRef] [PubMed]
90. Wang, W.; Pan, H.; Shi, Y.; Yu, B.; Pan, Y.; Liew, K.M.; Song, L.; Hu, Y. Sandwichlike coating consisting of alternating montmorillonite and β-FeOOH for reducing the fire hazard of flexible polyurethane foam. *ACS Sustain. Chem. Eng.* **2015**, *3*, 3214–3223. [CrossRef]
91. Agrawal, A.; Kaur, R.; Walia, R.S. Investigation on flammability of rigid polyurethane foam-mineral fillers composite. *Fire Mater.* **2019**, *43*, 917–927. [CrossRef]

92. Darder, M.; Matos, C.R.S.; Aranda, P.; Gouveia, R.F.; Ruiz-Hitzky, E. Bionanocomposite foams based on the assembly of starch and alginate with sepiolite fibrous clay. *Carbohydr. Polym.* **2017**, *157*, 1933–1939. [CrossRef]
93. Alis, A.; Majid, R.A.; Mohamad, Z. Morphologies and Thermal Properties of Palm-oil Based Rigid Polyurethane/Halloysite Nanocomposite Foams. *CET J. Chem. Eng. Trans.* **2019**, *72*, 415–420.
94. Pang, X.Y.; Chang, R.; Weng, M.Q. Halogen-free flame retarded rigid polyurethane foam: The influence of titanium dioxide modified expandable graphite and ammonium polyphosphate on flame retardancy and thermal stability. *Polym. Eng. Sci.* **2018**, *58*, 2008–2018. [CrossRef]
95. Salasinska, K.; Borucka, M.; Leszczyńska, M.; Zatorski, W.; Celiński, M.; Gajek, A.; Ryszkowska, J. Analysis of flammability and smoke emission of rigid polyurethane foams modified with nanoparticles and halogen-free fire retardants. *J. Therm. Anal. Calorim.* **2017**, *130*, 131–141. [CrossRef]
96. Bian, X.C.; Tang, J.H.; Li, Z.M. Flame retardancy of whisker silicon oxide/rigid polyurethane foam composites with expandable graphite. *J. Appl. Polym. Sci.* **2008**, *110*, 3871–3879. [CrossRef]
97. Yuan, Y.; Yu, B.; Shi, Y.; Ma, C.; Song, L.; Hu, W.; Hu, Y. Highly efficient catalysts for reducing toxic gases generation change with temperature of rigid polyurethane foam nanocomposites: A comparative investigation. *Compos. Part A Appl. Sci. Manuf.* **2018**, *112*, 142–154. [CrossRef]
98. Yuan, Y.; Wang, W.; Xiao, Y.; Yuen, A.C.Y.; Mao, L.; Pan, H.; Yu, B.; Hu, Y. Surface modification of multi-scale cuprous oxide with tunable catalytic activity towards toxic fumes and smoke suppression of rigid polyurethane foam. *Appl. Surf. Sci.* **2021**, *556*, 149792. [CrossRef]
99. Cheng, J.; Niu, S.; Kang, M.; Liu, Y.; Zhang, F.; Qu, W.; Guan, Y.; Li, S. The thermal behavior and flame retardant performance of phase change material microcapsules with modified carbon nanotubes. *Energy* **2022**, *240*, 122821. [CrossRef]
100. Jiang, Y.; Yan, P.; Wang, Y.; Zhou, C.; Lei, J. Form-stable phase change materials with enhanced thermal stability and fire resistance via the incorporation of phosphorus and silicon. *Mater. Des.* **2018**, *160*, 763–771. [CrossRef]
101. Niu, S.; Cheng, J.; Zhao, Y.; Kang, M.; Liu, Y. Preparation and characterization of multifunctional phase change material microcapsules with modified carbon nanotubes for improving the thermal comfort level of buildings. *Constr. Build. Mater.* **2022**, *347*, 128628. [CrossRef]
102. Hou, L.; Li, H.; Liu, Y.; Niu, K.; Shi, Z.; Liang, L.; Yao, Z.; Liu, C.; Tian, D. Synergistic effect of silica aerogels and hollow glass microspheres on microstructure and thermal properties of rigid polyurethane foam. *J. Non Cryst. Solids* **2022**, *592*, 121753. [CrossRef]
103. Huang, Y.; Zhou, J.; Sun, P.; Zhang, L.; Qian, X.; Jiang, S.; Shi, C. Green, tough and highly efficient flame-retardant rigid polyurethane foam enabled by double network hydrogel coatings. *Soft Matter* **2021**, *17*, 10555–10565. [CrossRef] [PubMed]
104. Wang, S.; Wang, X.; Wang, X.; Li, H.; Sun, J.; Sun, W.; Yao, Y.; Gu, X.; Zhang, S. Surface coated rigid polyurethane foam with durable flame retardancy and improved mechanical property. *Chem. Eng. J.* **2020**, *385*, 123755. [CrossRef]
105. Huang, Y.; Jiang, S.; Liang, R.; Sun, P.; Hai, Y.; Zhang, L. Thermal-triggered insulating fireproof layers: A novel fire-extinguishing MXene composites coating. *Chem. Eng. J.* **2020**, *391*, 123621. [CrossRef]
106. Zhang, C.; Xie, H.; Du, Y.; Li, X.; Zhou, W.; Wu, T.; Qu, J. Shelter Forest Inspired Superhydrophobic Flame-Retardant Composite with Root-Soil Interlocked Micro/Nanostructure Enhanced Mechanical, Physical, and Chemical Durability. *Adv. Funct. Mater.* **2023**, *33*, 2213398. [CrossRef]
107. Chen, Y.; Wang, W.; Qiu, Y.; Li, L.; Qian, L.; Xin, F. Terminal group effects of phosphazene-triazine bi-group flame retardant additives in flame retardant polylactic acid composites. *Polym. Degrad. Stab.* **2017**, *140*, 166–175. [CrossRef]

Disclaimer/Publisher's Note: The statements, opinions and data contained in all publications are solely those of the individual author(s) and contributor(s) and not of MDPI and/or the editor(s). MDPI and/or the editor(s) disclaim responsibility for any injury to people or property resulting from any ideas, methods, instructions or products referred to in the content.

Review

Multifunctional Textiles with Flame Retardant and Antibacterial Properties: A Review

Liping Jin, Chenpeng Ji, Shun Chen, Zhicong Song, Juntong Zhou, Kun Qian and Wenwen Guo *

Key Laboratory of Eco-Textiles, Ministry of Education, College of Textile Science and Engineering, Jiangnan University, 1800 Lihu Avenue, Wuxi 214122, China; 6223016009@stu.jiangnan.edu.cn (L.J.); 6223017060@stu.jiangnan.edu.cn (C.J.); 6213011001@stu.jiangnan.edu.cn (S.C.); 1091210212@stu.jiangnan.edu.cn (Z.S.); 1091210214@stu.jiangnan.edu.cn (J.Z.); qiankun_8@163.com (K.Q.)
* Correspondence: guoww@jiangnan.edu.cn

Abstract: It is well known that bacterial infections and fire-hazards are potentially injurious in daily life. With the increased security awareness of life and properties as well as the improvement of living standards, there has been an increasing demand for multifunctional textiles with flame retardant and antibacterial properties, especially in the fields of home furnishing and medical protection. So far, various treatment methods, including the spray method, the dip-coating method, and the pad-dry-cure method, have been used to apply functional finishing agents onto fabrics to achieve the functionalization in the past exploration stage. Moreover, in addition to the traditional finishing technology, a number of novel technologies have emerged, such as layer-by-layer (LBL) deposition, the sol-gel process, and chemical grafting modification. In addition, some natural biomasses, including chitin, chitosan (CS), and several synthetic functional compounds that possess both flame-retardant and bacteriostatic properties, have also received extensive attention. Hence, this review focuses on introducing some commonly used finishing technologies and flame retardant/antibacterial agents. At the same time, the advantages and disadvantages of different methods and materials were summarized, which will contribute to future research and promote the development and progress of the industry.

Citation: Jin, L.; Ji, C.; Chen, S.; Song, Z.; Zhou, J.; Qian, K.; Guo, W. Multifunctional Textiles with Flame Retardant and Antibacterial Properties: A Review. *Molecules* 2023, 28, 6628. https://doi.org/10.3390/molecules28186628

Academic Editors: Gaëlle Fontaine and Baljinder Kandola

Received: 31 May 2023
Revised: 7 August 2023
Accepted: 18 August 2023
Published: 14 September 2023

Copyright: © 2023 by the authors. Licensee MDPI, Basel, Switzerland. This article is an open access article distributed under the terms and conditions of the Creative Commons Attribution (CC BY) license (https://creativecommons.org/licenses/by/4.0/).

Keywords: flame retardant; antibacterial; functional textiles; finishing techniques

1. Introduction

Textiles, as a common product made of fiber materials, have been extensively utilized in all aspects of our daily lives and industries. However, due to the general tendency of textile fiber materials to burn and cause fires, it can result in damage to property and even human life. Therefore, the application expansion of textiles is highly limited. For decades, tremendous efforts have been made to enhance the flame retardancy of textiles by incorporating flame retardants into the fiber matrix or directly modifying the surface of textiles.

Surface modification is the most commonly used technique thanks to its simplicity and ease of operation for both synthetic and natural fabrics [1]. A wide variety of flame retardants are used to endow textiles with good flame retardancy, mainly involving inorganic or organic flame retardants. The common inorganic flame retardants, mainly inorganic phosphorus-containing, boron-containing, zinc-containing, iron-containing, and carbon-based materials, and the frequently used organic flame-retardants such as halogenated, phosphorus-containing, nitrogen-containing, and silicone-containing flame retardants [2–4]. As early-stage commercial flame-retardants, halogenated compounds perform outstanding functions by releasing halogen radicals to eliminate reactive radicals during combustion [2]. Unfortunately, it has been abandoned in its actual application since its severe toxicity to the environment and human safety [5]. In response to this challenge, halogen-free flame retardants have emerged, and phosphorus-containing flame retardants stand out. The high flame retardancy is due to the promotion of the substrate to form a char layer, which isolates the transfer of heat and combustible gases, thus preventing further combustion of the substrate [2,3]. In addition, the silicone-containing flame retardants usually form a vitrified layer on the polymer surface during combustion, which effectively hinders the transfer of oxygen, heat, and mass and reduces the flammability of the polymer [6,7]. In contrast, nitrogen-containing flame retardants release noncombustible gases to dilute combustible gases such as oxygen during combustion, and the flame retardant effect is relatively poor but friendly to the environment [8]. To effectively improve the flame retardant efficiency, the combination of different elements with flame retardant properties can provide synergistic flame retardant effects and impart additional thermal stability and mechanical properties to the composites. In general, the study of synergistic flame retardancy has gradually become the emphasis of flame retardancy research in recent years [9].

Since the outbreak of COVID-19 in 2019, its prevalence has had a terribly negative impact on human health and caused enormous panic in the public. Against this background, the demand for medical protective equipment and anti-bacterial textiles is on the rise. As we all know, textiles have been protecting humans from external environmental harm for a long time. However, some textiles can also serve as breeding grounds or carriers for bacteria and viruses due to their hygroscopicity [10], which may lead to the risk of inflammation, disease, and even death in the human body. Therefore, the antibacterial treatment of some fabrics has become extremely urgent and has received widespread attention from scholars. Notably, the surface treatment of fabrics with antibacterial agents is the most commonly used strategy, which is similar to the flame retardant surface treatment of fabrics. In addition, antibacterial agents are mainly divided into three categories: inorganic antibacterial agents (metal and oxide nanoparticles (NPs), carbon-based antibacterial materials along with their composites), organic antibacterial agents (quaternary ammonium salts, guanidine, halogenated amines, phenols, etc.), and natural antibacterial agents (chitin, CS), and then the antibacterial or bactericidal effect of the antibacterial substances is mainly exerted by either directly contacting the bacterial surface or releasing the antibacterial moiety onto the substrate [11,12]. With the emergence of inorganic and organic antibacterial agents, the industrial application of antibacterial products continues to deepen, and they are playing an irreplaceable role in the functional textile industry.

With the improvement of living conditions and the development of science and technology, the demand for multifunctional textiles in the market has grown in recent years. In particular, there is a large demand for flame retardant and antibacterial textiles in areas such as household products and medical protection. In order to develop novel functional fabrics with both flame-retardant and antibacterial properties, the relevant specific functional reagents and treatment methods have attracted considerable attention. Flame retardant and antibacterial dual-functional fabrics are usually achieved via a two-step or one-step method. The two-step method generally achieves the superposition of dual functions by introducing flame retardant and bacterial inhibitors in steps, which has the advantages of simplicity and wide applicability. However, there is a problem that some flame retardants or antibacterial agents may interact with each other, thus causing functional antagonism that is not conducive to simultaneously imparting excellent flame retardant or antibacterial properties to a fabric when the functional coating is applied in a stack. One-step methods normally treat fabrics by using synthetic agents with both flame retardant and antibacterial functions. However, the design of such multifunctional compounds is often difficult, and the synthesis process is complex. Therefore, it is a tremendous challenge to find the best processes and functional materials in the current direction of multifunctional textile research.

It is worth nothing that the specific finishing techniques have a pivotal influence on the processing efficiency and overall performance of the fabric. Then the traditional after-finishing technology mainly involves impregnation, padding, spraying, and pad-dry-cure techniques. These methods often do not require additional physical and chemical reactions but instead directly use functional solutions to treat the surface of fabrics, which have the advantages of being simple, effective, and easy to operate, but have the drawback of poor durability. In recent years, some promising environmentally friendly strategies have gradually attracted attention, such as LBL deposition, which usually uses deionized water as the solvent and positive and negative electrolytes as functional treatment agents [13]. Similarly, the sol-gel process is another environmentally friendly strategy that is favorable for the construction of functional surfaces for textile fibers by depositing thin organic-inorganic hybrid sol-gel films. In addition, this method has been selected as a simple and effective method to form a multifunctional protective coating since two or more siloxane precursors with different organic functions could be applied simultaneously [14,15]. Compared with the aforementioned techniques, chemical grafting modification exhibits unprecedented durability due to the stronger chemical bond linkage between functional agents and substrates [16]. Specifically, flame-retardant and antibacterial coatings are composed of inorganic nanomaterials, metal ions, or metal oxides. In situ modification technology has been extensively used for the construction of these coatings.

There is no systematic summary of research on flame-retardant antibacterial fabrics, especially the accompanying finishing techniques. This paper reviews the latest research results about the flame-retardant and antibacterial functional finishing of textiles over the past decade. Furthermore, it explores the nascent finishing agents as well as the adaptable post-finishing technique. Simultaneously, this review also discusses the advantages, disadvantages, and application scope of these techniques and briefly introduces the development of green environmental technologies.

2. Methods and Standards for Flame Retardant and Antibacterial Tests

2.1. Methods and Standards for Flame Retardant Tests

2.1.1. Cone Calorimetry

According to ASTM Standard E1354, the flame retardancy of the material was evaluated by the heat value of combustion released by subjecting a 10×10 cm^2 sample to a radiant heat stream ($\leq$100 kWm^{-2}) in the presence of an ignition source [17].

This paper mainly provides information such as the peak heat release rate (HRR), total heat release (THR), and peak heat release rate (PHRR) [18]. In general, the lower the value, the better the flame-retardant effect.

2.1.2. Limiting Oxygen Index (LOI)

According to ISO Standard 4589 and GB/T Standard 5454-1997, the lowest oxygen concentration to support the combustion of the material (limiting oxygen index) was obtained by igniting the tip of a vertically fixed sample ($\sim 7 \times 15$ cm^2), accompanied by a continuously decreasing oxygen concentration until the flame was extinguished.

A high oxygen index indicates that the material is not easily combustible, while a low oxygen index indicates that the material is easily combustible. Flammable Materials: <22%; Combustible Materials: 22~27%; Refractory Materials: >27%.

2.1.3. UL-94/Vertical Burning Test

According to the UL-94V standard, ASTM standard D3801, GB/T standard 17591-2006, and GB/T standard 5455-2014, the sample size was about 300×70 mm, which was exposed to a vertical flame for 12 s once or twice. Time to ignition, afterflame, afterglow, calculated residual mass, etc. were obtained.

There are two hierarchies that can be used for assessment: Firstly, V 0 (best grade), V-1 or V-2, wherein, V-2: after two 10 s combustion tests on the sample, the flame is extinguished within 60 s and combustibles are allowed to fall off; V-1: after two 10 s combustion tests on the sample, the flame is extinguished within 60 s and no combustibles can fall off; V-0: after two 10 s combustion tests on the sample, the flame is extinguished within 30 s and no combustibles can fall off [19]. Secondly, B1, B2, wherein, B1: time to ignition $\leq$ 5 s, time to deflagration $\leq$ 5 s, length of damage $\leq$ 150 mm; B2: time to ignition, time to deflagration $\leq$ 15 s, length of damage $\leq$ 200 mm.

2.2. Methods and Standards for Antibacterial Tests

2.2.1. The Shake Flask Method

According to GB/T standard 20944.3-2008, the rate of bacterial inhibition was obtained by determining the concentration of surviving bacteria in the specimen and control samples, respectively, after a period of shaking.

A sample has an antimicrobial effect if it inhibits *Staphylococcus aureus* and *Escherichia coli* by $\geq 70\%$.

2.2.2. The Inhibition Zone Method

According to SNV standard 195920–1992, the fabric with a diameter of 2 cm was placed on the agar medium and incubated at 37 °C for 24 h. The clear media around the disc indicates that the bacterial growth around the samples was inhibited, and the diameter of the area was measured to assess the antimicrobial activity of the samples.

In general, if the diameter of the suppression ring is less than or equal to 7 mm, it is judged as having no suppression; if the diameter of the suppression ring is greater than 7 mm, it is judged as having suppression.

2.2.3. AATCC Test Method

According to ASTM standard 147-2004, successive parallel lines were drawn on the agar medium inoculated with bacterial solution, and the sample was uniformly attached to the parallel stripes. After incubation for a period of time, clean areas will appear on the agar surface at the delineated areas due to the interruption of bacterial reproduction. The antibacterial activity against bacteria was examined by measuring the average clear inhibition zone using the relevant equation.

The size of the zone of inhibition cannot be used as a quantitative assessment of antimicrobial activity. However, antimicrobial-treated materials are comprehensively evaluated by comparing them to untreated materials and to sample materials with known inhibitory activity and including observations of the zone of inhibition.

3. Recent Advances in Multifunctional Textiles with Flame Retardant and Antibacterial Properties

The functional finishing of traditional textiles is mainly divided into two strategies. One is to mix the functional agents with textile raw materials and prepare functional textiles after the spinning process. This method has the advantage of good washing durability, but the cost is relatively high and usually has a significant influence on the mechanical properties of textile fibers. The other method is to perform surface modification treatments on textiles. Currently, surface modification technology is commonly used for the fabrication of functional fabrics such as flame retardant, antibacterial, and hydrophobic fabrics, which is a simple, convenient, and efficient way to endow traditional fibers/fabrics with specific functions. In the process of functional finishing, such as flame retardancy and bacteriostasis, the finishing techniques have a great influence on the final performance of the fabrics. Generally speaking, the finishing technology for fibers/fabrics mainly includes traditional finishing methods such as dip-coating [20] and spraying techniques [15] and some novel finishing strategies, including chemical grafting modification [21,22], layer-by-layer self-assembly [23,24], sol gel [25,26], and in situ deposition [27].

Ulteriorly, the traditional finishing technology of fabrics mainly includes the traditional impregnation method, pad-dry-cure method, coating method, and spray method. These methods have low finishing costs but unsatisfactory washability, and long-term use of flame-retardant effects will be affected by water, light, and other conditions. Presently, various surface-modifying technologies, such as the sol-gel method, nanoparticle adsorption, layer-by-layer self-assembly method, plasma treatment, and the graft copolymerization modification method, have been utilized for preparing flame-retardant, anti-bacterial, hydrophobic, UV-resistant, self-cleaning, multi-functional textiles on the basis of synergistic flame-retardant technology.

3.1. The Traditional Finishing Techniques

3.1.1. The Spray Method

The spraying method is one of the traditional flame-retardant finishing technologies. The finishing agents were dissolved into a certain solvent and then introduced onto the surface of fabric by simple spraying, which easily forms a thin functional coating on fabric surfaces [28]. For instance, Attia et al. developed the novel nanocomposites (DPHM-Ag NP) based on diphosphate malonate as organic phosphates and silver NPs, and then the nanocomposites coatings were sprayed on the surface of polyester (PS) and cotton-polyester (CB) blend fabrics. The treated fabric meets the standards of high-class flame-retardant textiles with a 0 mm/min rate of burning. Furthermore, the antibacterial properties were enhanced with the clear bacterial inhibition zone reaching 4.48 mm for *Staphylococcus aureus* (*S. aureus*) [29].

3.1.2. The Dip-Coating Method

Dip-coating is a finishing method that involves immersing fabrics in a functional agent solution and is accompanied by a drying treatment [30]. This method is simple to operate; the crosslinking reaction between the fabric and the flame retardant is weak during the process of finishing, and most of the flame retardant is just attached to the surface of the fabric, so the durability of the flame-retardant fabric is generally poor. To achieve the flame-retardant antibacterial properties of textiles, a novel agent (tetramethylcyclosiloxyl-piperazin) tetra guanidine (GNCTSi) was designed and successfully applied to form a functional coating on the surface of cotton fabrics. The treated cotton fabrics have enhanced properties, with LOI reaching 30.1% and char length remaining at 6.5 cm after burning. To a certain extent, the coating improves washing durability and thus has less impact on the flame retardancy of cotton fabrics. Furthermore, it also exhibits improved antibacterial effects with inhibition zones of 2.5 mm and 2.3 mm against *Escherichia coli* (*E. coli*) and *S. aureus*, respectively [31]. In addition, Atousa et al. prepared a suspension with ZrO_2 NPs along with cetyltrimethylammonium bromide (CTAB), maleic acid (MA), sodium hypophosphite (SHP), and urea by using an impregnation bath. MA was used as a cross-linking agent, while SHP acted as the catalyst to stabilize NPs on the fabric surface and prevent fabric from creasing. In the end, the test results showed improved flame-retardant properties, antibacterial activities, and self-cleaning properties of the treated cotton fabrics [32]. In order to enhance the coating's durability, some polymeric coatings, such as polyvinyl alcohol (PVA) and polyurethane (PU), were used as binder for flame retardant components. In the work of Ghada et al., nano chitosan (n CS), melamine phosphate (MP), and melamine salt of CS phosphate (MCSP) were prepared and then mixed with PVA to construct PVA/MCSP and PVA/n CS/MP coatings on cotton fabrics by the dip-coating method (as shown in Figure 1). The PVA/MCSP30 coating displayed the optimum flame resistance with self-extinguished behavior and a very high LOI of 58.2%, while the LOI for the original fabric was only 17.2%. In addition, it also exhibited good coating durability as well as better antibacterial properties for both *E. coli* (inhibition zone diameter of 27.6 mm) and *S. aureus* (inhibition zone diameter of 30.5 mm) [33]. Similarly, the condensed tannin extract from Dioscorea cirrhosa tubers was also used as the functional agent for silk fabric, and the treated fabric has good antibacterial properties and flame retardancy [34].

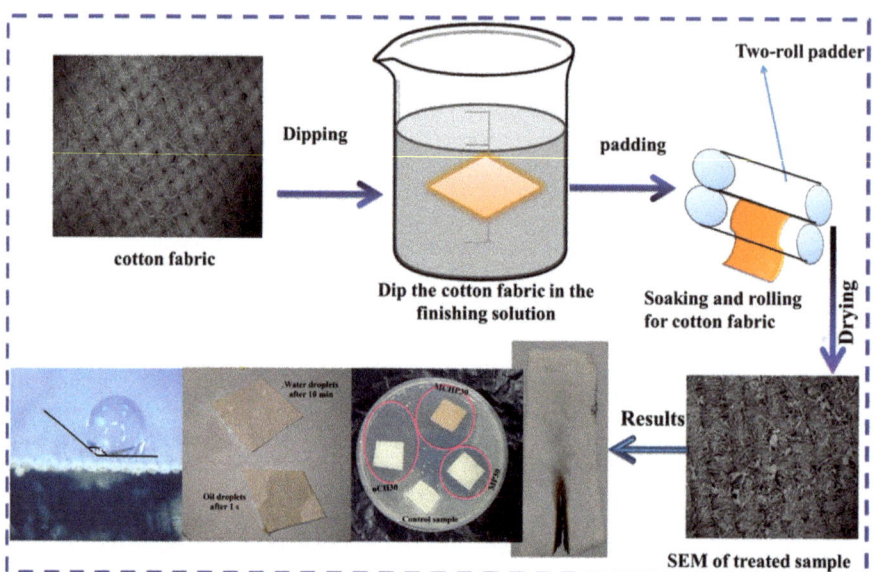

Figure 1. A schematic diagram for preparing cotton fabrics coated with CS-based flame retardant coatings [33] (Copyright 2022, Elsevier).

3.1.3. The Pad-Dry-Cure Method

The pad-dry-cure method treats the fabric by padding, drying, and curing it after it has absorbed enough functional solution [35]. In these processes, high-temperature drying makes flame-retardant or antibacterial functional materials chemically cross-linked with fabrics to obtain durable flame retardancy. Ahmed et al. first fabricated blended fabrics with different compositions, weaving structures, grams per square meter, thicknesses, and thread densities. In the following step, a three-dimensional tetrakis (hydroxymethyl) phosphonium chloride (THPC)-urea polymer coating was synthesized and deposited onto the blended fabrics. The finished fabric showed excellent antibacterial properties (99.9%) and excellent flame retardancy (LOI~36.8%); additionally, it also possessed superior water repellence properties (151.5°) [36]. Singh et al. prepared a thyme oil-embedded functional microcapsule via in situ synthesis of CS phosphate as the shell material, and then the microcapsules were introduced onto linen fabrics via the pad-dry method. The finished fabric presented excellent antibacterial properties (>98%), flame retardancy (LOI > 28), antioxidant activity (96%), mosquito repellency (100%), and an excellent fragrance. Moreover, the functional properties were durable after at least 20 washes [37].

The pad-dry-cure method was also applied to deposit nanocomposite coatings on fabrics. The TiO_2 NPs were prepared by the sol-gel method using titanium tetraisopropoxide. The development of nano TiO_2 onto cotton fabric was accomplished when nano TiO_2 was coated onto cotton fabric by the traditional pad-dry-cure method in the presence of polycarboxylic acid [1,2,3,4-butane tetracarboxylic acid], SHP, and CS phosphate. The results confirmed that 1,2,3,4-butane tetracarboxylic acid, TiO_2, and CS phosphate are helpful in increasing the flame resistance and antibacterial properties of cotton fabrics [38]. Similarly, Dhineshbabu et al. prepared colloidal methyl silicate and MgO nanoparticle-embedded methyl silicate solutions through the sol-gel method. Subsequently, cotton fabrics were separately modified with silica and MgO/methyl silicate composites via an optimized pad-dry-cure method. The MgO/methyl silicate composite-coated fabrics showed enhanced burning performance, significant water-repellent properties, and better antibacterial activity against *S. aureus* and *E. coli* than methyl silicate-coated and uncoated fabrics [39]. In another work, an equimolar sol mixture of the precursors P, P-diphenyl-N-(3-(trimethoxysilyl)propyl) phosphinic amide (SiP) and 1H,1H,2H,2H-perfluorooctyltriethoxysilane (SiF) was employed to form a two-component sol-gel inorganic–organic hybrid coating on the fabric by the pad-dry-cure method. It leads to good flame retardancy and antibacterial properties, with an inhibition rate of 92.9% against E. coli and 80.4% against S. aureus [40].

3.2. LBL Self-Assembly Technology

Whether in textile flame retardant finishing or other fields, LBL self-assembly technology has been widely used because of its simple operation, ease of control, and environmental friendliness [41]. It is a simple and versatile technology for preparing multifunctional coatings. These coatings are formed by repeatedly depositing alternate layers of oppositely charged materials, which experience attraction and undergo self-regulation within individual layers by electrostatic action. LBL self-assembly can be uniformly coated on the surface of textiles to achieve flame retardant, antibacterial, and other multifunctional properties, which have been applied to PS, cotton, polyamide, and silk fabrics [42]. At present, the commonly used positively charged compounds mainly include CS, polyetherimide (PEI), DL-arginine (DL), etc. The negatively charged electrolytes mainly include phytic acid (PA), sodium alginate (SA), ammonium polyphosphate (APP), and so on. Except for flame retardancy, some of these charged materials can also endow base materials with antibacterial properties, while others need to be combined with antibacterial materials to achieve multi-functionalization.

Herein, the bio-based materials PA and CS are often selected to impart flame retardant as well as antibacterial properties to fabrics through LBL technology. PA and its salts, such as ammonium phytate (AP), are generally composed of a large number of phosphoric acid groups and ammonium ions. They can catalyze the degradation of fabric substrate to form more char residues and release incombustible gases such as NH_3 or N_2 to dilute combustible gases during combustion. Therefore, the flame-retardant properties of fabrics in both condensed and gaseous phases are ultimately improved [1,43]. CS contains NH_2 groups that can generate -NH^{3+} groups under acidic conditions, so it applies as a positive charge in LBL self-assembly. Moreover, the -NH^{3+} group can destroy the cell wall or interfere with the normal physiological activities of cells, thus exerting both flame retardant and bacteriostatic effects [44]. Liu et al. deposited CS and AP on cotton fabric to manufacture fully bio-based flame-retardant and antibacterial cotton fabrics using LBL technology. The modified textile with a low weight gain (8 wt%) performed perfect self-extinguishing behavior in the vertical burning test. Moreover, the CS/AP coating has an effective antimicrobial rate of 99.83% against E. coli, which mainly depends on the introduction of CS [44]. In their other work, a fully bio-based CS/AP coating was also applied to a viscose fabric (as shown in Figure 2B,C)). The 2BL/Viscose showed sharp improvements in thermal stability in the higher temperature zone, accompanied by a LOI of 29% and self-extinguishing behavior in the combustion test. Additionally, 2BL/Viscose possessed a high bacteriostatic function of 99.99% for both E. coli and S. aureus [45]. However, the antibacterial ability of the system using PA/CS alone is limited and generally requires synergistic use with other antibacterial agents. In order to enhance the antimicrobial ability and improve the broad-spectrum antibacterial effect, it is necessary to introduce other antibacterial agents into the CS/AP coatings. For instance, Eva et al. soaked the LBL-treated samples in a 2% Cu^{2+} solution after depositing PA and CS-urea on cotton. The VFT results showed that 12 BL of PA/CS-urea-Cu^{2+} could stop the burning flame, and the PHRR and THR were reduced by 61% and 54%, respectively. Moreover, for the antibacterial test, 100% reduction of S. aureus and Klebsiella pneumoniae can be achieved (as shown in Figure 2A) [13].

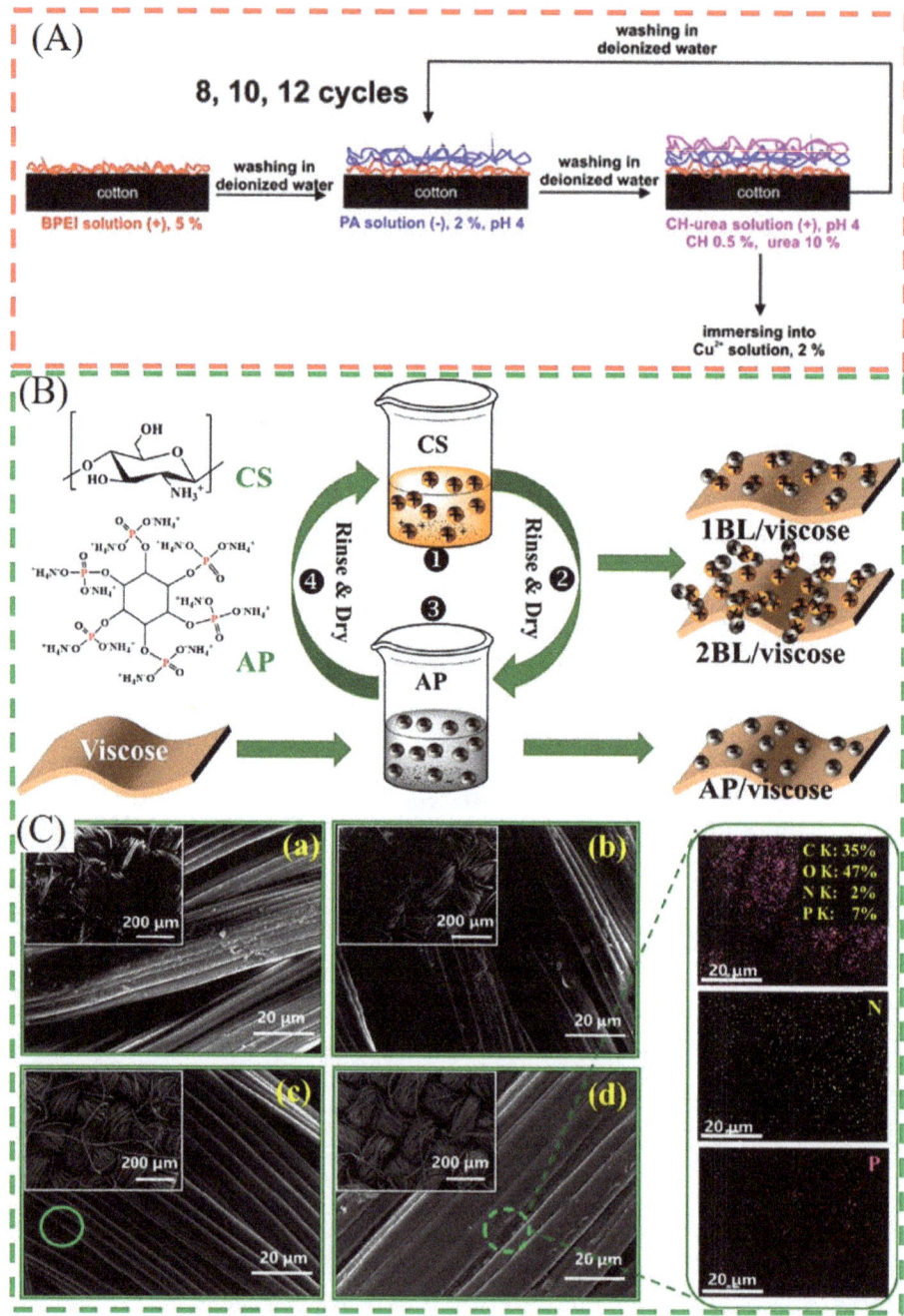

Figure 2. (**A**) The LBL deposition process of FR/antimicrobial nanocoating on cotton [13] (Copyright 2020, Elsevier). (**B**) The preparation process of 1BL/viscose, 2BL/viscose, and AP/viscose. (**C**) SEM pictures (×200 and ×4000) of the pure and coated viscose fabrics: (**a**) viscose, (**b**) AP/viscose, (**c**) 1BL/viscose, (**d**) 2BL/viscose, and EDS mappings of 2BL/viscose [45] (Copyright 2021, Elsevier).

Arginine, a kind of novel renewable material, has an alkaline cationic chemical group that possesses high antibacterial properties [46]. In addition, non-combustible gases are released during combustion, acting as a gas-phase flame retardant [47]. Recently, many researchers have proven that the combination of arginine and CS can have a good inhibitory effect on cell growth and hence exert an antibacterial function. Moreover, arginine-functionalized CS also exhibited good flame retardancy and smoke suppression [48–50]. Except for CS, PA can also chelate with the ammonium ions of DL through hydrogen bonds, and they were deposited on CB fabric using LBL assembly. The finishing CT fabric with 20 bilayers showed enhanced fire safety and an efficient inhibition diameter of 4.0 mm against *S. aureus* compared to the untreated CT fabric (0 mm) [51]. In addition, bio-based riboflavin sodium phosphate (vitamin B2, VB2) is an anionic phosphorus-containing compound that has been extensively used in disease treatment. The negatively charged VB2 can cooperate with CS to be used as the LBL assembly agents for the modification of silk fabric, and the prepared colored silk fabric with 10 assembly bilayers exhibited great flame retardancy and antibacterial performance (>90% inhibition rate against both *S. aureus* and *E. coli*). The employment of LBL technology to prepare fully bio-based flame-retardant and antibacterial coatings follows the concept of environmentally friendly development and has been widely used for the functional modification of fabrics [52].

Among numerous antibacterial agents, N-halamines have attracted extensive attention due to their broad-spectrum antibacterial activity, long-term efficacy, and renewability [53,54]. With the great demand for multifunctional materials, a novel nitrogen/silicon-containing N-halamine cationic polymer (PCQS) containing both flame-retardant and antibacterial components has been designed and used as the positive electrolyte, which then interacts with negatively charged PA molecules to coat cotton fabric. The cotton-PEI/(PCQS/PA)$_{30}$-Cl exhibited an increased LOI of 28.5% with a lower char length of 7.9 cm in the vertical flammability test. In addition, the treated fabric could inactivate 6.01 logs of *S. aureus* and 6.00 logs of *E. coli* within 1 min of contact time, demonstrating effective antimicrobial activity [55].

3.3. Chemical Grafting Modification

Chemical grafting modification is a technique that introduces functional groups to fibers or fabrics by forming covalent bonds. Flame retardant and antibacterial functionalization can be realized via the chemical grafting technique, which generally shows higher durability than other finishing methods [56]. As a typical scale inhibitor, diethylene triamine penta methylene phosphonic acid (DTPMPA) is rich in N and P elements and has become a potential flame retardant. In addition, the phosphonic acid groups in the structure of DTPMPA provide sufficient binding sites that can easily chelate with metal ions, such as silver ions. The lyocell fabric with flame-retardant properties (FR-lyocell) has been accomplished by grafting a novel flame-retardant ammonium salt of diethylene triamine penta methylene phosphonic acid (ADTPMPA). Subsequently, the FR-lyocell was further treated with Ag nanoparticles (Ag NPs) to develop antibacterial properties. The flame retardant and antibacterial lyocell fabric (FRAg-lyocell) exhibits unbelievable flame retardancy (LOI of 44.8%) and good washing durability (LOI has still been maintained at 31.3% after nearly 20 laundering cycles). Moreover, pHRR and THR values were suppressed effectively. At the same time, it also possesses excellent antimicrobial ability against both *S. aureus* and *E. coli* [56]. Xu et al. synthesized a water-soluble N-halamine precursor based on s-triazine (TIAPC) by introducing iminodiacetic acid. After grafting modification with TIAPC, the treated cotton fabric was then chlorinated with NaClO solution and chelated with metal Al^{3+} ions. Cotton-TIAPC-Cl-Al presented a high-efficacy and rapid bactericidal effect against *S. aureus* and *E. coli*, with 100% bacterial reduction in 1 min. In addition, the hydrophobic property of cotton-TIAPC-Cl-Al was greatly improved after chlorination. However, it cannot self-extinguish in the vertical burning test, implying limited flame retardancy [57].

Recently, nanogels, with large specific surface areas and higher functional efficiency, have been widely exploited as antimicrobial materials [58]. However, to date, there are very few reports in the literature on the use of nanogels for antimicrobial and flame retardant applications. In the work of Li et al., novel nanogels (NG3) were synthesized by Michael addition, which contain both the flame retardant elements of phosphorus, nitrogen, and silicon and the antibacterial component of N, N'-dimethyl-N-(3-(trimethoxysilyl) propyl) dodecane-1-chloroamine. By grafting on the cotton fibers and fabrics, the treated cotton fabric has self-extinguishing behavior, implying improved fire safety. Because NG3 easily destroys cell membranes and causes cell lysis, the grafted cotton fabrics can eliminate nearly 99% of bacteria for both *S. aureus* and *E. coli*. In addition, the NG3 with good biocompatibility and antibacterial properties plays a positive role in preventing wound infections, and anti-infection experiments of healing efficiency reaching 97.7% after 14 days' treatment have confirmed it. More importantly, functional cotton fabrics modified with nanogels exhibit relatively low mechanical and comfort properties (Figure 3) [16]. In another work, a binary mixture of acrylonitrile and 4-vinyl pyridine under the condition of ceric ammonium nitrate as initiator was copolymerized and grafted on the cotton fabrics by chemical induction. Taking advantage of the flame-retardant properties of synergistic nitrogen and phosphorus elements, the modified cotton fabrics have self-extinguishing abilities with a slow propagation rate. Meanwhile, the treated samples possess excellent antibacterial properties, with 41~96% antibacterial activity [59].

The guanidyl-based organic compounds can be used as antibacterial agents, which have the advantages of nontoxicity, high-temperature resistance, outstanding antibacterial effects, and long action periods. Commonly, the nitrogen-containing guanidyl group in the guanidyl-based antibacterial agent also exerts a better flame-retardant effect that can be combined with a phosphorus-containing group to form a phosphorus-nitrogen synergistic flame-retardant and antibacterial agent. Two novel and efficient antibacterial and flame-retardant guanidine-based compounds, N, N-di (ethyl phosphate) biguanide (DPG) and mono chlorotriazine triethyl phosphite guanidine (MCTPG), were successfully synthesized and then grafted onto cotton fabric by generating covalent bonds. The treated cotton fabric obtained good flame retardancy and antibacterial properties; the former DPG-treated cotton fabric obtained an LOI of 31.2%; and the antibacterial ratios of *S. aureus* and *E. coli* were 96.4% and 99.2%, respectively. The latter MCTPG-treated cotton fabric gained a LOI value of 31.2%, and the char length decreased to 8.5 cm. In addition, its inhibition zone base for *E. coli* and *S. aureus* reached 2.9 mm and 2.8 mm, respectively [10,60].

3.4. In Situ Deposition of Inorganic Metal Materials

In recent years, due to its excellent thermal and catalytic properties. Additionally, metal materials have excellent antibacterial properties with free radical capture abilities that have a wide application prospect in the functional fields of antibacterial and UV resistance. These metal materials were deposited on the surface of fabrics by direct reduction or synthesis [61].

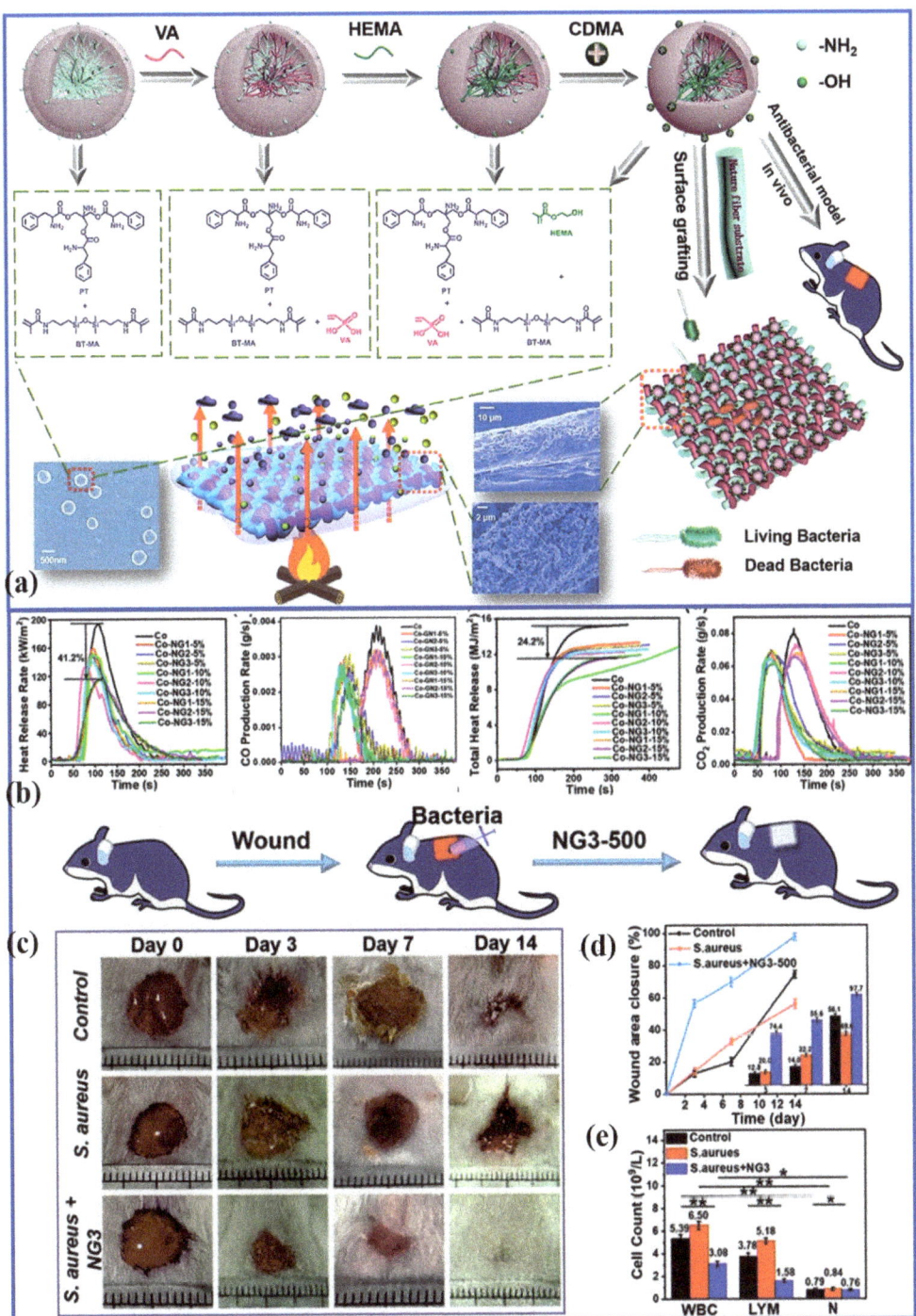

Figure 3. (**a**) Graphical abstract. (**b**) Cone calorimeter curves of Co-NG1, Co-NG2, and Co-NG3. (**c**) Photographs of wounds treated by the control, *S. aureus* and *S. aureus* + NG3-500 nanogels (the

minimum scale is 1 mm). (**d**) Assessment of the wound size reduction. (**e**) Expression level of WBC, LYM, and N in the blood of mice wounds after 3 d of treatment. Results are presented as mean ± standard deviation, and * indicates a significant difference ($p < 0.05$); ** indicates a significant difference compared with all other conditions ($p < 0.01$) [16] (Copyright 2021, Elsevier).

3.4.1. In Situ Deposition of Neat Metallic Oxide

The biomedical applications of NPs, especially metal oxides, have attracted great interest. ZnO NPs are the most famous type of NPs that inhibit the growth of Escherichia coli. Moreover, ZnO NPs with low cost, non-toxicity, and recyclability also acted as efficient photocatalysts. However, their easy deactivation and low acid resistance severely limit the development of ZnO NPs. To solve the problem, Bahare et al. synthesized ZnO@SiO2 NPs and coated them on the PET fabric by applying zinc acetate and sodium silicate as two precursors in an aqueous ammonia solution at 90 °C. The silica-supported ZnO improved these problems by limiting the size of the NPs (approximately 28.29 mm) and protecting them from acid solution corrosion. The treated samples have enhanced anti-dripping properties because of the inherent thermal resistance of Si. Furthermore, the treated PET fabrics eradicated almost 100% of *E. coli* [62]. However, some synthetic fibers, such as PS fibers, lack chemically active groups and have compact structures, which makes it difficult to absorb and carry functional reagents. In order to realize compatibility between these fiber products and functional components, the solvent crazing technique is applied to these fiber structures [63].

3.4.2. In Situ Deposition Assisted by Polymer Coatings

Unfortunately, metal oxides often have insufficient adhesion to fabrics when used alone. Polymer coating is helpful to improve the adhesion of metal compounds to the surface of fabrics. Commonly, pyrrole and aniline are used to prepare polymer coatings on fabrics that could provide active sites for metallic or other inorganic materials. Mahmoud et al. produced a polypyrrole-silver composite (Ppy-Ag) coating on the cotton/PS substrate through vapor phase polymerization (VPP) and redox reactions [64]. Polypyrrole can act as an effective stabilizer for Ag NP to solve the instability problem of treated fabrics after washing steps. The coated textile displayed an inhibition zone diameter of 25 mm and 28 mm for *E. coli* and *S. aureus*, respectively, verifying a supreme antibacterial property. It also had superior conductivity features with a low electrical resistance of 0.0218 kΩ. Furthermore, the treated fabrics show good washing fastness, implying improved stability of silver-containing coatings for textiles [64]. Polyaniline (PANI), a popular conducting polymer, has been regarded as a flame retardant for polymers due to its better char-forming ability. Cai et al. fabricated a PANI @ TS-silk fabric electrode that exhibits good charging and discharging cycle stability and high area-specific capacitance. Furthermore, the treated fabric electrode has good flame retardance and excellent antibacterial properties (99%) [65]. In another work, polyaniline was polymerized to form polyaniline chains on the surface of nanotubes in the presence of dispersed halloysite nanotubes. The polyaniline chains were decorated onto the fabric successfully due to the nanotubes aligned on the fabric's surface. Compared with the untreated fabric, the burning rate of the coated fabric decreased by 72%, and the clear antibacterial inhibition zone was recorded at 6 mm. Furthermore, the tensile strength of coated textile fabrics was maintained due to the alignment of nanotubes on the surface of the fabrics [66].

3.4.3. In Situ Deposition Assisted by Complexation Reaction

Many organic compound molecules (ions) containing unsaturated or active groups such as amino, carboxyl, and hydroxyl groups are prone to interact with metal ions, and organometallic complexes are usually obtained through certain coordination, complexation, and redox reactions. At present, this method is widely used in fabric functional finishing. Owing to the abundant nitrogen source, guanidine salts (e.g., guanidine carbonate, nitrate, and phosphate) exhibited intriguing flame retardancy. These compounds are able to improve the flame retardancy of the polymer matrix and promote the formation of the carbonized layer, which can act as a physical barrier. Guanazole, a low-cost compound with the chemical structure of 3,5-diamino-1,2,4-triazole, can coordinate with metal centers, so it has become an excellent flame retardant ligand. Hu and Wang et al. formed guanazole-zinc and guanazole-silver in aqueous solutions and then deposited them on cotton fabric surfaces by a dipping process. As expected, the cotton fabrics modified with guanazole-zinc and guanazole-silver exhibit outstanding flame retardancy with 29.5% and 27.5% LOI, respectively. Additionally, the samples possess self-extinguished behavior after removing the igniter during the vertical burning test and have reached the UL-94 V-0 level of flame retardant after the vertical burning test. In the micro-scale combustion calorimeter test, the HRR of guanazole-zinc and guanazole-silver modified cotton fabrics, respectively, reached 64.4% and 59.1%, while the THR of guanazole-zinc and guanazole-silver modified cotton fabrics reached 26.4% and 14.8%. More than this, the treated cotton fabrics also showed augmented antibacterial capacity against *S. aureus* and *E. coli*. Notably, the guanazole-silver-coated cotton fabrics also reflect the antifungal effect on Penicillium, Aspergillus niger, and Fusarium chlamydosporum [67].

As we all know, water and fire are incompatible. However, it is difficult to directly apply water as a fire-resistant material due to its high mobility. Hydrogel polymers with water as the main component can be used as fire-retardant materials to form a fire-resistant layer and reduce water loss, thereby improving the flame-retardancy of the coated fabric. Therefore, in the work of Yu et al., a novel fire-preventing triple-network (TN) hydrogel composed of poly (N-isopropylacrylamide) (PNIPAAm)/SA/PVA was prepared and laminated on cotton fabric, which was then put through the ionic coordination crosslinking process in $CaCl_2$ solution to form a stable structure. During the process, Ag NPs were also embedded into the hydrogel. Compared to neat fabric, the hydrogel-fabric laminates were nearly undamaged after being exposed to fire for 12 s, which is attributed to energy absorption as the water in the hydrogel is heated and evaporates. At the same time, outstanding antibacterial functions (>96%) against *E. coli* and *S. aureus* were achieved. Moreover, the introduction of a hydrogel layer also improves the mechanical strength of fabrics. Thus, the results demonstrated that the TN hydrogel, as a fire-resistant polymer, has potential for life-saving (Figure 4) [68].

Metal phenolic networks (MPNs), which consist of a variety of phenolic compounds and metals, have been promising candidates for the surface functionalization of substrates. For decades, naturally occurring compounds and their derivatives as eco-friendly antibacterial and flame-retardant agents for fabrics have attracted extensive attention from scholars. Researchers [69,70] have applied MPNs on the surface of silk fabrics, wherein Zhang et al. combined tannic acid (TA) with ferrous ions to form FR and antibacterial materials and applied them to silk fabrics that possess durable increasing FR with a 27.5% LOI value and almost no decrease even after 20 washes. In the vertical burning test, the treated sample shows a damaged length of only 11.2 cm but 30.0 cm of the original. The antibacterial activity significantly increased from 22% to 95% and maintained over 90% of its properties even after 20 washes [69]. In another work, Cheng et al. extracted polyphenols under alkaline conditions to develop flame-retardant macromolecular polyphenols through oxidative polymerization. The silk was dyed with the aforementioned extracted natural dyes and post-mordanted with metallic salts. The results not only showed improved flame-retardant and antibacterial properties but also antioxidant behaviors, washing fastness, perspiration, and wet rubbing fastness [70]. The MPNs can be applied to wood fibers as well to solve the problem of limiting application caused by their poor flame retardancy and antibacterial behavior. In the study of Jiang et al., wood fibers were immersed in a single solution of TA and ferrous salt successively to form a TA-Fe-wood complex and then further modified with silver nanoparticles (Ag NPs) to structure an Ag NPs layer. The TA/Fe/Ag NPs@wood fibers were successfully prepared (as shown in Figure 5). In the test of cone calorimetry, the TA/Fe/Ag NPs@wood fibers displayed enhanced flame retardancy, with the peak heat release rate and the peak smoke production rate reducing by 71.5% and 56.5%, respectively. Not only that, it also increases antibacterial activity for both *E. coli* and *S. aureus*. At the same time, the problem of the matrix being darkened by MPNs was solved [71].

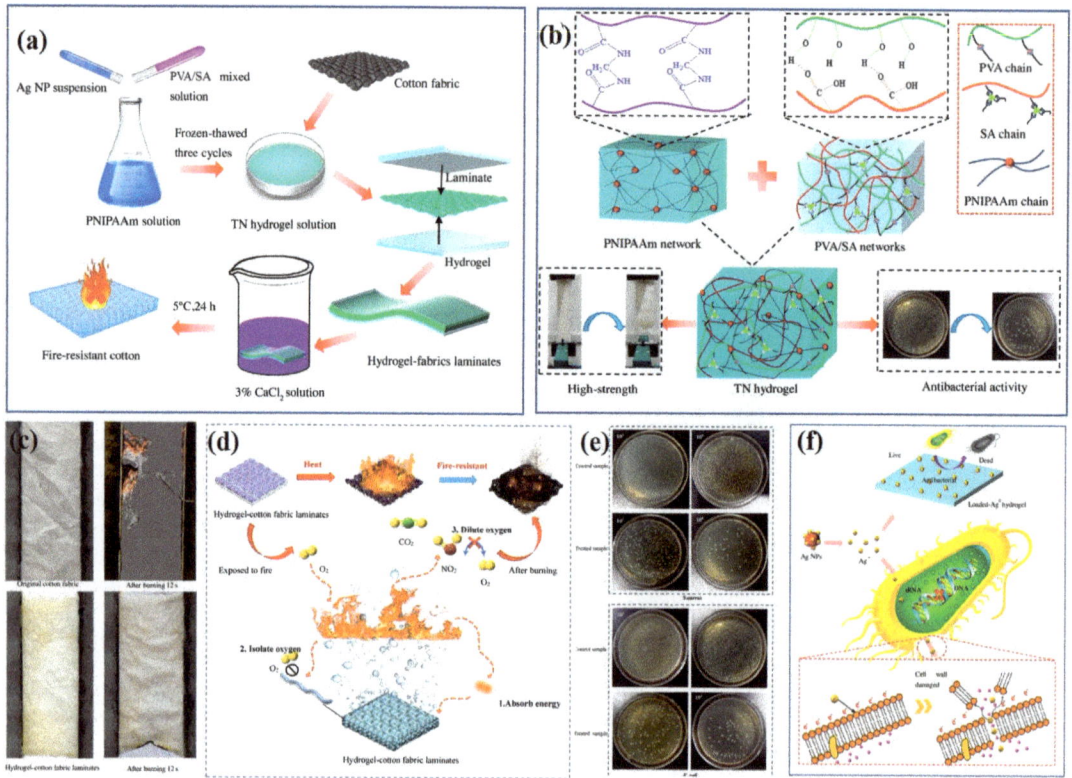

Figure 4. (**a**) Schematic diagram of the preparation of fire-resistant hydrogel-fabric laminates. (**b**) Interpenetrating polymer network structure and antibacterial activity of triple-network composite hydrogels. (**c**) Vertical fire-retardant testing of untreated cotton fabric and hydrogel–fabric laminates. (**d**) Schematic illustration of the mechanism of fire resistance of hydrogel-cotton fabric laminates. (**e**) the picture of bacteria recovered from the samples against *S. aureus* and *E. coli*. (**f**) schematic illustration of the antibacterial mechanism of Ag NPs [68] (Copyright 2021, Elsevier).

Apart from TA (Figure 6a), some other flavonoids, including Catechin (Figure 6b), Proanthocyandins (Figure 6c), Rutin (Figure 6d)), Quercetins (Figure 6f), Baicalin (Figure 6g), have attracted great attention in the fields of dyeing and functionalization of textiles simultaneously. Guo et al. applied grape seed proanthocyanidins (GSPs), which are a kind of recycled low-value byproduct rich in polyphenolic compounds, to the coloration of silk with a flame-retardant and antibacterial functionalized treatment. The dyed silk performs progressive flame retardancy due to the condensed phase flame retardancy mechanism of GSPs. In addition to enhancing the antibacterial properties effectively, washing, rubbing, perspiration, and light color fastness have also been improved to a certain extent [72]. Three flavonoids (baicalin, quercetin, and rutin) have been utilized in silk fabrics under the action of two metal salts (ferrous sulfate and titanium sulfate) mordanting as well. The results of the vertical burning test indicate improved flame-retardancy (the detailed data are shown in Table 1) and smoke suppression due to the good char formation ability of the silk fabrics in the process of combustion [73].

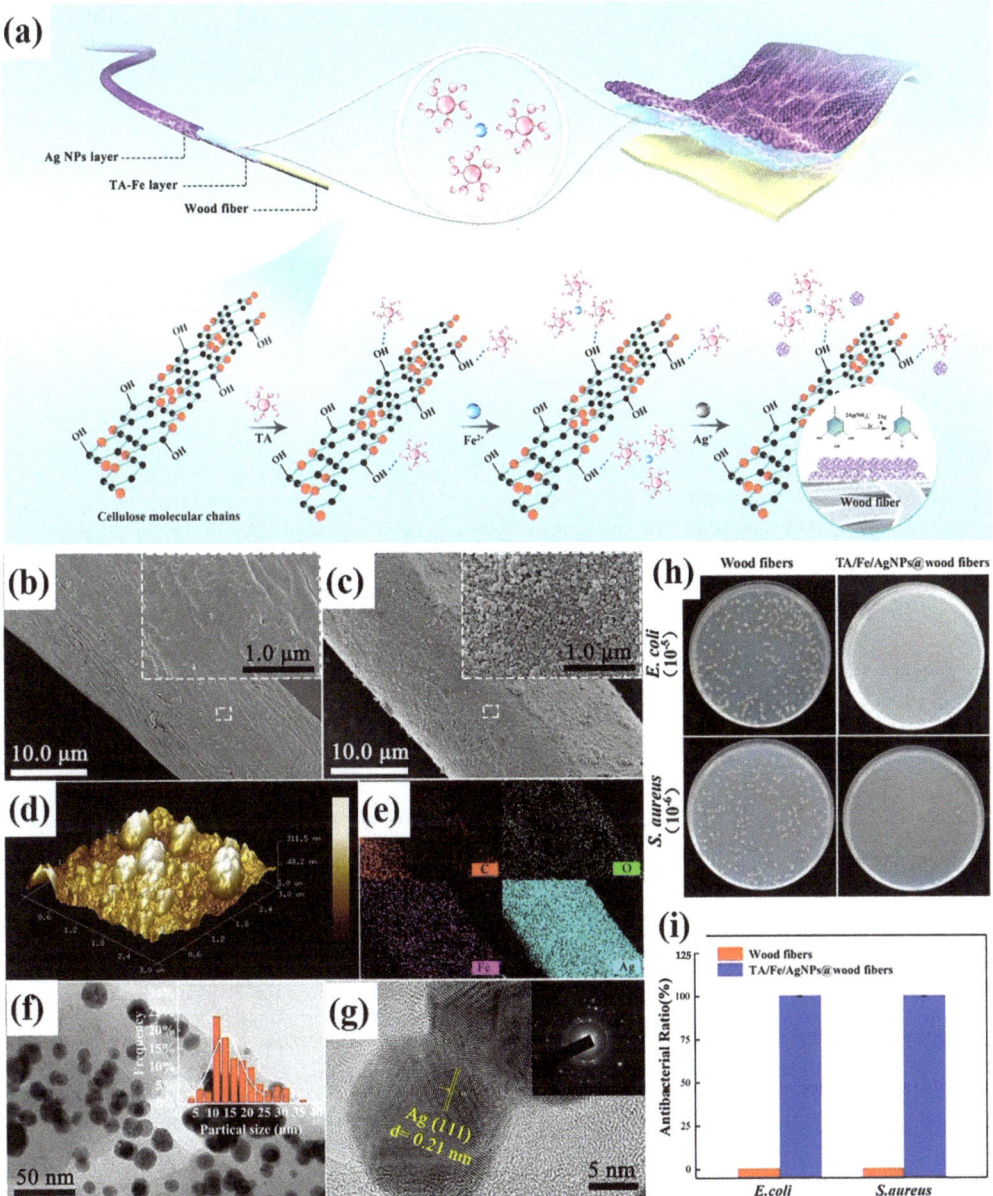

Figure 5. (**a**) Synthesis of TA/Fe/Ag NPs @ wood fibers. (**b**,**c**) SEM images of wood fibers: before and after modification. (**d**) AFM images of the modified fibers. (**e**) Element maps of C, O, and Ag for the sample shown. (**f**) TEM image of the modified fibers; inset: Ag NP size distribution. (**g**) HRTEM image of a single Ag NP showing the (111) lattice plane of Ag; inset: corresponding SAED image. (**h**) Impacts of TA/Fe/Ag NP modification on the antibacterial activity of wood fibers: Plate count tests show the effects of wood fibers with and without TA/Fe/ Ag NP modification on the growth of *E. coli* and *S. aureus*, including the appearance of counting plates and (**i**) antibacterial ratio [71] (Copyright 2021, Elsevier).

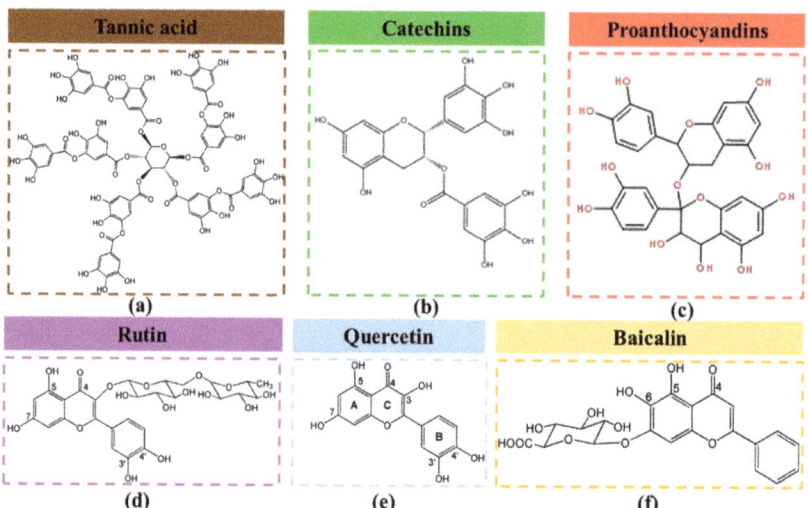

Figure 6. (**a**) Chemical structure of tannic acid [69] (Copyright 2020, Elsevier). (**b**) Chemical structure of Catechins. (**c**) Chemical structure of Proanthocyandins. (**d–f**) Chemical structures of three flavonoids [73] (Copyright 2019, Elsevier).

Table 1. The flame retardancy and antibacterial effect of phenolic compounds with metal mordanting.

MPNs	Seif-Extinguish	Vertical Combustion Test	LOI (%)	*S. aureus* (%)	*E. coli* (%)	Reference
TA-Fe^{2+}		B1	27.5	~98	~97	[48]
Tea Stem Extract-Fe^{2+}	Yes	B1	~26.7	97	93	[49]
Tea Stem Extract-Fe^{3+}	Yes	B1	~26.7	95	80	[49]
Tea Stem Extract-Ti^{4+}	Yes	B1	~26.3	96	83	[49]
baicalin-Fe^{2+}	No	B1	~27	~90	~93	[52]
baicalin-Ti^{4+}	No	B1	~27	~91	~91	[52]
quercetin-Fe^{2+}	No	B1	~27	~87	~90	[52]
quercetin-Ti^{4+}	No	B1	~27.2	~86	~92	[52]
Rutin-Fe^{2+}	No	B1	~26.8	~85	~90	[52]
Rutin-Ti^{4+}	No	B1	~27	~87	~87.8	[52]
GSPs-Fe^{2+}		B1	~27.8		99	[51]
GSPs-Fe^{3+}		B1	~28		91	[51]
GSPs-Ti^{4+}			~26.8		96	[51]

3.5. Sol-Gel Method

The sol-gel technique involves hydrolysis and condensation reactions using siloxane or metal alkoxide as precursors. First, a sol system is formed through the hydrolysis process; subsequently, a micro-nanoscale organic or inorganic coating will be formed on the surface of the fabric through a condensation reaction [74]. This method has the advantages of a simple process, mild reaction conditions, high efficiency, and good film-forming properties. It plays an important role in the functionalization of textiles, such as wrinkle resistance, dyeing, UV protection, antistatic, antibacterial, flame retardant, and hydrophobic properties. The durable antibacterial and flame-retardant cotton fabrics were developed via simultaneous hydrolytic condensation of $N_3P_3[NH(CH_2)_3Si(OC_2H_5)_3]_6$ and polymerization of dopamine (PDA) on cotton fabric. Ag NPs were then introduced onto fabrics via in situ reactions with PDA. Considerable flame retardancy can be observed for treated cotton fabric even at a low loading (7.2%) of the hybrid coating. In addition, the antibacterial activity of the treated fabric reached 99.99% for both *S. aureus* and *E. coli*. The modification showed excellent durability and nearly intact antimicrobial properties, and flame retardancy was maintained after 30 washing cycles [75]. In another work, a multifunctional composite coating (APP @ SiO$_2$-PDA @ Ag) composed of APP, PDA, PDMS-silica (PDMS-SiO$_2$), and Ag NPs was constructed on the surface of PET fabrics. The APP @ SiO$_2$-PDA @ Ag PET fabric showed an LOI of 29.0% and could self-extinguish in the VFT, and its PHRR and THR were 34% and 26% lower than those of the pure PET fabric. Notably, the multifunctional PET fabrics also exhibited excellent antibacterial activity against *E. coli* and *S. aureus* and superhydrophobicity (>150°). More importantly, the APP@SiO$_2$-PDA@Ag-coated PET fabrics still maintained good flame retardant and antibacterial performances after multiple washing cycles [76] (Figure 7).

The introduction of functionalized trialkoxysilane as the sol-gel agent has made significant progress in the chemical modification process of textiles, which is beneficial for producing unique surface properties. A three-component equimolar sol mixture of SiF, 3-(trimethoxysilyl)-propyldimethyloctadecyl ammonium chloride (SiQ), P, and SiP was constructed on cotton fabric by the sol-gel method. The treated fabrics simultaneously achieved flame-retardant, antibacterial (bacterial reduction of 100%), and water-repellent properties due to the thermal stability of SiP, the antibacterial properties of SiQ, and the hydrophobicity of SiF [77]. Following this work, the same group further optimized the structure of the multifunctional coating to increase the washing speed of treated cotton fabrics. They applied the prepared Stöber silica particles onto cotton fibers to form a particle-containing polysiloxane layer, which is based on tetraethyl orthosilicate, in the preparation work before the process of sol-gel. Eventually, the results showed enhanced washing fastness under the influence of the deposition of the silica particles. At the same time, it still exhibits excellent antibacterial activity, with R values of 81.6 and 100% for *E. coli* and *S. aureus* [14].

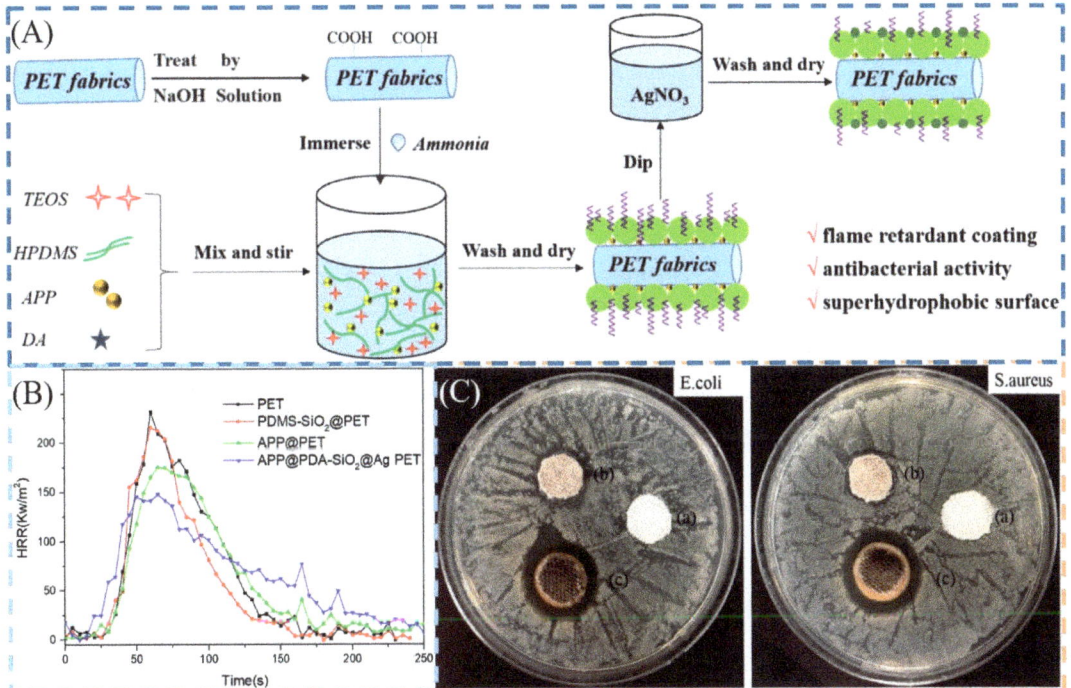

Figure 7. (**A**) Schematic illustration of the fabrication process for APP @ SiO$_2$-PDA @ Ag PET fabrics. (**B**) HRR curves of the pristine and coated PET fabrics. (**C**) (a) Antibacterial activities of the pristine PET fabrics; (b) APP @ SiO$_2$-PDA PET fabrics; and (c) APP @ SiO$_2$-PDA @ Ag PET fabrics [76]. [Copyright 2021 Elsevier].

4. Conclusions and Perspectives

This review summarizes the different treatments and adaptable finishing agents to obtain flame-retardant, antibacterial, multi-functional fabrics. Through the latest decade of research, it was found that the treatment methods not only affect processing efficiency but also have strong ties with the final functional properties and wearability of fabric. At the same time, the relevant functional materials were further introduced here, including their source, characteristics, and mechanism. Moreover, the advantages and disadvantages of these treatment methods and finishing agents were briefly introduced, which can provide a basic reference for relevant research in this field.

Although finishing methods and agents have been developing rapidly, many shortcomings remain and need to be solved. (1) Improving durability. The functional-coating textile products inescapably undergo friction and water-washing during daily applications. It will lead to bad results from a damaged coating and decrease or even eliminate the effect of the functional fabrics. Therefore, it is proposed that more research focus be laid on the durability of the coating. (2) Developing environmentally friendly but high-performance finishing agents. It is an eternal task for the whole of humanity to promote sustainable development. However, the effect of most green functional treatments is greatly limited. Therefore, it is integral to spend effort researching innocuous technology and chemicals while developing high-performance fabrics. (3) Increasing yields and realizing industrialization earlier. Plenty of research just stays in the laboratory stage, and it can hardly be applied in practice. Increasing production could promote the industrialization process, and realizing industrialization earlier will improve the quality of life and accelerate social progress. (4) It is well known that dual-functional fabrics can be realized by a two-step or one-step method. The two-step method may suffer from functional antagonism when flame retardants and antimicrobial agents are utilized simultaneously, which is detrimental to the construction of bifunctional coatings. In addition, the one-step method is important in the design and synthesis of multifunctional compounds. Therefore, to achieve the multifunctionalization of textiles, the best efforts are needed to find the optimal process and functional materials.

The above challenges will be gradually overcome with the continuous development of science and technology as well as the appearance of innovative technologies and materials. Owing to their powerful functionality and portability, multifunctional fabrics have been increasing in popularity in markets and will have good development prospects for a long time in the future.

Author Contributions: Conceptualization, K.Q.; investigation, L.J., Z.S. and J.Z.; writing—original draft, L.J., C.J. and S.C.; writing—review and editing, W.G.; supervision, W.G. All authors have read and agreed to the published version of the manuscript.

Funding: The authors would like to acknowledge the financial support by the Open Project of State Key Laboratory of Environment-friendly Energy Materials (Grant No. 21kfhg13), the Textile Light Applied Basic Research Project (Grant No. J202107), the Hong Kong Scholars Program (Grant No. XJ2020003), the Doctor Project of Innovation and Entrepreneurship in Jiangsu Province (Grant No. JSSCBS20210821), the National Natural Science Foundation of China (Grant No. 22205082), and the Natural Science Foundation of Jiangsu Province (Grant No. BK20221098).

Institutional Review Board Statement: Not applicable.

Informed Consent Statement: Written informed consent has been obtained from the patient(s) to publish this paper.

Conflicts of Interest: The authors declare no conflict of interest.

Abbreviations

ASTM	American Society for Testing and Materials
GB/T	Guobiao Standards (Chinese National Standards)
IEC	International Electrotechnical Commission
ISO	International Organization for Standardization
NFPA	National (US) Fire Prevention Association
UL-94	Standard released by the Underwriters Laboratories (USA)

References

1. Li, P.; Wang, B.; Xu, Y.-J.; Jiang, Z.; Dong, C.; Liu, Y.; Zhu, P. Ecofriendly Flame-Retardant Cotton Fabrics: Preparation, Flame Retardancy, Thermal Degradation Properties, and Mechanism. *ACS Sustain. Chem. Eng.* **2019**, *7*, 19246–19256. [CrossRef]
2. Costes, L.; Laoutid, F.; Brohez, S.; Dubois, P. Bio-based flame retardants: When nature meets fire protection. *Mater. Sci. Eng. R Rep.* **2017**, *117*, 1–25. [CrossRef]
3. Qiu, X.; Li, Z.; Li, X.; Zhang, Z. Flame retardant coatings prepared using layer by layer assembly: A review. *Chem. Eng. J.* **2018**, *334*, 108–122. [CrossRef]
4. Liang, F.; Xu, Y.; Chen, S.; Zhu, Y.; Huang, Y.; Fei, B.; Guo, W. Fabrication of Highly Efficient Flame-Retardant and Fluorine-Free Superhydrophobic Cotton Fabric by Constructing Multielement-Containing POSS@ZIF-67@PDMS Micro-Nano Hierarchical Coatings. *ACS Appl. Mater. Interfaces* **2022**, *14*, 56027–56045. [CrossRef]
5. Sykam, K.; Hussain, S.S.; Sivanandan, S.; Narayan, R.; Basak, P. Non-halogenated UV-curable flame retardants for wood coating applications: Review. *Prog. Org. Coat.* **2023**, *179*, 107549. [CrossRef]
6. Gao, M.; Wu, W.; Xu, Z.-Q. Thermal degradation behaviors and flame retardancy of epoxy resins with novel silicon-containing flame retardant. *J. Appl. Polym. Sci.* **2013**, *127*, 1842–1847. [CrossRef]
7. Wang, X.; Hu, Y.; Song, L.; Xing, W.; Lu, H. Thermal degradation behaviors of epoxy resin/POSS hybrids and phosphorus-silicon synergism of flame retardancy. *J. Polym. Sci. Part B Polym. Phys.* **2010**, *48*, 693–705. [CrossRef]
8. Lu, Y.; Zhao, P.; Chen, Y.; Huang, T.; Liu, Y.; Ding, D.; Zhang, G. A bio-based macromolecular phosphorus-containing active cotton flame retardant synthesized from starch. *Carbohydr. Polym.* **2022**, *298*, 120076. [CrossRef]
9. Song, T.; Li, Z.S.; Liu, J.G.; Yang, S.Y. Novel phosphorus–silicon synergistic flame retardants: Synthesis and characterization. *Chin. Chem. Lett.* **2012**, *23*, 793–796. [CrossRef]
10. Zhang, J.; Chen, B.; Liu, J.; Zhu, P.; Liu, Y.; Jiang, Z.; Dong, C.; Lu, Z. Multifunctional antimicrobial and flame retardant cotton fabrics modified with a novel N,N-di(ethyl phosphate) biguanide. *Cellulose* **2020**, *27*, 7255–7269. [CrossRef]
11. Lin, J.; Chen, X.; Chen, C.; Hu, J.; Zhou, C.; Cai, X.; Wang, W.; Zheng, C.; Zhang, P.; Cheng, J.; et al. Durably Antibacterial and Bacterially Antiadhesive Cotton Fabrics Coated by Cationic Fluorinated Polymers. *ACS Appl. Mater. Interfaces* **2018**, *10*, 6124–6136. [CrossRef] [PubMed]
12. Wu, M.; Ma, B.; Pan, T.; Chen, S.; Sun, J. Silver-Nanoparticle-Colored Cotton Fabrics with Tunable Colors and Durable Antibacterial and Self-Healing Superhydrophobic Properties. *Adv. Funct. Mater.* **2016**, *26*, 569–576. [CrossRef]
13. Magovac, E.; Vončina, B.; Budimir, A.; Jordanov, I.; Grunlan, J.C.; Bischof, S. Environmentally Benign Phytic Acid-Based Nanocoating for Multifunctional Flame-Retardant/Antibacterial Cotton. *Fibers* **2021**, *9*, 69. [CrossRef]
14. Vasiljević, J.; Zorko, M.; Štular, D.; Tomšič, B.; Jerman, I.; Orel, B.; Medved, J.; Kovač, J.; Simončič, B. Structural optimisation of a multifunctional water- and oil-repellent, antibacterial, and flame-retardant sol–gel coating on cellulose fibres. *Cellulose* **2017**, *24*, 1511–1528. [CrossRef]
15. Guo, W.; Wang, X.; Huang, J.; Zhou, Y.; Cai, W.; Wang, J.; Song, L.; Hu, Y. Construction of durable flame-retardant and robust superhydrophobic coatings on cotton fabrics for water-oil separation application. *Chem. Eng. J.* **2020**, *398*, 125661. [CrossRef]
16. Li, N.; Han, H.; Li, M.; Qiu, W.; Wang, Q.; Qi, X.; He, Y.; Wang, X.; Liu, L.; Yu, J.; et al. Eco-friendly and intrinsic nanogels for durable flame retardant and antibacterial properties. *Chem. Eng. J.* **2021**, *415*, 129008. [CrossRef]
17. Schartel, B.; Hull, T.R. Development of fire-retarded materials—Interpretation of cone calorimeter data. *Fire Mater.* **2007**, *31*, 327–354. [CrossRef]
18. Li, K.; Mao, S.; Feng, R. Estimation of Heat Release Rate and Fuel Type of Circular Pool Fires Using Inverse Modelling Based on Image Recognition Technique. *Fire Technol.* **2019**, *55*, 667–687. [CrossRef]
19. Laoutid, F.; Bonnaud, L.; Alexandre, M.; Lopez-Cuesta, J.M.; Dubois, P. New prospects in flame retardant polymer materials: From fundamentals to nanocomposites. *Mater. Sci. Eng. R Rep.* **2009**, *63*, 100–125. [CrossRef]
20. Lin, D.; Zeng, X.; Li, H.; Lai, X. Facile fabrication of superhydrophobic and flame-retardant coatings on cotton fabrics via layer-by-layer assembly. *Cellulose* **2018**, *25*, 3135–3149. [CrossRef]
21. Reddy, P.R.S.; Agathian, G.; Kumar, A. Ionizing radiation graft polymerized and modified flame retardant cotton fabric. *Radiat. Phys. Chem.* **2005**, *72*, 511–516. [CrossRef]
22. Liu, Z.; Xu, M.; Wang, Q.; Li, B. A novel durable flame retardant cotton fabric produced by surface chemical grafting of phosphorus- and nitrogen-containing compounds. *Cellulose* **2017**, *24*, 4069–4081. [CrossRef]

23. Jiang, Z.; Wang, C.; Fang, S.; Ji, P.; Wang, H.; Ji, C. Durable flame-retardant and antidroplet finishing of polyester fabrics with flexible polysiloxane and phytic acid through layer-by-layer assembly and sol-gel process. *J. Appl. Polym. Sci.* **2018**, *135*, 46414. [CrossRef]
24. Pan, Y.; Liu, L.; Zhang, Y.; Song, L.; Hu, Y.; Jiang, S.; Zhao, H. Effect of genipin crosslinked layer-by-layer self-assembled coating on the thermal stability, flammability and wash durability of cotton fabric. *Carbohydr. Polym.* **2019**, *206*, 396–402. [CrossRef] [PubMed]
25. Bentis, A.; Boukhriss, A.; Grancaric, A.M.; El Bouchti, M.; El Achaby, M.; Gmouh, S. Flammability and combustion behavior of cotton fabrics treated by the sol gel method using ionic liquids combined with different anions. *Cellulose* **2019**, *26*, 2139–2153. [CrossRef]
26. Lin, D.; Zeng, X.; Li, H.; Lai, X.; Wu, T. One-pot fabrication of superhydrophobic and flame-retardant coatings on cotton fabrics via sol-gel reaction. *J. Colloid Interface Sci.* **2019**, *533*, 198–206. [CrossRef]
27. Ling, C.; Guo, L. An eco-friendly and durable multifunctional cotton fabric incorporating ZnO and a branched polymer. *Cellulose* **2021**, *28*, 5843–5854. [CrossRef]
28. Yu, R.; Tian, M.; Qu, L.; Zhu, S.; Ran, J.; Liu, R. Fast and simple fabrication of SiO_2/poly(vinylidene fluoride) coated cotton fabrics with asymmetric wettability via a facile spray-coating route. *Text. Res. J.* **2018**, *89*, 1013–1026. [CrossRef]
29. Attia, N.F.; Morsy, M.S. Facile synthesis of novel nanocomposite as antibacterial and flame retardant material for textile fabrics. *Mater. Chem. Phys.* **2016**, *180*, 364–372. [CrossRef]
30. Zheng, C.; Sun, Y.; Cui, Y.; Yang, W.; Lu, Z.; Shen, S.; Xia, Y.; Xiong, Z. Superhydrophobic and flame-retardant alginate fabrics prepared through a one-step dip-coating surface-treatment. *Cellulose* **2021**, *28*, 5973–5984. [CrossRef]
31. Wei, D.; Dong, C.; Liu, J.; Zhang, Z.; Lu, Z. A Novel Cyclic Polysiloxane Linked by Guanidyl Groups Used as Flame Retardant and Antimicrobial Agent on Cotton Fabrics. *Fibers Polym.* **2019**, *20*, 1340–1346. [CrossRef]
32. Moazami, A.; Montazer, M. A novel multifunctional cotton fabric using $ZrO2NPs$/urea/CTAB/MA/SHP: Introducing flame retardant, photoactive and antibacterial properties. *J. Text. Inst.* **2015**, *107*, 1253–1263. [CrossRef]
33. Makhlouf, G.; Abdelkhalik, A.; Ameen, H. Preparation of highly efficient chitosan-based flame retardant coatings with good antibacterial properties for cotton fabrics. *Prog. Org. Coat.* **2022**, *163*, 106627. [CrossRef]
34. Yang, T.-T.; Guan, J.-P.; Tang, R.-C.; Chen, G. Condensed tannin from Dioscorea cirrhosa tuber as an eco-friendly and durable flame retardant for silk textile. *Ind. Crops Prod.* **2018**, *115*, 16–25. [CrossRef]
35. Okeil, A.A. Citric Acid Crosslinking of Cellulose Using TiO_2 Catalyst by Pad-Dry-Cure Method. *Polym.-Plast. Technol. Eng.* **2008**, *47*, 174–179. [CrossRef]
36. Ahmed, M.T.; Morshed, M.N.; Farjana, S.; An, S.K. Fabrication of new multifunctional cotton–modal–recycled aramid blended protective textiles through deposition of a 3D-polymer coating: High fire retardant, water repellent and antibacterial properties. *New J. Chem.* **2020**, *44*, 12122–12133. [CrossRef]
37. Singh, N.; Sheikh, J. Sustainable development of mosquito-repellent, flame-retardant, antibacterial, fragrant and antioxidant linen using microcapsules containing Thymus vulgaris oil in in-situ generated chitosan-phosphate. *Cellulose* **2021**, *28*, 2599–2614. [CrossRef]
38. El-Shafei, A.; ElShemy, M.; Abou-Okeil, A. Eco-friendly finishing agent for cotton fabrics to improve flame retardant and antibacterial properties. *Carbohydr. Polym.* **2015**, *118*, 83–90. [CrossRef]
39. Dhineshbabu, N.R.; Manivasakan, P.; Karthik, A.; Rajendran, V. Hydrophobicity, flame retardancy and antibacterial properties of cotton fabrics functionalised with MgO/methyl silicate nanocomposites. *RSC Adv.* **2014**, *4*, 32161–32173. [CrossRef]
40. Vasiljević, J.; Tomšič, B.; Jerman, I.; Orel, B.; Jakša, G.; Kovač, J.; Simončič, B. Multifunctional superhydrophobic/oleophobic and flame-retardant cellulose fibres with improved ice-releasing properties and passive antibacterial activity prepared via the sol–gel method. *J. Sol-Gel Sci. Technol.* **2014**, *70*, 385–399. [CrossRef]
41. Alongi, J.; Carosio, F.; Malucelli, G. Current emerging techniques to impart flame retardancy to fabrics: An overview. *Polym. Degrad. Stab.* **2014**, *106*, 138–149. [CrossRef]
42. Alongi, J.; Carosio, F.; Frache, A.; Malucelli, G. Layer by Layer coatings assembled through dipping, vertical or horizontal spray for cotton flame retardancy. *Carbohydr. Polym.* **2013**, *92*, 114–119. [CrossRef] [PubMed]
43. Cheng, X.-W.; Tang, R.-C.; Guan, J.-P.; Zhou, S.-Q. An eco-friendly and effective flame retardant coating for cotton fabric based on phytic acid doped silica sol approach. *Prog. Org. Coat.* **2020**, *141*, 105539. [CrossRef]
44. Li, P.; Wang, B.; Liu, Y.Y.; Xu, Y.J.; Jiang, Z.M.; Dong, C.H.; Zhang, L.; Liu, Y.; Zhu, P. Fully bio-based coating from chitosan and phytate for fire-safety and antibacterial cotton fabrics. *Carbohydr. Polym.* **2020**, *237*, 116173. [CrossRef] [PubMed]
45. Li, P.; Liu, C.; Wang, B.; Tao, Y.; Xu, Y.-J.; Liu, Y.; Zhu, P. Eco-friendly coating based on an intumescent flame-retardant system for viscose fabrics with multi-function properties: Flame retardancy, smoke suppression, and antibacterial properties. *Prog. Org. Coat.* **2021**, *159*, 106400. [CrossRef]
46. Hou, F.; Zhu, M.; Liu, Y.; Zhu, K.; Xu, J.; Jiang, Z.; Wang, C.; Wang, H. A self-assembled bio-based coating with phytic acid and DL-arginine used for a flame-retardant and antibacterial cellulose fabric. *Prog. Org. Coat.* **2022**, *173*, 107179. [CrossRef]
47. Liu, Y.; Zhang, J.; Ren, Y.; Zhang, G.; Liu, X.; Qu, H. Biomaterial arginine encountering with UV grafting technology to prepare flame retardant coating for polyacrylonitrile fabric. *Prog. Org. Coat.* **2022**, *163*, 106599. [CrossRef]

48. Song, J.; Feng, H.; Wu, M.; Chen, L.; Xia, W.; Zhang, W. Preparation and characterization of arginine-modified chitosan/hydroxypropyl methylcellose antibacterial film. *Int. J. Biol. Macromol.* **2020**, *145*, 750–758. [CrossRef]
49. Tang, H.; Zhang, P.; Kieft, T.L.; Ryan, S.J.; Baker, S.M.; Wiesmann, W.P.; Rogelj, S. Antibacterial action of a novel functionalized chitosan-arginine against Gram-negative bacteria. *Acta Biomater.* **2010**, *6*, 2562–2571. [CrossRef]
50. Su, Z.; Han, Q.; Zhang, F.; Meng, X.; Liu, B. Preparation, characterization and antibacterial properties of 6-deoxy-6-arginine modified chitosan. *Carbohydr. Polym.* **2020**, *230*, 115635. [CrossRef]
51. Jiang, Z.; Hu, Y.; Zhu, K.; Li, Y.; Wang, C.; Zhang, S.; Wang, J. Self-assembled bio-based coatings for flame-retardant and antibacterial polyester–cotton fabrics. *Text. Res. J.* **2021**, *92*, 368–382. [CrossRef]
52. Lv, Z.; Hu, Y.-T.; Guan, J.-P.; Tang, R.-C.; Chen, G.-Q. Preparation of a flame retardant, antibacterial, and colored silk fabric with chitosan and vitamin B2 sodium phosphate by electrostatic layer by layer assembly. *Mater. Lett.* **2019**, *241*, 136–139. [CrossRef]
53. Dong, A.; Wang, Y.J.; Gao, Y.; Gao, T.; Gao, G. Chemical Insights into Antibacterial N-Halamines. *Chem. Rev.* **2017**, *117*, 4806–4862. [CrossRef]
54. Ren, H.; Du, Y.; Su, Y.; Guo, Y.; Zhu, Z.; Dong, A. A Review on Recent Achievements and Current Challenges in Antibacterial Electrospun N-halamines. *Colloid Interface Sci. Commun.* **2018**, *24*, 24–34. [CrossRef]
55. Li, S.; Lin, X.; Liu, Y.; Li, R.; Ren, X.; Huang, T.-S. Phosphorus-nitrogen-silicon-based assembly multilayer coating for the preparation of flame retardant and antimicrobial cotton fabric. *Cellulose* **2019**, *26*, 4213–4223. [CrossRef]
56. Xiao, M.; Guo, Y.; Zhang, J.; Liu, Y.; Ren, Y.; Liu, X. Diethylene triamine penta methylene phosphonic acid encountered silver ions: A convenient method for preparation of flame retardant and antibacterial lyocell fabric. *Cellulose* **2021**, *28*, 7465–7481. [CrossRef]
57. Xu, D.; Wang, S.; Hu, J.; Liu, Y.; Jiang, Z.; Zhu, P. Enhancing antibacterial and flame-retardant performance of cotton fabric with an iminodiacetic acid-containing N-halamine. *Cellulose* **2021**, *28*, 3265–3277. [CrossRef]
58. Keskin, D.; Zu, G.; Forson, A.M.; Tromp, L.; Sjollema, J.; van Rijn, P. Nanogels: A novel approach in antimicrobial delivery systems and antimicrobial coatings. *Bioact. Mater.* **2021**, *6*, 3634–3657. [CrossRef] [PubMed]
59. Kaur, I.; Bhati, P.; Sharma, B. Antibacterial, flame retardant, and physico-chemical properties of cotton fabric graft copolymerized with a binary mixture of acrylonitrile and 4-vinylpyridine. *J. Appl. Polym. Sci.* **2014**, *131*, 40415. [CrossRef]
60. Dong, C.; He, P.; Lu, Z.; Wang, S.; Sui, S.; Liu, J.; Zhang, L.; Zhu, P. Preparation and properties of cotton fabrics treated with a novel antimicrobial and flame retardant containing triazine and phosphorus components. *J. Therm. Anal. Calorim.* **2017**, *131*, 1079–1087. [CrossRef]
61. Jiang, T.; Liu, L.; Yao, J. In situ deposition of silver nanoparticles on the cotton fabrics. *Fibers Polym.* **2011**, *12*, 620–625. [CrossRef]
62. Nozari, B.; Montazer, M.; Mahmoudi Rad, M. Stable ZnO/SiO$_2$ nano coating on polyester for anti-bacterial, self-cleaning and flame retardant applications. *Mater. Chem. Phys.* **2021**, *267*, 124674. [CrossRef]
63. Kale, R.D.; Soni, M.; Potdar, T. A flame retardant, antimicrobial and UV protective polyester fabric by solvent crazing route. *J. Polym. Res.* **2019**, *26*, 189. [CrossRef]
64. Abu Elella, M.H.; Goda, E.S.; Yoon, K.R.; Hong, S.E.; Morsy, M.S.; Sadak, R.A.; Gamal, H. Novel vapor polymerization for integrating flame retardant textile with multifunctional properties. *Compos. Commun.* **2021**, *24*, 100614. [CrossRef]
65. Cai, H.; Liu, Z.; Xu, M.; Chen, L.; Chen, X.; Cheng, L.; Li, Z.; Dai, F. High performance flexible silk fabric electrodes with antibacterial, flame retardant and UV resistance for supercapacitors and sensors. *Electrochim. Acta* **2021**, *390*, 138895. [CrossRef]
66. Elsayed, E.M.; Attia, N.F.; Alshehri, L.A. Innovative Flame Retardant and Antibacterial Fabrics Coating Based on Inorganic Nanotubes. *Chem. Sel.* **2020**, *5*, 2961–2965. [CrossRef]
67. Nabipour, H.; Wang, X.; Rahman, M.Z.; Song, L.; Hu, Y. An environmentally friendly approach to fabricating flame retardant, antibacterial and antifungal cotton fabrics via self-assembly of guanazole-metal complex. *J. Clean. Prod.* **2020**, *273*, 122832. [CrossRef]
68. Yu, Z.; Liu, J.; He, H.; Ma, S.; Yao, J. Flame-retardant PNIPAAm/sodium alginate/polyvinyl alcohol hydrogels used for fire-fighting application: Preparation and characteristic evaluations. *Carbohydr. Polym.* **2021**, *255*, 117485. [CrossRef]
69. Zhang, W.; Yang, Z.-Y.; Tang, R.-C.; Guan, J.-P.; Qiao, Y.-F. Application of tannic acid and ferrous ion complex as eco-friendly flame retardant and antibacterial agents for silk. *J. Clean. Prod.* **2020**, *250*, 117485. [CrossRef]
70. Cheng, T.-H.; Liu, Z.-J.; Yang, J.-Y.; Huang, Y.-Z.; Tang, R.-C.; Qiao, Y.-F. Extraction of Functional Dyes from Tea Stem Waste in Alkaline Medium and Their Application for Simultaneous Coloration and Flame Retardant and Bioactive Functionalization of Silk. *ACS Sustain. Chem. Eng.* **2019**, *7*, 18405–18413. [CrossRef]
71. Jiang, P.; Zhu, Y.; Wu, Y.; Lin, Q.; Yu, Y.; Yu, W.; Huang, Y. Synthesis of flame-retardant, bactericidal, and color-adjusting wood fibers with metal phenolic networks. *Ind. Crops Prod.* **2021**, *170*, 113796. [CrossRef]
72. Guo, L.; Yang, Z.-Y.; Tang, R.-C.; Yuan, H.-B. Grape Seed Proanthocyanidins: Novel Coloring, Flame-Retardant, and Antibacterial Agents for Silk. *ACS Sustain. Chem. Eng.* **2020**, *8*, 5966–5974. [CrossRef]
73. Zhou, Y.; Tang, R.-C.; Xing, T.; Guan, J.-P.; Shen, Z.-H.; Zhai, A.-D. Flavonoids-metal salts combination: A facile and efficient route for enhancing the flame retardancy of silk. *Ind. Crops Prod.* **2019**, *130*, 580–591. [CrossRef]
74. Malucelli, G. Surface-Engineered Fire Protective Coatings for Fabrics through Sol-Gel and Layer-by-Layer Methods: An Overview. *Coatings* **2016**, *6*, 33. [CrossRef]
75. Li, Y.; Wang, B.; Sui, X.; Xie, R.; Xu, H.; Zhang, L.; Zhong, Y.; Mao, Z. Durable flame retardant and antibacterial finishing on cotton fabrics with cyclotriphosphazene/polydopamine/silver nanoparticles hybrid coatings. *Appl. Surf. Sci.* **2018**, *435*, 1337–1343. [CrossRef]

76. Li, Q.; Zhang, S.; Mahmood, K.; Jin, Y.; Huang, C.; Huang, Z.; Zhang, S.; Ming, W. Fabrication of multifunctional PET fabrics with flame retardant, antibacterial and superhydrophobic properties. *Prog. Org. Coat.* **2021**, *157*, 106296. [CrossRef]
77. Vasiljević, J.; Tomšič, B.; Jerman, I.; Orel, B.; Jakša, G.; Simončič, B. Novel multifunctional water- and oil-repellent, antibacterial, and flame-retardant cellulose fibres created by the sol–gel process. *Cellulose* **2014**, *21*, 2611–2623. [CrossRef]

Disclaimer/Publisher's Note: The statements, opinions and data contained in all publications are solely those of the individual author(s) and contributor(s) and not of MDPI and/or the editor(s). MDPI and/or the editor(s) disclaim responsibility for any injury to people or property resulting from any ideas, methods, instructions or products referred to in the content.

Review

Synthesis and Applications of Supramolecular Flame Retardants: A Review

Simeng Xiang [†], Jiao Feng [†], Hongyu Yang * and Xiaming Feng *

College of Materials Science and Engineering, Chongqing University, Shapingba, Chongqing 400044, China; 202109021141t@stu.cqu.edu.cn (S.X.); 202109021177t@cqu.edu.cn (J.F.)
* Correspondence: yhongyu@cqu.edu.cn (H.Y.); fengxm@cqu.edu.cn (X.F.)
† The authors contributed equally to this work.

Abstract: The development of different efficient flame retardants (FRs) to improve the fire safety of polymers has been a hot research topic. As the concept of green sustainability has gradually been raised to the attention of the whole world, it has even dominated the research direction of all walks of life. Therefore, there is an urgent calling to explore the green and simple preparation methods of FRs. The development of supramolecular chemistry in the field of flame retardancy is expanding gradually. It is worth noting that the synthesis of supramolecular flame retardants (SFRs) based on non-covalent bonds is in line with the current concepts of environmental protection and multi-functionality. This paper introduces the types of SFRs with different dimensions. SFRs were applied to typical polymers to improve their flame retardancy. The influence on mechanical properties and other material properties under the premise of flame retardancy was also summarized.

Keywords: supramolecular flame retardants; sustainability; fire safety; polymers; mechanical properties

Citation: Xiang, S.; Feng, J.; Yang, H.; Feng, X. Synthesis and Applications of Supramolecular Flame Retardants: A Review. *Molecules* 2023, 28, 5518. https://doi.org/10.3390/molecules28145518

Academic Editor: Gaëlle Fontaine

Received: 29 May 2023
Revised: 12 July 2023
Accepted: 13 July 2023
Published: 19 July 2023

Copyright: © 2023 by the authors. Licensee MDPI, Basel, Switzerland. This article is an open access article distributed under the terms and conditions of the Creative Commons Attribution (CC BY) license (https://creativecommons.org/licenses/by/4.0/).

1. Introduction

With the development of science and technology, the application of polymer materials has penetrated all aspects of our life and production, making existence more convenient [1,2]. At the same time, there are potential risks. Because most polymer materials are rich in carbon, hydrogen, and other elements, their intrinsic molecular structure determines the combustibility or flammability. They may decompose and burn at high temperatures, causing fires [3,4]. In the new situation of consumption upgrading and the rapid-development of emerging industries, higher requirements are put forward for the performance of polymers. Improving the flame retardancy of polymers can improve the reliability and application fields of their products (such as new energy vehicles [5], electronics and electrical products [6,7], and aerospace products [8]). In addition, Figure 1 shows the number of scientific research publications on flame retardancy of some typical polymers in the past decade. The increasing trend also reflects that the research of flame-retardant polymers is challenging and developmental.

Flame retardants (FRs), as additives, are applied to polymer materials. They achieve flame retardancy mainly in the condensed phase and/or gas phase. In the past, halogenated FRs were the mainstay. However, in practical application, some halogen FRs will release harmful substances (such as corrosive hydrogen halide gas and toxic carcinogens dioxins and furans) during thermal decomposition, which will undoubtedly cause great harm to human health and the ecological environment [9,10]. Meanwhile, some laws and regulations also put forward explicit requirements for the use of FRs. In 2013, the global ban on hexabromocyclododecane was proposed by the United Nations Environment Programme in the Stockholm Convention on Persistent Organic Pollutants [11]. In 2019, the European Union issued regulations on the prohibition of halogen FRs in electronic displays [12]. In 2022, New York State amended the control content of FRs in upholstered furniture, mattresses, electronic display housings, supports, etc., in the Bill (S4630B/A5418B) [13].

Today, with the gradual deepening of environmental awareness, environmental-friendly halogen-free FRs with no (low) toxicity, low smoke, and low corrosion [9,14–16] are being developed. Statistically, halogen-free flame-retardant products dominated the market in 2020 (up from 59% share) [17]. Among them, phosphorus FRs [18,19], nitrogen FRs [20,21], phosphorus–nitrogen FRs [22–25], and metal-compound FRs [26–28] have been widely studied. In the synthesis and development of FRs, it is found that a particular status does not conform to the concept of green and sustainable development. Because some synthesis conditions are relatively harsh (such as high temperature, high pressure, and inert environment) and the synthesis procedure is relatively complex, some involve toxic and harmful organic solvents, such as trichloromethane, ether, acetone, tetrahydrofuran, and acetonitrile [29–32]. This brings lots of trouble to the subsequent processing. It also burdens the environment greatly. Therefore, preparing the high-efficiency FRs in a simple, safe, and environmentally friendly manner has become an urgent focus.

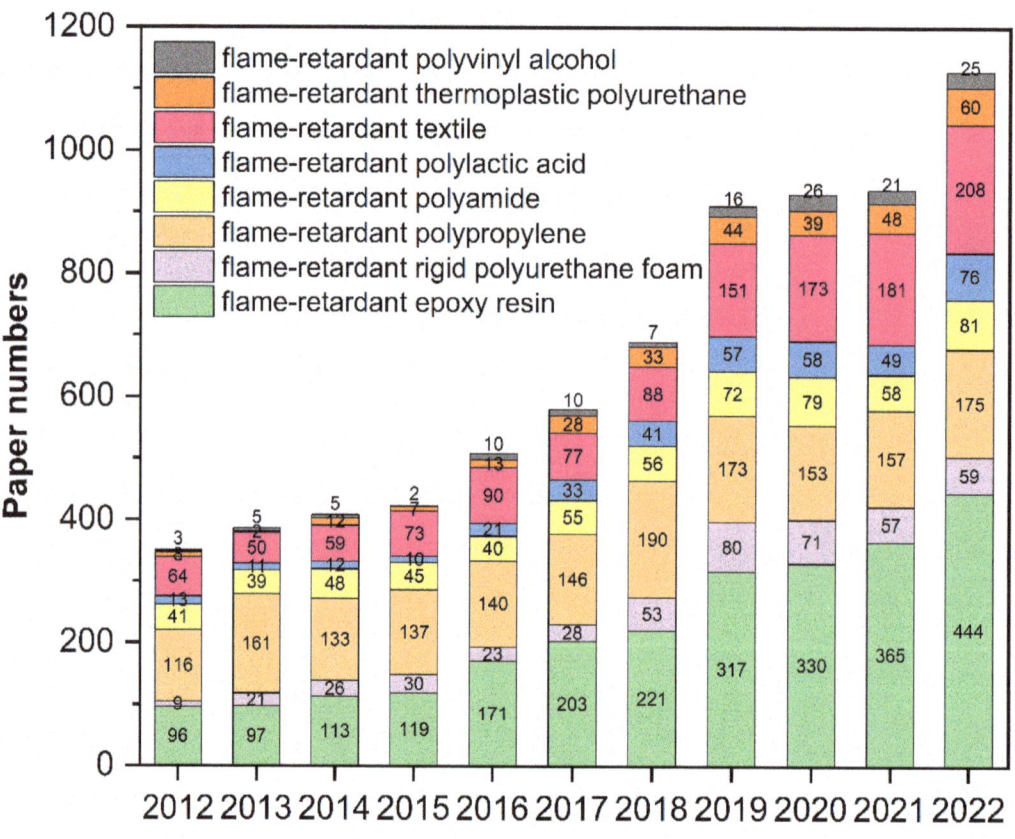

Figure 1. Paper publications of flame-retardant typical polymers from 2012 to 2022. (Data from Web of Science, as of December 2022).

Supramolecular chemistry usually refers to the combination of two (or more) molecules with specific structures and properties by non-covalent intermolecular interactions (such as ion attraction, ion-dipole interaction, dipole-dipole interaction, hydrogen bonding, and electrostatic interaction) [33]. Supramolecular chemistry is at the forefront of scientific development, which has been extensively developed in biomedicine [34], photoelectric materials [35], self-healing materials [36], and binders [37]. In the field of flame retardancy, supramolecular chemistry is also gradually expanding. It is noteworthy that supramolecular-assembly has the characteristics of easy synthesis and greenness. At present, there is no systematic review on the research progress and application of supramolecular flame retardants (SFRs) in the field of flame retardancy. This work mainly introduces the types of SFRs and their applications in the typical polymer materials. Under the premise of flame retardancy, the improvement of polymer materials in other properties by SFRs is also concerned.

2. Synthesis of Supramolecular Flame Retardants

According to the morphological structure of FRs, FRs synthesized based on supramolecular self-assembly are classified into one-dimensional supramolecular flame retardants (1D SFRs), two-dimensional supramolecular flame retardants (2D SFRs) and three-dimensional supramolecular flame retardants (3D SFRs) (Figure 2).

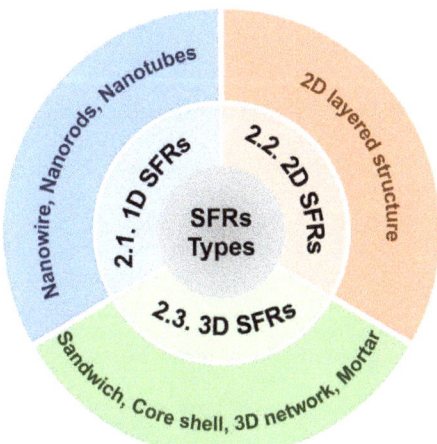

Figure 2. Classification of SFRs and corresponding structural types.

2.1. One-Dimensional Supramolecular Flame Retardants

Some inorganic materials with 1D nanostructures, such as halloysite nanotubes (HNTs) [38,39], multiwalled carbon nanotubes (MWCNTs) [40], various whiskers [41,42], nanowire materials [43], etc., have certain flame-retardant and smoke-suppressing effects. But there are problems with agglomeration, which deteriorates the mechanical properties of the matrix due to excessive addition. How to design and optimize the more advantageous 1D organic-inorganic hybrid FRs has certain research significance for the development of efficient flame-retardant systems and improving the mechanical properties of the matrix. Currently, some studies have used such 1D inorganic nanomaterials as the building blocks and selected some classical flame-retardant phosphorus–nitrogen sources for controllably encapsulating the blocks by self-assembly to construct 1D organic-inorganic hybrid SFRs [44–46].

Ting Chen et al. [44] reported a supramolecular nanorod with a core-shell structure. Covalent polymers (named HP) of flame-retardant phosphorus and nitrogen sources were firstly assembled by the Kabachnik-Fields reaction. β-FeOOH can play an effective role in smoke suppression. β-FeOOH has a tetragonal crystal system structure, and Fe^{3+} is located in the voids of octahedra. Using this unique structure, it formed an organic-inorganic hybrid by coordination with HP containing polyphenolic structures (Figure 3a). SEM showed that HP was encapsulated on the surface of spindle-like β-FeOOH nanorods (Figure 3b).

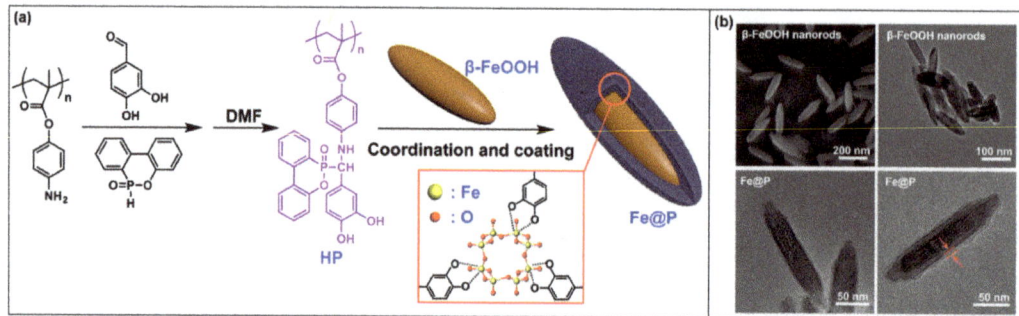

Figure 3. (a) The process of modifying β-FeOOH. (b) SEM images before and after modification of β-FeOOH [44].

Sheng Shang et al. [45] used HNTs with surface rich in active sites (e.g., Al-OH, Si-O functional groups) as the building blocks. Firstly, melamine (MEL) was modified on the surface of HNTs in aqueous solution based on hydrogen bonding. The supramolecular self-assembly was then performed by means of hydrogen bonding and ionic attraction between the phosphate group of phytic acid (PA) and the amino group of MEL (Figure 4a). A flame-retardant functionalized modified nanotube structure was successfully prepared (Figure 4b). In addition to HNTs as substrates for self-assembly, MWCNTs have also been chosen as the building blocks [46]. MEL and PA were grafted onto the surface of MWCNTs using ionic interactions and π-π stacking by successive ultrasonic stirring in aqueous solution at 80 °C (Figure 4c). It was observed that the grafted MWCNTs changed from smooth to rough (Figure 4d). Moreover, the polarity of MEL-PA makes its water contact angle smaller, indicating that its wettability improves (Figure 4e).

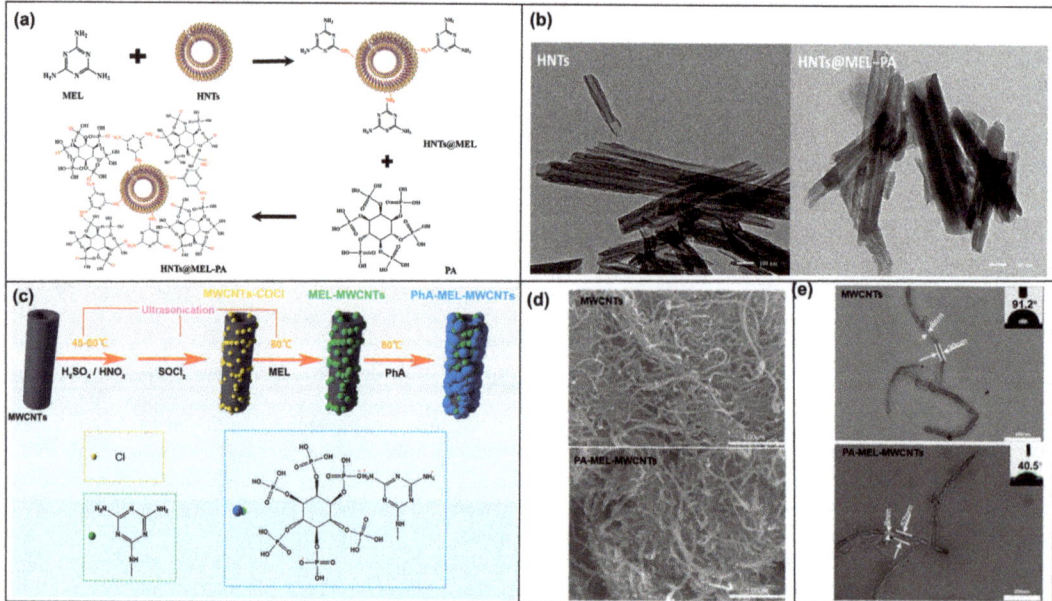

Figure 4. (a) The process of modifying HNTs. (b) SEM images before and after modification of HNTs [45]. (c) Grafting procedure of MWCNTs. (d) SEM images before and after grafting of MWCNTs. (e) Water contact Angles before and after grafting of MWCNTs [46].

2.2. Two-Dimensional Supramolecular Flame Retardants

Currently, the construction of 1D SFRs mainly relies on the 1D structure of the selected building blocks. The diversity and additivity of supramolecular self-assembly interactions should be fully utilized. More options in assembly materials can be available to expand supramolecular assemblies from 1D to 2D. 2D SFRs exhibit a 2D lamellar form, which can affect the transfer of air, combustible volatile substances and heat, thus achieving a barrier effect. In addition, such structures tend to have a high aspect ratio and can be used to enhance the mechanical properties of the matrix [47–50].

PA is a biomass-based flame-retardant monomer, mainly extracted from plant seeds. The structure of PA is that there are six phosphate groups in the inositol ring. In the process of thermal decomposition, it can trap and burn free radicals and catalyze carbon formation [51]. MEL is a triazine compound with a nitrogen heterocyclic structure (nitrogen content up to 68%), which is a kind of typical bulk-additive nitrogen-based FR. During thermal decomposition, non-flammable gases (e.g., NH_3 and N_2) are released, which play the roles of dilution, heat absorption, and cooling [52,53]. MEL and PA are actively studied because of their modifiable chemical structures. The following summarizes the types of 2D SFRs designed using MEL and PA as the basic building units (Figure 5). The supramolecular self-assembly of MEL and PA occurs in the aqueous phase. There is a double synergistic effect of ion attraction (phosphate anion and -NH_3^+) and hydrogen bonding between PA and MEL. And the special triazine ring structure of MEL leads to π-π stacking, thus forming a 2D nanolayered structure (Figure 5a) [54,55]. The MEL-PA assembly can be obtained by simple filtration, washing and drying. Based on MEL-PA, the idea of grafting metal ions (e.g., Cu^{2+}, Zn^{2+}, Ni^{2+}, Mg^{2+}, and Mn^{2+}, Figure 5b) was developed, which can improve the synergistic flame-retardant effects of MEL-PA such as cross-linking and catalytic carbonization and smoke release inhibition [56–58]. The metal ions mainly chelate strongly with the phosphate group structure of PA. Xiaodong Qian [57] found that Cu^{2+}, Zn^{2+}, and Ni^{2+} doping into MEL-PA made the 2D lamellar structure smoother. Wen Xiong Li [56] found that doping with Mn^{2+} made MEL-PA nanosheets thicker and the surface rougher. This is because during the self-assembly process, Mn^{2+} was added to the MEL-PA supramolecular structure, which affected the strength of internal interaction forces and made the assembly skeleton expand outward. In addition, based on MEL and PA raw materials, other organic or inorganic compounds were selected for multi-component self-assembly, demonstrating the flexibility of supramolecular self-assembly (Figure 5c) [58–61]. Introduced components, such as sulfanilic acid [60] and amine-functionalized AL_2O_3 [61], have multiple active sites and are capable of self-assembly with MEL-PA in the aqueous phase through multiple synergies such as ion attraction and hydrogen bonding. Some of the morphology and structure of the multi-component modified MEL-PA will change, and some will still show 2D sheet structure.

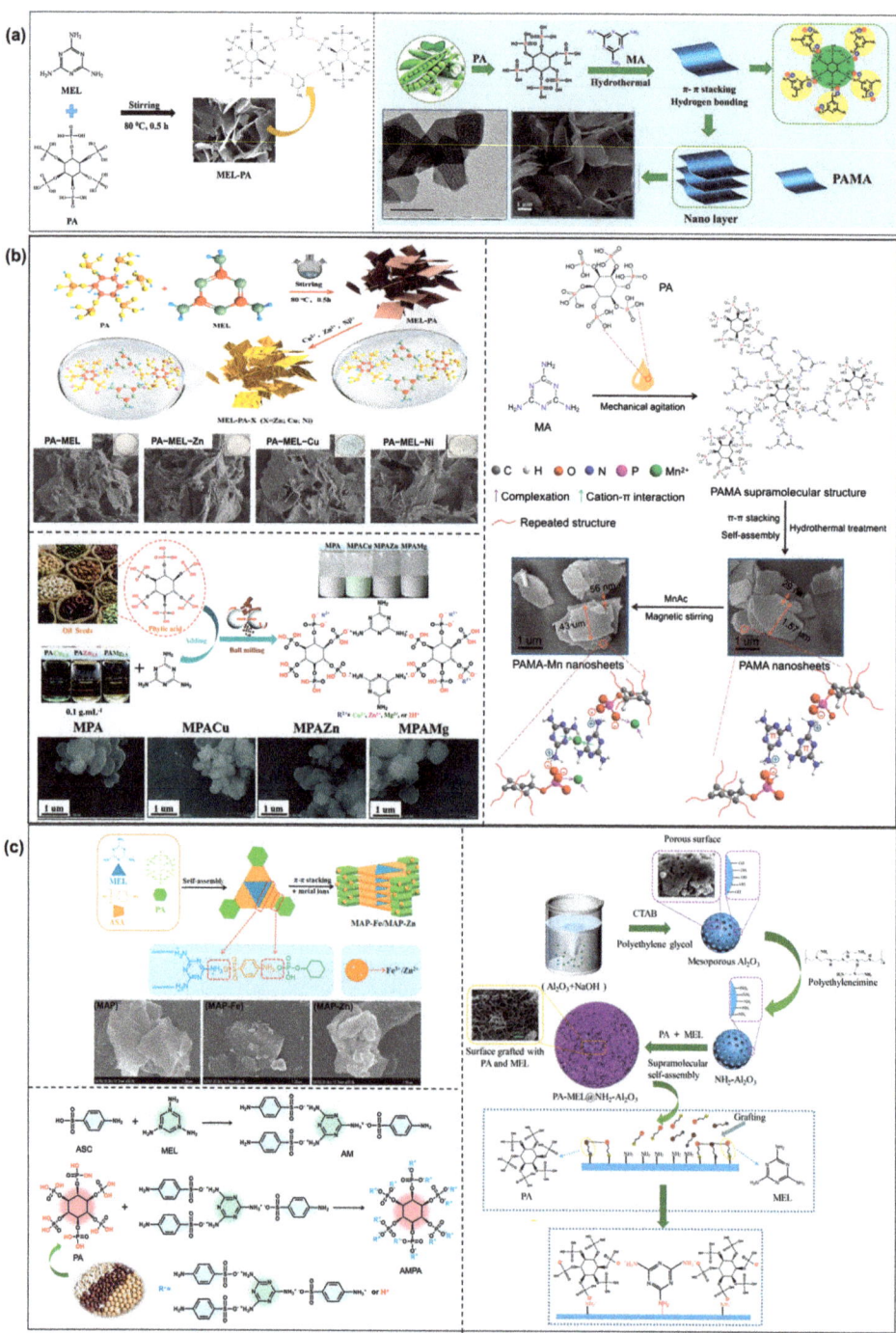

Figure 5. (**a**) Supramolecular self-assembly process of MEL and PA [54,55]. (**b**) Different metal ions were grafted on the basis of MEL-PA [56,57,62]. (**c**) Modify different organic or inorganic compounds on the basis of MEL-PA [58,60,61].

In addition to MEL-PA, other components containing flame-retardant elements have been selected for the self-assembly of 2D SFRs [63–66]. Most of them take advantage of multiple interactions such as electrostatic interaction, hydrogen bonding, π-π interaction, coordination between metal ions and chelating groups among different assembly units in the aqueous phase. Peifan Qin [65] selected MEL and sodium trimetaphosphate (STMP) with hydrogen bond association groups as the assembly units to construct a SFR (named MAP), based on multiple hydrogen bond interactions. MAP showed a ribbon-shaped layered structure (Figure 6a). Kuruma Malkappa [64] made full use of MEL and cyanuric acid for their phenyl-like ring structure of triazine to synthesize a 2D FR (named MCA). MCA nanosheets were obtained based on the conjugation of large π electron clouds on a benzoid ring and multiple hydrogen bonds. Dimethyl sulfone (DMSO) with high polarity was used as the solvent. Using triethylamine (TEA) as the acid-binding agent, PZS was synthesized by the substitution reaction between cyclotriphosphazene and 4,4′-sulfonyl diphenol. Then PZS first aggregated into nuclei based on hydrogen bonding and then attached to the surface of MCA. It was found that the surface of MCA nanosheets became smooth after hybridization with PZS (Figure 6b).

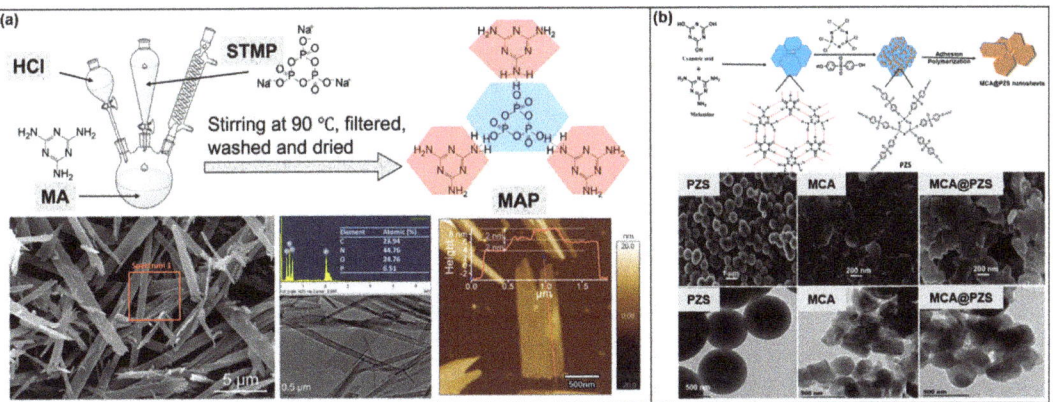

Figure 6. (a) The preparation of MAP, the corresponding micromorphology and EDS analysis [65]. (b) The hybridization process of MCA and PSZ, and the corresponding SEM images [64].

2.3. Three-Dimensional Supramolecular Flame Retardants

The design of 3D SFRs is mainly divided into two categories. The first category is supramolecular self-assembly into sandwich structures or core-shell structures on the basis of some 2D lamellar inorganic materials. Some 2D materials, such as black phosphorus (BP) [67], layered double hydroxides (LDH) [68], graphene [69], MoS$_2$ [70], etc., may face problems of poor stability, compatibility, and uneven dispersion. Therefore, 3D SFRs are assembled from these materials as templates, using some substances containing flame-retardant effects to modify them. Liang Cheng [71] used Si$_3$N$_4$ nanosheet as the template and introduced PA in the aqueous solution. The Si$_3$N$_4$ template is modified by PA based on intermolecular forces. Then, MEL was introduced, which continued to be assembled with the PA (as described earlier, the Mel-PA assembly presented a 2D lamellar structure) to build a sandwich structure. The surface of Si$_3$N$_4$ modified by PA-MEL changed from smooth and porous to rough, but the crystal structure remained unchanged (Figure 7a). Metakaolinite also uses this idea [72]. Due to the abundant hydroxyl groups on the surface and between layers of metakaolinite, PA was introduced based on hydrogen bonding. And then a layered stacking structure was constructed by electrostatic assembly with MEL (Figure 7b). Xiaming Feng [73] and Shuilai Qiu [74] utilized melamine cyanurate as a bridge. Using MoS$_2$ sheets (Figure 7c) and aminated-BP nanosheets (Figure 7d) as the templates, respectively, the sandwich structure was self-assembled. Besides, there are core-shell structures. Yanlong Sui [75] took -NH$_2$-modified SiO$_2$ nanoparticles as the core, coated PA as the shell by strong electrostatic action in the mixed solution of ethanol and water, and finally introduced Ni^{2+} based on coordination. The addition of Ni^{2+} made the surface of SiO_2@PA microspheres coarser than the smooth surface of SiO_2 nanospheres. The microspheres were all independently dispersed. The thickness of the composite shell was about 40 nm (Figure 7e).

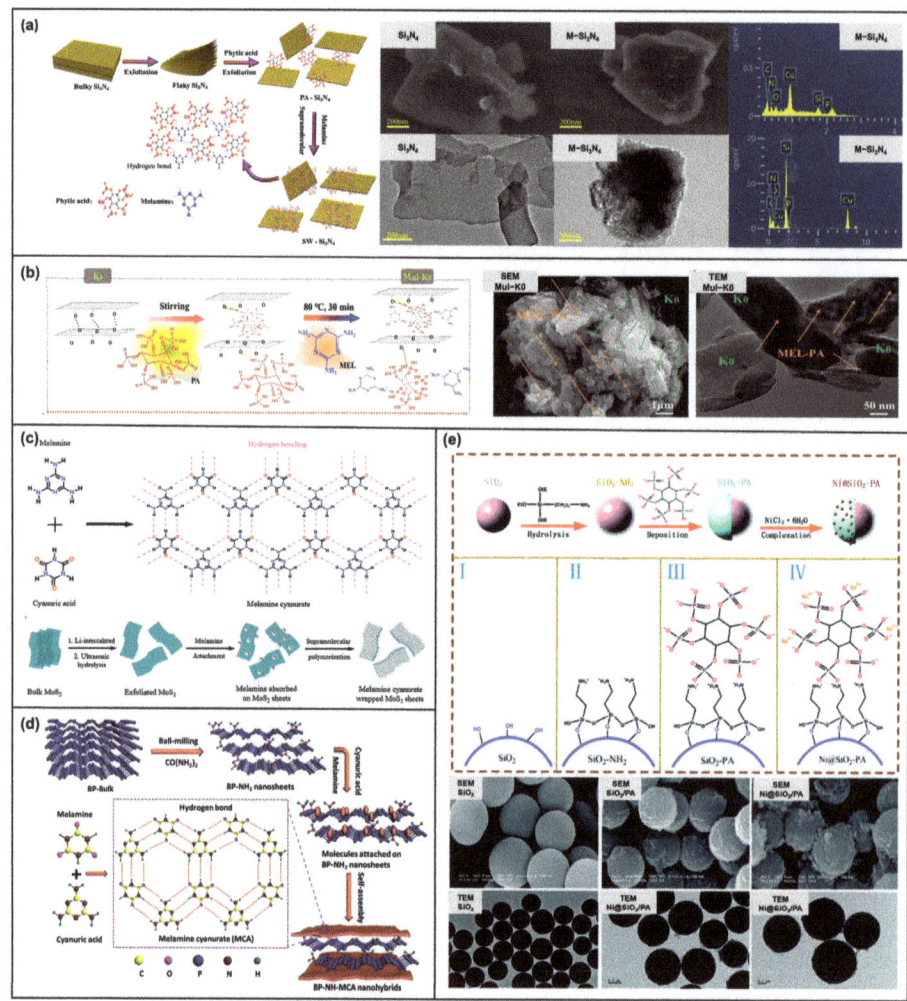

Figure 7. (**a**) Modification diagram of Si_3N_4 nanosheet; SEM, TEM and element spectra before and after modification [71]. (**b**) Multi-dimensional modification diagram of metakaolinite; SEM and TEM images after modification [72]. (**c**) The sandwich structure assembly diagram of MoS_2 sheet and melamine cyanurate [73]. (**d**) The sandwich structure assembly diagram of aminated-BP nanosheets and melamine cyanurate [74]. (**e**) The assembling process of SiO_2 nanosphere core-shell structure. SEM and TEM images of SiO_2 nanospheres before and after modification [75].

The second category is supramolecular self-assembly among small molecules, showing a 3D network structure. Most of these small molecules contain flame-retardant elements with abundant active sites. Shuitao Gao [76] used PA and branched polyethylenimide (b-PEI) as the basic units. A 3D network was assembled through hydrogen bonding and electrostatic interaction between the phosphonic acid group of PA and the amino and imino groups of b-PEI (Figure 8a). Shuo-ping Chen [77] assembled 1-aminoethyl diphosphonic acid (AEDPH4) and ethylenediamine (En) by ion attraction in aqueous solution. The specific assembly mode was: AEDPH4 transferred two hydrogen cations (H⁺) to En. At this time, there was a strong hydrogen bond between the AEDPH4 units to form a 1D zigzag chain. After receiving 2 H⁺, the En units had multiple N-H ·· O hydrogen bonds with these 1D chains. Finally, a 3D Mosaic network structure was formed (Figure 8b).

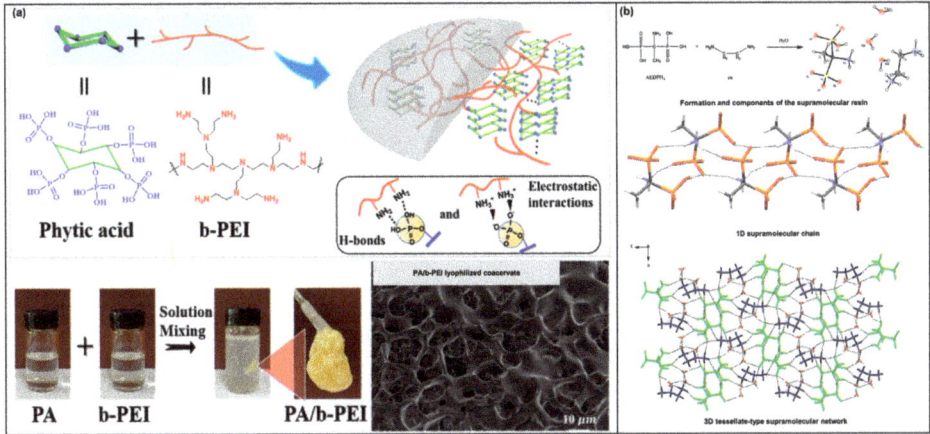

Figure 8. (**a**) Schematic diagram and physical diagram of PA and b-PEI assembly; SEM image of the assembled sample after lyophilization [76]. (**b**) The chemical structure of two units for assembly, and the structure diagram in the assembly process [77].

3. Applications of Supramolecular Flame Retardants

3.1. Flame Retardancy

Polymers are now used in all areas of human production and life. Polymers can be divided into plastics, fibers, and rubber, etc. [78]. Plastics are widely used due to their strong adhesion, light weight, excellent mechanical properties, and durability [79–81]. The typical ones are epoxy resins (EP) [74], rigid polyurethane foam (RPUF) [82], polypropylene (PP) [83], polyvinyl alcohol (PVA) [84], thermoplastic polyurethane (TPU) [85], Polyamide 6 (PA 6) [86], and poly lactic acid (PLA) [87]. In addition, textiles also penetrate all aspects of human life [88]. However, most polymers are composed of two elements, C and H. They are highly flammable. It limits the scope of their application and poses a significant fire risk. Therefore, it is very important to try various methods to make polymers stable at high temperatures and nonflammable [89–91]. SFRs have shown excellent effects in this regard. And they are more environmentally friendly and easier to synthesize [46,63].

3.1.1. EP

Peifan Qin et al. [65] used MAP (Figure 5a) to increase the flame retardancy of EP. MAP was added to EP by mechanical stirring at 140 °C and cured at 180 °C to produce an EP-MAP composite. The whole synthesis route and curing process are shown in Figure 9a. The flame retardancy of the EP-MAP composite was found in the vertical combustion test (UL-94) and the limiting oxygen index (LOI) test. 4 wt% MAP was added to make EP-MAP-4% composites reach a V-0 rating and the LOI value reach 30% (Figure 9b). It showed better flame retardancy than pure EP. In addition, the fire resistance of the material was further verified by the cone calorimeter test (CCT). With the addition of MAP, the peak heat release rate (PHRR) value of EP composites decreased from 1076 Kw/m² to 370 Kw/m², and the total heat release (THR) value decreased from 90 MJ/m² to 72.9 MJ/m². Figure 9c shows that the decreasing range is 65.6% and 17.7%, respectively. It proves that EP-MAP has good flame retardancy. Compared with pure EP, the peak smoke production rate (PSPR) and total smoke production (TSP) of EP-MAP composites decreased by 63.7% and 45.4%, respectively, which also proves that EP-MAP has good smoke suppression performance. The possible flame-retardant mechanism was speculated (Figure 9d), which is divided into four steps to flame retardant EP. In the first two steps, water is formed by endothermic condensation by P-O-H-N hydrogen bonds. And the vaporized MEL ring dilutes the combustible gas concentration. Inhibit the diffusion of combustible gas in the gas phase. At the same time, MAP undergoes ring-opening and cracking reactions, generating acid sources for catalytic dehydrogenation and expanding coke to promote the formation of a dense expanded carbon layer. In the last two steps, many aromatic crosslinked structures contain heterocycles of P, N, and O elements, which make the diffusion path of combustible gas tortuous, thus slowing down the heat and mass transfer rate and achieving the flame-retardant effect.

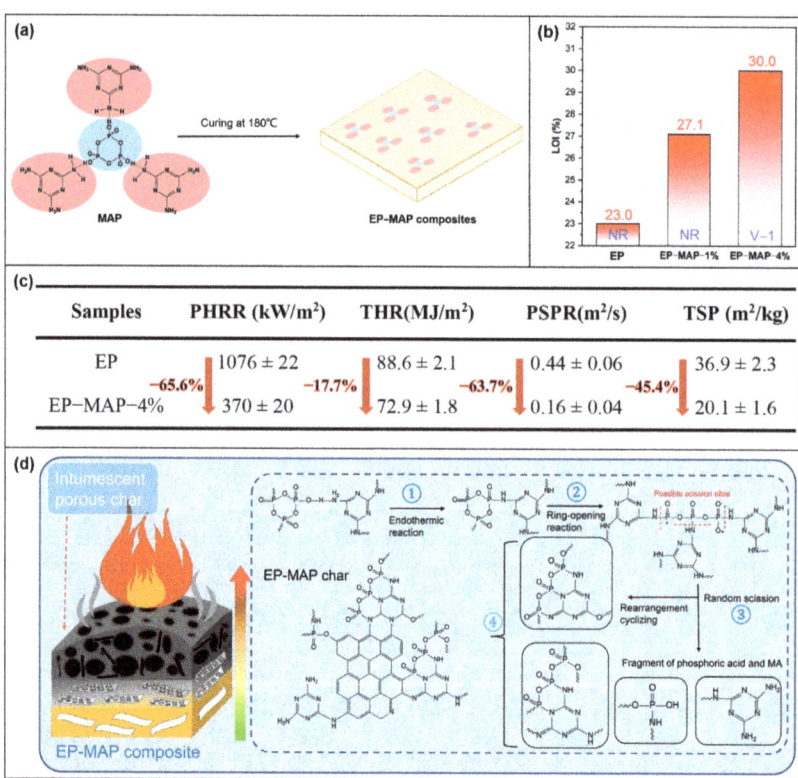

Figure 9. (**a**) The production route of composite materials. (**b**) Results of LOI and UL-94 tests. (**c**) CCT data: PHRR, THR, PSPR, TSP. (**d**) EP-MAP-4% carbon residue photograph. [65].

The low loading of SFRs into the EP not only greatly improves its flame retardancy but also its thermal stability. Shuilai Qiu and co-workers [74] applied BP-NH-MCA (Figure 7d) to EP. The authors used thermogravimetry (TGA) to analyze the activation energy of thermal decomposition of pure EP and EP/BP-NH-MCA composites at different heating rates. From the data (Figure 10a), it is clear that the addition of 2 wt% BP nanosheets and 2 wt% BP-NH-MCA hybrid particles can increase the thermal degradation activation energy of EP by 11.00% and 22.85%, respectively. While the activation energy of EP/BP-NH-MCA 2.0 is 34.81% higher than that of pure EP at the maximum mass loss rate of the composites, indicating a good synergistic flame-retardant effect between BP and MCA. There is also a great improvement in the thermal stability of the composites. Compared to pure EP, the addition of 0.5 wt% to 2.0 wt% BP-NH-MCA reduces the PHRR of the composites by 21.9 to 47.2%

and the THR values by 26.7 to 42.3% (Figure 10b,c). The catalytic carbonization is facilitated by the highly efficient phosphorus–nitrogen synergistic flame-retardant system constructed by BP-NH$_2$ together with MCA supramolecules, resulting in a denser continuous carbon layer than EP and EP/BP2.0 (Figure 10d). The possible flame-retardant mechanism of EP/BP-NH-MCA composites can be divided into two stages (Figure 10e). In the first stage, the BP nanosheet acts as a physical barrier, effectively preventing the escape of combustible volatiles from the EP matrix. Besides, MCA is broken down and volatilizes non-flammable gases (HOCN, CO_2, and NH_3) to dilute the fuel and interrupt the combustion process. In the second stage, most BP is oxidized into a series of PO_x and phosphoric acid derivatives in the air and combined with the nitrogen structure to form a heat-stable carbon layer. At the same time, the coke network and the cross-linked phosphorus oxynitride form a physical barrier, which can delay the escape of flammable volatiles from the EP matrix.

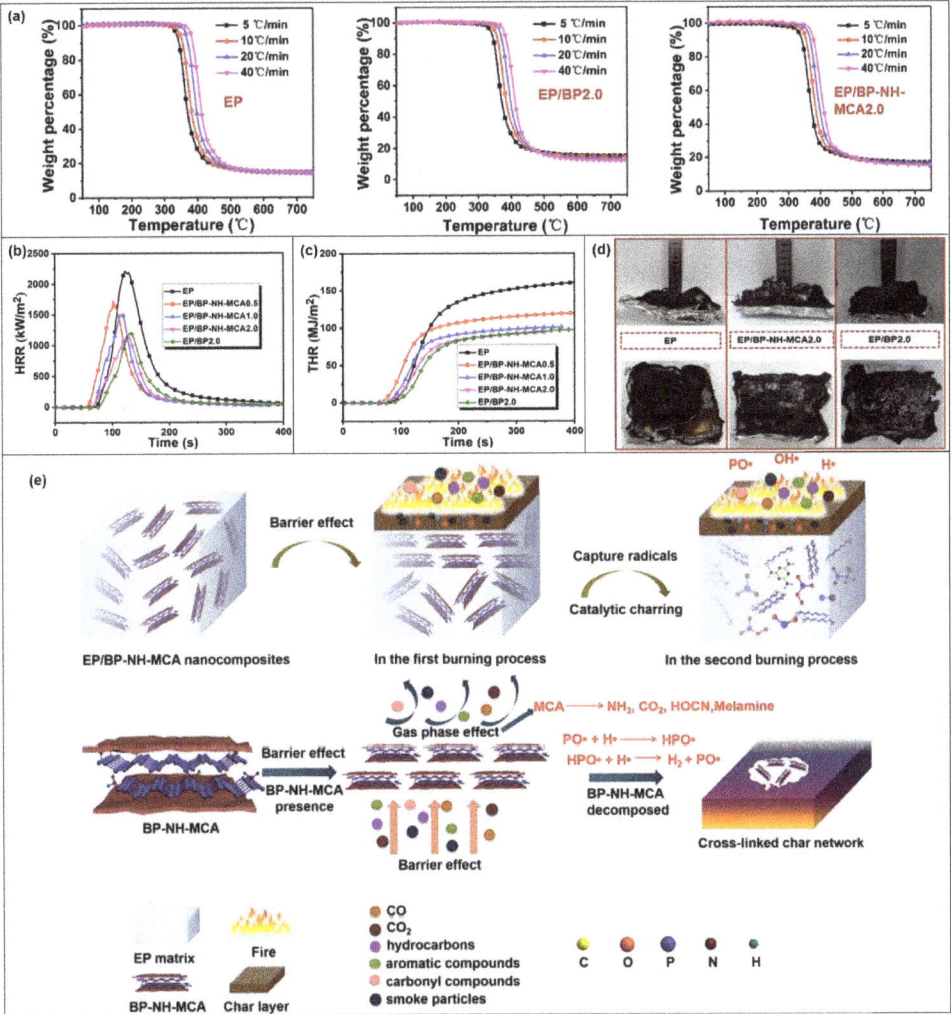

Figure 10. (**a**) TGA curves of EP, EPBP2.0 and EPBP-NH-MCA2.0 at different heating rates. (**b**) HRR curve of EP composites. (**c**) THR. (**d**) SPR. (**e**) TSP, Schematic diagram of flame-retardant mechanism [74].

3.1.2. Rigid Polyurethane Foam

Engin Burgaz [92] found that compared with the effects of -COOH modified MWCNTs or nanosilica applied to RPUF alone, their effects were not as good as those of binary complexes assembled based on multiple hydrogen bonding (Figure 11a). The effects of modified MWCNTs and nanosilica assemblies of different proportions on the properties of RPUF were also investigated. The results show that modified MWCNTs (0.4 wt%) and nanosilica (0.1 wt%) assemblies can improve the thermal decomposition of RPUF (Figure 11b). The thermal decomposition temperature reached its highest at 5%, 15% weightlessness and maximum mass loss (Figure 11c).

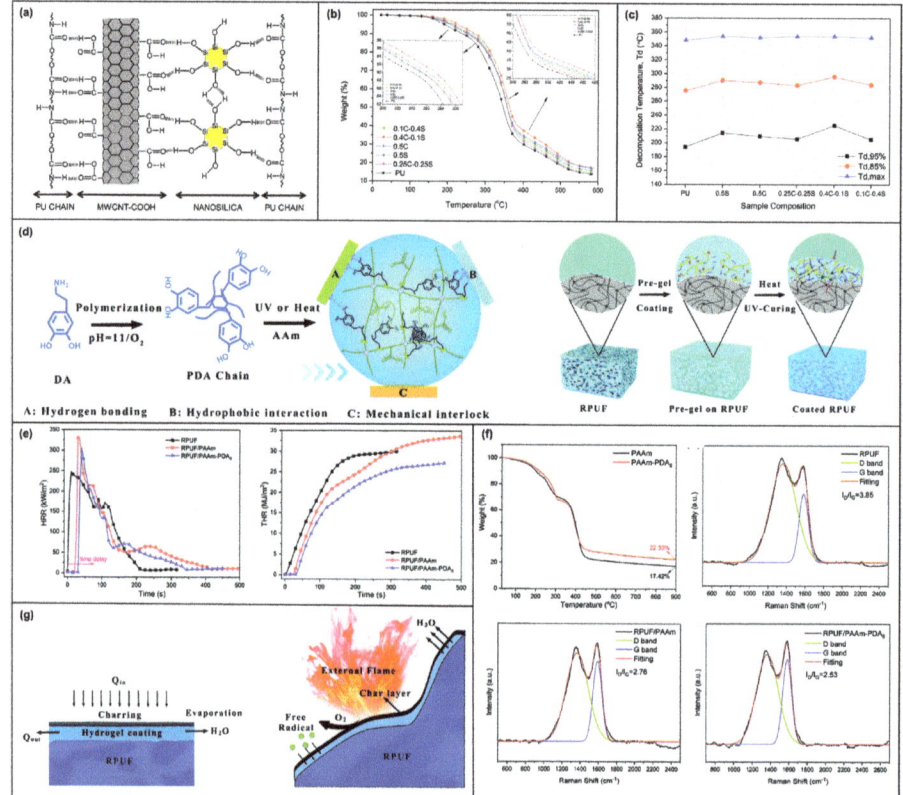

Figure 11. (**a**) Schematic representation of the forces between modified MWCNTs and nanosilica assemblies and RPUFs. (**b**) TGA of RPUF under different assembly conditions. (**c**) Comparison of decomposition temperature under different weightlessness. (**d**) Diagram of dual network hydrogel applied to RPUF [92]. (**e**) The HRR and THR curves of the RPUF before and after the flame-retardant coating treatment. (**f**) Raman spectra of RPUF carbon residues before and after the flame-retardant coating treatment. (**g**) Schematic diagram of the flame-retardant mechanism of RPUF treated with the coating [82].

Yubin Huang [82] proposed a dual network hydrogel based on the combination of covalent and non-covalent interactions. Use the hydrogel as a flame-retardant coating for RPUF. Supramolecular-assembly was mainly manifested by multiple hydrogen bonds and π-π stacking between polyacrylate and polydopamine chains (Figure 11d). In CCT, the ignition time of pure RPUF is only 6 s. The ignition time of RPUF after the double network hydrogel coating is increased to 36 s, indicating that the fire resistance is improved. The mean HRR of RPUF with a dual network hydrogel coating is reduced by 39.7% (Figure 11e). According to Raman spectrum analysis, the residual carbon quality of RPUF coated with dual network hydrogel is the densest (Figure 11f), which can effectively play a flame-retardant role in the condensed phase. In addition, the aerogel coating evaporates water when it burns, creating a concentration that dilutes combustible gases and heat in the gas phase. It also has the function of quenching free radicals (Figure 11g).

3.1.3. PP

Congrui Qi and colleagues [93] used melamine-trimesic acid (MEL-TA) supramolecular aggregates to modify the surface of ammonium polyphosphate (APP). The modified APP and carbonated foaming agent (CFA) were then blended into the PP matrix. The combustion properties of PP composites and the corresponding flame-retardant mechanism were investigated. The cross-sectional SEM images of the PP composites are shown in Figure 12a. The dispersion effect of modified APP in PP is obviously better than that of unmodified APP, which improves the compatibility in the matrix. The TGA image (Figure 12b) also shows that the maximum thermal decomposition rate of the PP composite is significantly lower than that of pure PP, indicating that the addition of APP@MEL-TA has improved the thermal stability of PP. The ratio of PP composites and the corresponding flame-retardant performance data are shown in Figure 12c. When the mixing ratio of APP@MEL-TA and CFA is 4:1, the LOI value of PP-5 reaches its maximum value of 34.8%. In addition, PP-5 can reach the V-0 rating in the UL-94 test without dripping, while pure PP can not reach the rating with dripping. Pure PP releases a lot of heat after ignition, with a PHRR of 937 kW/m^2 and a THR of 87.8 MJ/m^2. The PHRR and THR values of PP-5 decreased to 97 kW/m^2 and 38.5 MJ/m^2, respectively (Figure 12 d,e), which further proves that the modified FR has an excellent flame-retardant effect.

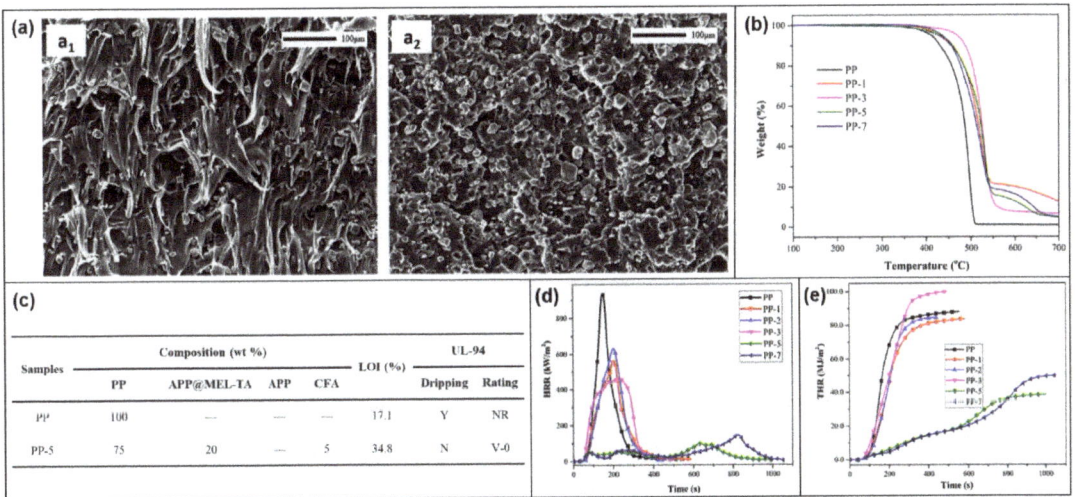

Figure 12. (a) SEM image of fracture surface: (a_1) PP-7 and (a_2) PP-5. (b) TGA curves of PP composites. (c) The ratio and flame-retardant test results of PP composites (d) HRR. (e) THR [93].

3.1.4. PVA

Lei Liu et al. [94] designed and synthesized a polyamine small molecule (named HCPA) that can be used as a FR for PVA. The flame-retardant effects of different HCPA concentrations on PVA were investigated (Figure 13a). UL-94 and LOI measurements were used to evaluate the flame retardancy of the film. Pure PVA film burns rapidly within 12 s of ignition, accompanied by heavy dripping, resulting in failure to reach the rating (Figure 13b). However, the PVA/5.0HCPA film rapidly self-extinguishes after the first ignition and continues to burn for only 3 s after the second ignition, during which no molten drops are observed, thus achieving the required UL-94 VTM V-0 rating (Figure 13c). The addition of 5.0 wt% HCPA increases the LOI of the original PVA from 19.0% to 24.3%. Similarly, with the increase in HCPA content to 10 wt%, the LOI value will increase to 25.6% (Figure 13c).

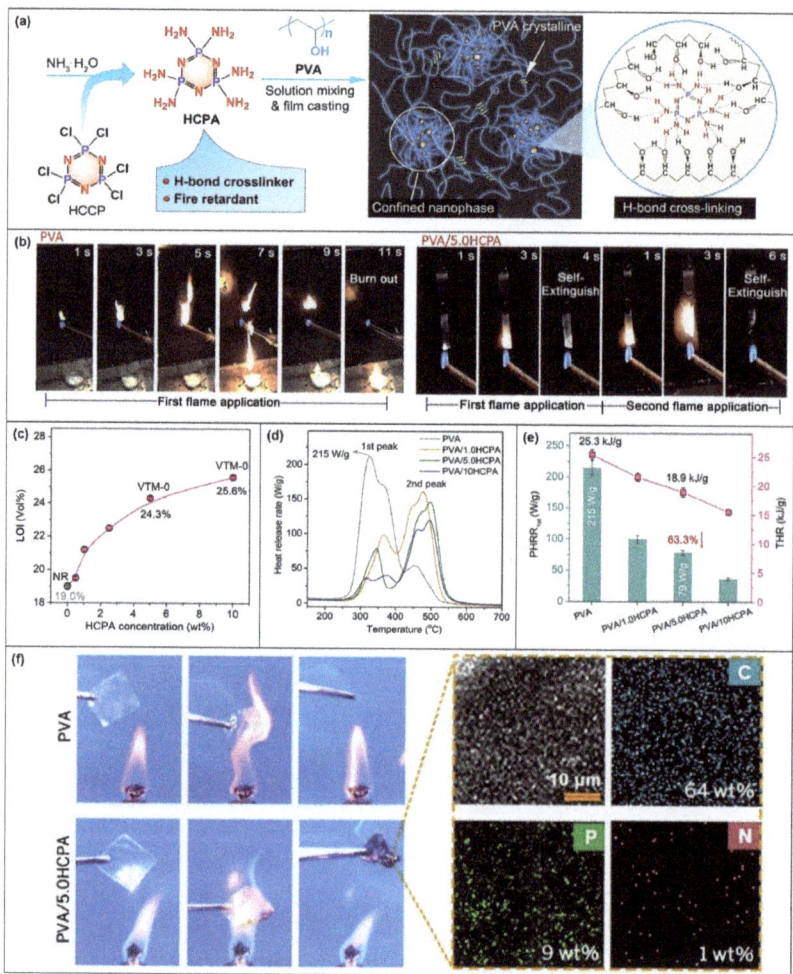

Figure 13. (a) Preparation route of PVA/HCPA composite membrane. (b) Digital images of PVA and PVA/5.0HCPA film vertical combustion test. (c) LOI test results. (d) HRR of MCC. (e) PHRR and THR values. (f) Composite film combustion image and SEM-EDS image of residual carbon after PVA/5.0HCPA film combustion. [94].

Figure 13d,e shows the test data from the Micro Cone Calorimetry (MCC) method. The first PHRR value of PVA/5.0HCPA film decreases by 63% compared with the original PVA. The THR value decreases from 25.3 kJ/g to 18.9 kJ/g. Furthermore, adding PVA/10HCPA can reduce the THR value by 38.5%. The decrease in these flame-retardant parameters confirms that HCPA improves the flame retardancy of PVA. After the PVA/5.0HCPA composite film is ignited by an alcohol lamp, it leaves a complete and dense carbon residue, which is in sharp contrast with the original PVA with no residue (Figure 13f), reflecting the good carbonization ability of HCPA. Therefore, PVA/HCPA can achieve a synergistic flame-retardant effect in the gas phase and condensed phase.

3.1.5. TPU

Liang Cheng et al. [71] proposed SW-Si_3N_4 (Figure 7a), which can effectively reduce the fire risk of TPU materials. The TGA analysis curves are shown in Figure 14a. The addition of 5 wt% SW-Si_3N_4 generates 4.97 wt% of residual carbon at 800 °C in SW-Si_3N_4/TPU, which is higher than the 0.30 wt% of residual carbon in pure TPU. In addition, the SW-Si_3N_4 hybrid nanosheets in the TPU matrix are able to suppress heat transfer more effectively compared to the Si_3N_4 nanomaterials. The combustion behavior of the TPU composites was investigated by CCT. Figure 14b,c shows the HRR and THR curves. Figure 14d,e shows the data curves of SPR and TSP. It is obvious that the smoke release is inhibited by each doping amount of TPU composites, which also confirms the good smoke suppression property of SW-Si_3N_4.

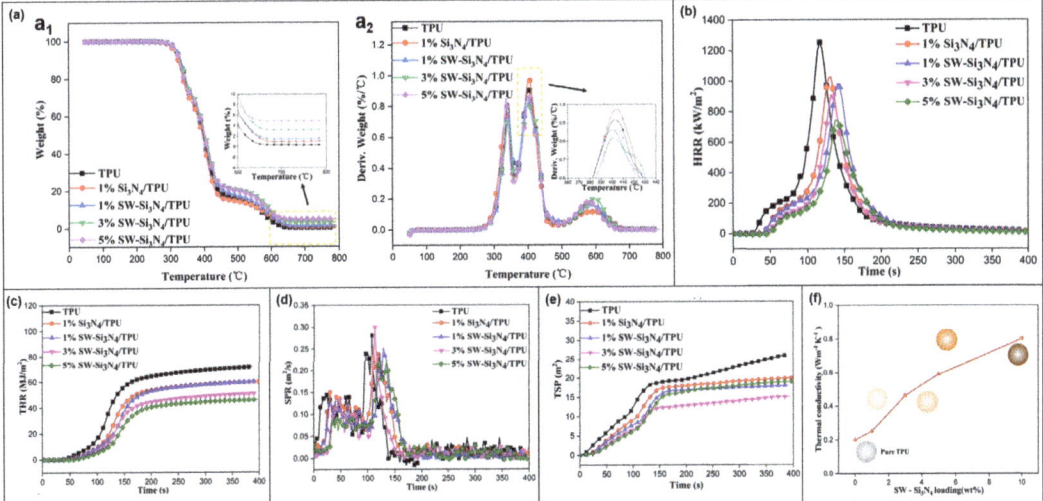

Figure 14. (a) Data curves: (a_1) TGA and (a_2) DTG. (b) HRR. (c) THR. (d) SPR. (e) TSP. (f) Thermal conductivity of TPU composites [71].

The thermal conductivity of SW-Si_3N_4/TPU (Figure 14f) shows that the thermal conductivity increases with increasing filler content and reaches approximately a fourfold increase in 10% SW-Si_3N_4/TPU. The TPU composite has an internal heat transfer path. It is conducive to heat diffusion and can reduce the risk of fire due to heat buildup locally in the matrix.

3.1.6. PA6

Xiaodong Qian et al. [57] studied the effect of PA-MEL with metal ions grafted (Figure 5b) on the properties of PA6. The LOI value of the composite increases from 21.5% to 30.0% after the addition of PA-MEL-phosphate/transition metal (Figure 15a), demonstrating the good flame-retardant effect. The flame-retardant properties of the composites were further evaluated by CCT. Pure glass fiber reinforced PA6 (GFPA) burns rapidly after ignition and has a PHRR value of up to 739.97 kW/m^2. The addition of PA-MEL reduces the PHRR of the composite to 612.38 kW/m^2. In particular, the addition of PA-MEL-Cu leads to a reduction of up to 32.37% in the PHRR value (Figure 15b). Moreover, the THR value of the PA-MEL-Cu composite decreases by 28.04% compared with GFPA (Figure 15c). All SPR peaks for the composites are lower than GFPA after the addition of the PA-MEL-phosphate or transition metal (Figure 15d). The TSP of the GFPA/PA-MEL-Cu composites is reduced by up to 36.7% (Figure 15e). Compared with unmodified PA-MEL, the addition of PA-MEL-Zn will increase the number of carbon layers (Figure 15f) and the degree of graphitization of the residual char (Figure 15g). A possible flame-retardant mechanism was proposed. PA promotes the formation of pyrophosphate or polyphosphate and catalyzes the formation of a stable carbon layer. MEL releases non-combustible gases such as NH_3 and H_2O to achieve dilution. In addition, the incorporation of transition metals has a synergistic flame-retardant effect, which can promote the formation of dense carbon layers.

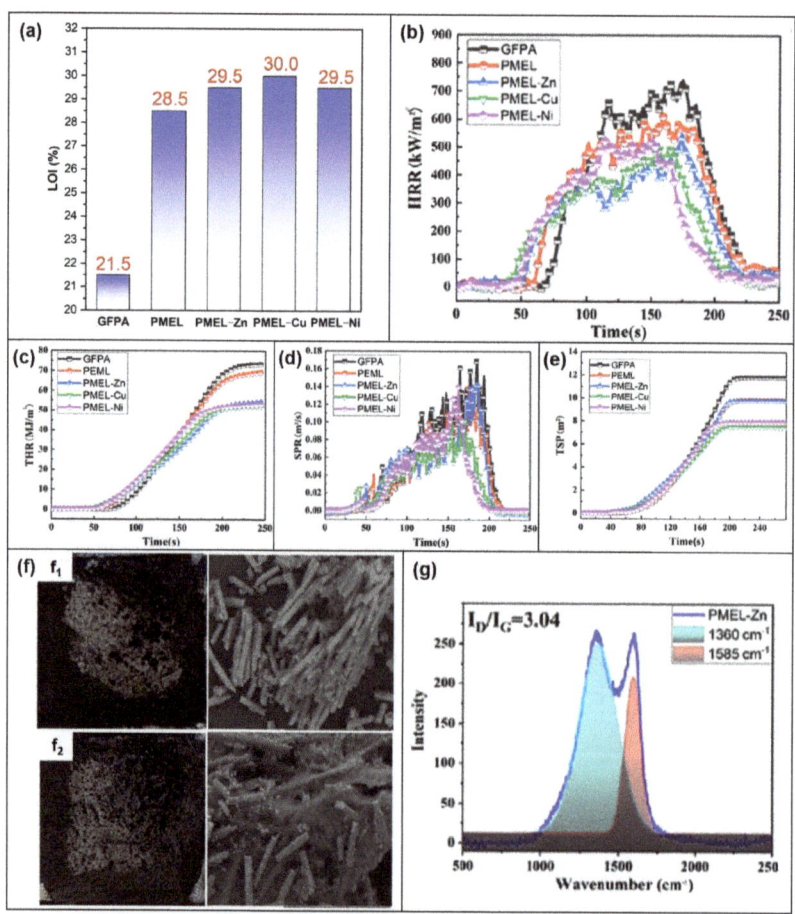

Figure 15. (a) Ratio of PA composite and LOI values. (b) HRR. (c) THR. (d) SPR. (e) TSP. (f) SEM image of carbon residue: (f_1) PA-MEL and (f_2) PA-MEL-Cu. (g) Raman image of PA-MEL-Zn. [57].

3.1.7. PLA

Qinyong Liu et al. [58] self-assembled MEL, paminobenzene sulfonic acid (ASA), and PA into a new SFR (named MAP) by a simple green method. The product after chelation with Fe^{3+} or Zn^{2+} was named MAP-Fe and MAP-Zn (Figure 5c) and was applied to PLA. The LOI value of pure PLA is only 20.4%, which not only fails to achieve a rating in the UL-94 test but is accompanied by severe dripping. In contrast, the PLA composite achieves a V-2 rating for PLA-2% MAP while also shortening the burning time after two ignitions. PLA-3% MAP-Zn even achieves a V-0 rating, minimizing the burning time and increasing the LOI value to 29.2% (Figure 16a). Interestingly, the authors used two thermocouples (T1 and

T2) to record the temperature changes at two locations during the UL-94 test (Figure 16b). The internal temperature of the PLA composite significantly reduces. Figure 16c shows the temperature profiles for each of the two ignitions. At the end of the first ignition for 10 s, T1 and T2 are reduced to 139 °C and 113 °C for 3.0% MAP-Zn, respectively (483 °C for pure PLA). The values for pure PLA are 490.1 kW/m^2 and 73.3 MJ/m^2, respectively, while PLA-3.0% MAP-Zn shows the lowest PHRR and THR values of 398.5 kW/m^2 and 65.6 MJ/m^2 (Figure 16d,e). The possible flame-retardant mechanism of PLA composites was summarized. Firstly, MAP releases NH_3 and phosphorus-containing radicals in the gas phase to dilute oxygen with combustible volatiles while trapping H· and OH· during combustion, interrupting the chain reaction of combustion. In the condensed phase, further cross-linking of MEL and ASA will form a stable carbon layer, while Fe^{3+} and Zn^{2+} will optimize the existence of the carbon layer.

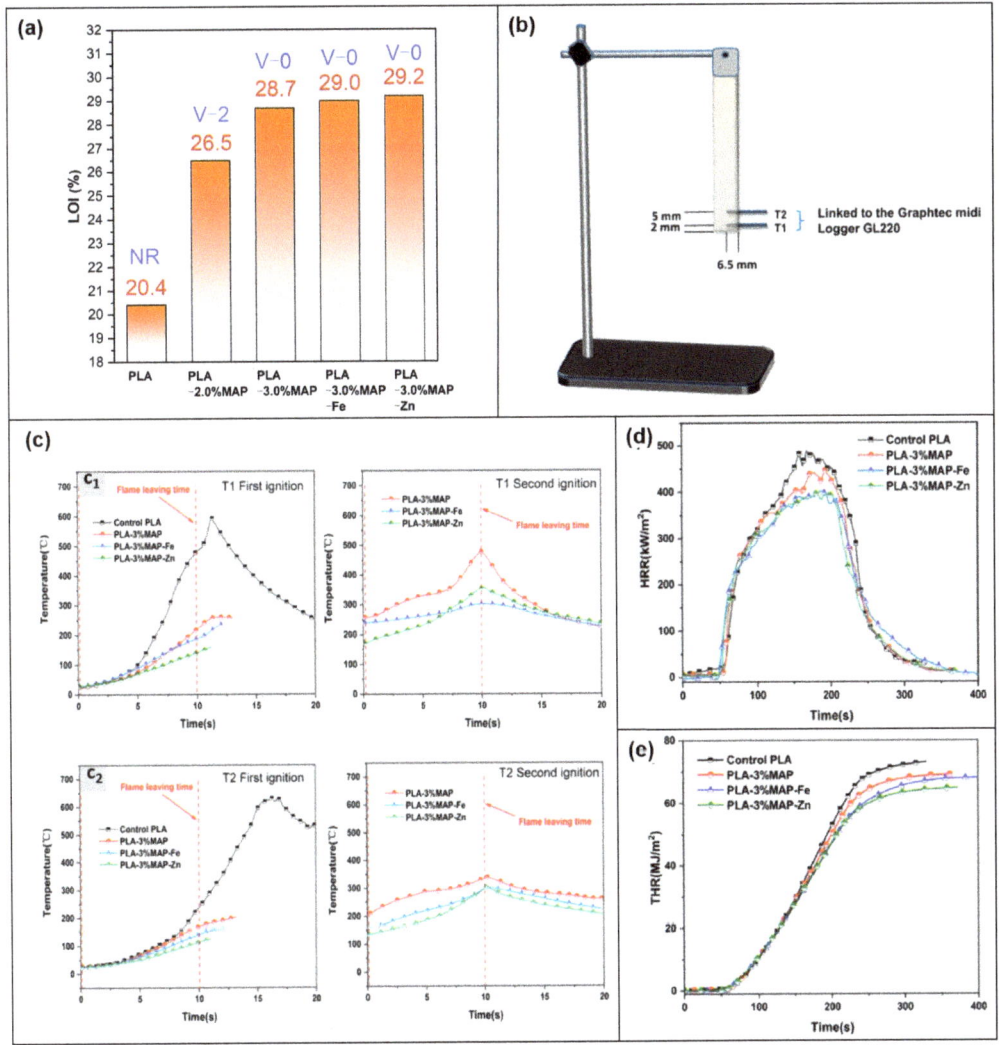

Figure 16. (**a**) Test results of LOI and UL-94. (**b**) Diagram of internal temperature test. (**c**) Temperature data recorded by thermocouple: (**c$_1$**) T1 and (**c$_2$**) T2, (**d**) HRR. (**e**) THR. [58].

3.1.8. Cotton Fabrics

Cotton fabrics are considered to be one of the most popular natural fibers due to their perspiration absorption, renewable nature, and comfortable feel [95,96]. However, cotton fabric is flammable. And the flame will spread rapidly after being ignited, thus causing a fire [97,98]. In the method of flame-retardant treatment of textiles, the most typical method is to assemble the coating layer by layer (LBL). The self-assembly of LBL makes the layers have attractive forces such as electrostatic attraction, hydrogen bonding and coordination bonds, forming supramolecular coatings, thus giving textiles flame retardancy [99,100].

Wen An et al. [101] constructed flame-retardant and antistatic fabrics by LBL assembly, using pure cotton fabric as the backbone and dipping the cotton fabric alternately into cationic casein (CA) and anionic graphene oxide (GO) solutions (Figure 17a). The LBL-assembled coated cotton fabrics form the carbon layer earlier in the process of combustion. A comparison of the carbon residues at 600 °C shows that the thermal stability of the cotton fabric increases with the number of assembled layers on the surface of the fabric (Figure 17b). The progressively higher LOI values (Figure 17c) also indicate that the flame resistance of the coated cotton fabric increases with the assembled layer number of rGO/CA.

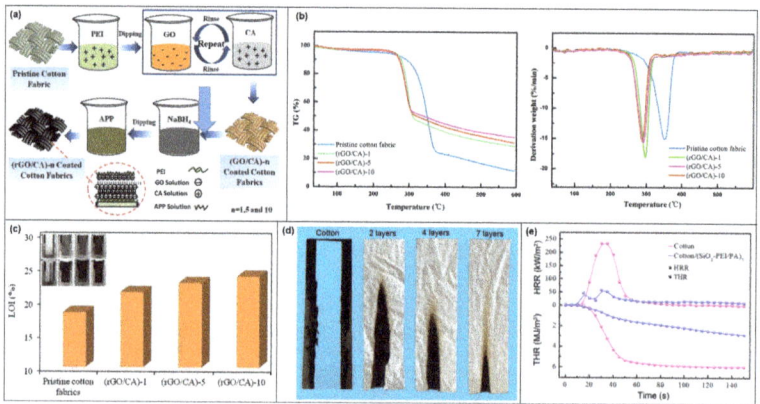

Figure 17. (a) Diagram of LBL assembly coating cotton fabric. (b) TGA and DTG curves of pure cotton fabrics and coated cotton fabrics with different cycles. (c) LOI test values [101]. (d) Digital images of vertical flammability test. (e) HRR and THR curves of cotton and cotton/(SiO$_2$-PEI/PA)$_7$ [102].

Shanshan Li et al. [102] designed a novel organic-inorganic hybrid intumescent flame-retardant coating by using nanosilica (SiO$_2$) covered with polyethyleneimine (PEI) and PA. It was applied to the surface of cotton fabric by LBL assembly. It achieved a good flame-retardant effect with only 7 bilayers. The UL-94 test shows that compared with pure cotton fabric, the combustion of coated cotton fabric is relatively slow. The spread of flame from the bottom of the coated cotton fabric is gradually weakened (Figure 17d). The PHRR of Cotton/(SiO$_2$-PEI/PA)$_7$ is 58 kW/m^2. It is 75% lower than that of untreated cotton fabric. The coating assembled by LBL within 150 s reduced the THR value of cotton fabric from 5.94 MJ/m^2 to 2.83 MJ/m^2 (Figure 17e).

3.2. Mechanical Properties

As we all know, in order to achieve efficient flame retardancy, the loss of mechanical properties is inevitable [103,104]. The successful preparation of SFRs has made great contributions to achieving flame retardancy and minimizing the loss of mechanical properties [46,105].

A new green FR, Ni@SiO$_2$-PA (Figure 7e), has been applied to flame retardant EP. The strength of EP nanocomposites has been significantly improved. This core-shell FR was made by Yanlong Sui et al. [75] The tensile modulus of EP and its composites is shown in Figure 18a. Due to the strong hydrogen bonding, there is a strong interface interaction between Ni@SiO$_2$PA and the EP matrix, which makes the tensile modulus of EP/Ni@SiO$_2$-PA3.0 higher than that of pure EP by 22.2%. Compared with the smooth surface of pure EP, large pores form in the EP/SiO$_2$1.0 composite with the obvious agglomeration of SiO$_2$ particles. EP/Ni@SiO$_2$-PA1.0 and EP/Ni@SiO$_2$-PA5.0 composites show a rough and inhomogeneous fracture surface (Figure 18b), demonstrating the extremely high supramolecular shell-matrix compatibility.

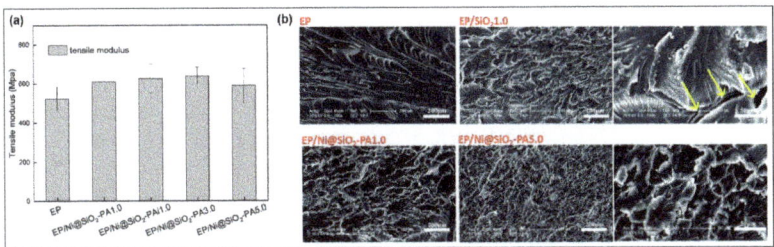

Figure 18. (a) Tensile modulus diagram. (b) SEM images of EP and EP composites [75].

Melamine cyanorate/α-ZRP nanosheets (MCA @ α-ZRP) [106], also made by supramolecular assembly technology, were applied to TPU (Figure 19a). And H· was used to strengthen the interface with the TPU matrix and limit the fluidity of the polymer chain, thus successfully avoiding the decrease of mechanical properties with the addition of FRs. As shown in Figure 19b, TPU composites all show better mechanical properties than pure TPU. TPU/MCA@α-ZrP increases the tensile strength by 43.1% at the maximum (Figure 19c). There is a synergistic effect in the MCA@α-ZrP blend, which reduces the influence of MCA alone on the flexibility of TPU (Figure 19d) and increases the fracture strain of TPU composites from 629% to 664%. Besides, this study calculates the fracture energy of the TPU composites in order to indirectly estimate their toughness (Figure 19e). The hydrogen bond breakage between the MCA@α-ZrP blend and TPU, leads to the reconstruction of the H-bond, inducing stress transfer and ductility, which consume more fracture energy. Therefore, TPU/MCA@α-ZrP shows the highest fracture energy of 162 MJ/m^3.

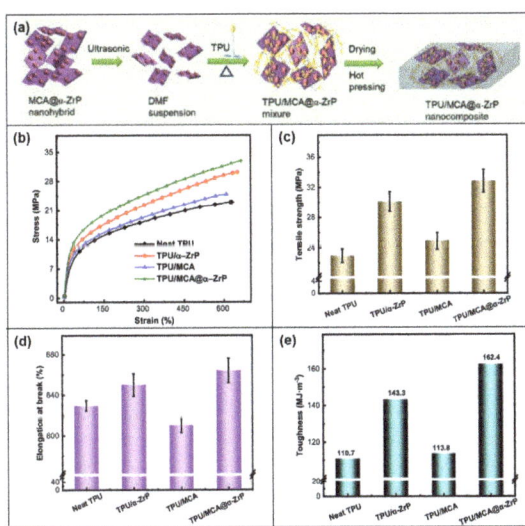

Figure 19. (a) Preparation of TPU/MCA@α-ZrP nanocomposites. (b) Stress-strain curves of TPU and its composites. (c) Tensile strength histogram. (d) Elongation at break histogram. (e) Fracture energy data image [106].

Kuang Li et al. [107] prepared soy protein (SP)-based films (PVP@LS) with supramolecular network structures including dynamic H, π-π interactions and interconnected water transport interactions, which can effectively dissipate energy when the films are stretched. These non-covalent interactions are beneficial to improve the adhesion and cohesion of the films so that they can have high tenacity and tensile strength along with high flame retardancy. As shown in Figure 20a,b, the tensile strength and toughness of SP/PVP@LS-2 film are 111.39% and 386.54% higher than those of pure SP film, respectively. The peak values are as high as 16.15 MPa and 23.50 MJ/m^3. In order to further prove the enhancement of the mechanical properties of the SP composite film, the author compared the strength and elongation at break of the SP composite film with other reports. SP/PVP@LS shows better tensile strength (Figure 20c). Figure 20d shows a possible toughening mechanism for SP/PVP@LS films, showing the microscopic and macroscopic mechanisms of the SP composite film under external loading stress, respectively. This in turn demonstrates the toughening effect of a strong and stable supramolecular crosslinking network on the SP composite film.

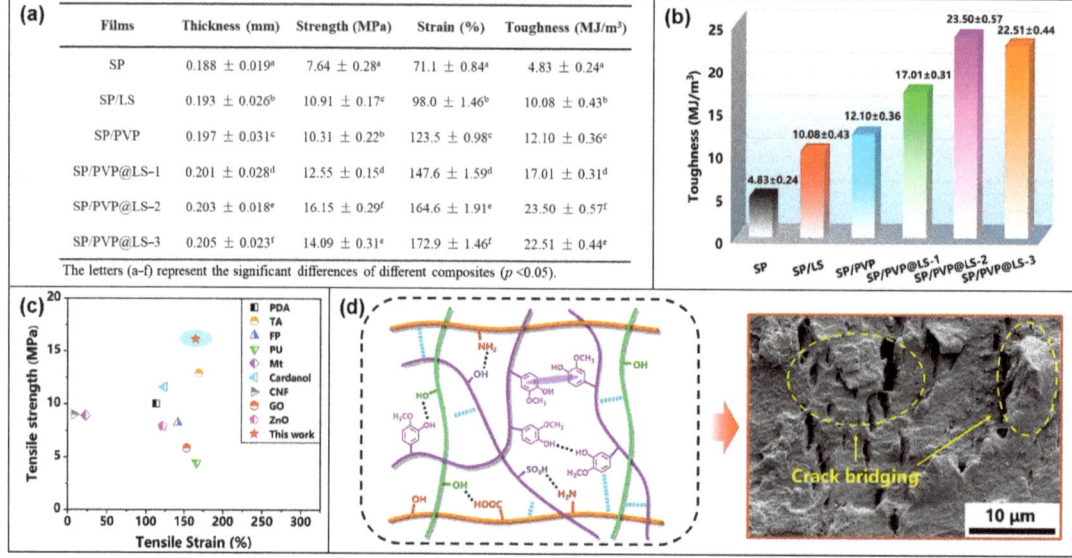

Figure 20. (a) Data on the mechanical properties of SP and its composites. (b) Toughness of the materials. (c) Comparison of the strength and flexibility of SP/PVP@LS films with other films. (d) Illustration of possible toughening mechanism analysis [107].

3.3. Other Properties

In fact, all kinds of electronic equipment, furniture, clothing, etc. used in our lives are easy to burn after encountering high temperatures or an open flame [108]. As a guarantee of fire safety, FRs can solve the burning problem. But it will greatly reduce the excellent performance of the original material [109]. The advantage of supramolecular structures formed by various non-covalent interactions is their high compatibility with the material. SFRs can play a flame-retardant role without destroying other special properties of the material. And they can even play a role in gaining. Therefore, researchers focus on the effects of supramolecular structure on the thermal insulation, self-healing, and UV-blocking of flame-retardant materials. In addition, there are also studies on improving the electrochemical properties of batteries [110–112], the viscoelasticity of thermoplastic materials [113] and the flame-retardant durability of fibrous materials [114].

3.3.1. Thermal Insulation Performance

Comfortable indoor temperatures have always been a concern for people. A lot of energy is needed to achieve this goal. So heat insulation materials with high heat insulation performance are very important [115]. However, most insulation materials are flammable, and the addition of FRs will destroy the insulation performance. But the addition of SFRs can reduce the reduction in insulation performance of many combustible insulation materials while achieving a high level of flame retardancy [60,105].

Xueyong Ren et al. [105] modified the gas coagulation of cellulose nanofibril (CNF) by in-situ supramolecular assembly of MEL-PA. A simple simulation experiment was designed to test the thermal insulation ability of the MEL-PA/CNF composite aerogel. As shown in Figure 21a, when the match head was placed on the top of the modified MEL-PA/CNF composite aerogel, neither the composite aerogel nor the match head ignited, but the match head on the top of the unmodified CNF aerogel ignited. When the aerogel burns for 28 s, which not only shows the flame-retardant effect of the MEL-PA/CNF composite aerogel but also shows the thermal insulation effect. To further demonstrate the thermal insulation properties of the composite aerogels, the authors tested their thermal conductivity and observed their heat transfer using an infrared camera. The thermal conductivity of the MEL-PA CNF aerogel is relatively higher due to the enhanced interfacial interaction between the MEL-PA and CNF aerogels (Figure 21b). But the thermal propagation curve (Figure 21c) shows that the heat transfer could reach a relatively stable state after 60 min. And the infrared thermography (Figure 21d) similarly shows that the heat transfer can be effectively blocked before and after the modification. From the above analysis, this work concluded that the supramolecularly-assembled MEL-PA does not significantly deteriorate the thermal insulation of CNF aerogels.

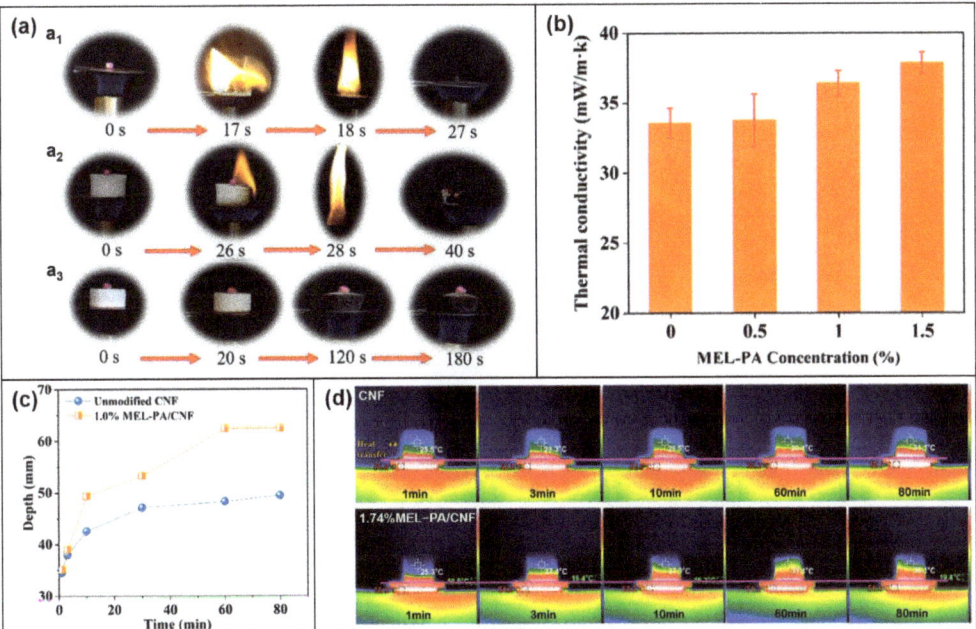

Figure 21. (a) Screenshot of insulation inspection test: (a_1) match end, (a_2) Place the match head in the unmodified CNF aerogel and (a_3) Match heads placed in 1.0%MEL-PA/CNF composite aerogel. (b) Thermal conductivity of MEL-PA/CNF composite aerogels. (c) Thermal penetration depth of aerogel before and after modification. (d) Thermal infrared image of CNF and 1.74%MEL-PA/CNF [105].

3.3.2. Self-Healing Property

Self-healing behavior contributes to the recycling of materials, helping to achieve sustainability and the rapid development of recycled polymers [116,117]. The presence of reversible non-covalent bonds imparts dynamic structural properties to the material, hence the self-healing properties of supramolecular materials [63,118].

The cyclic phosphonitrile-based polymer electrolyte (CPSHPE) prepared by Binghua Zhou et al. [118] has self-healing properties. Figure 22a shows the self-healing mechanism of CPSHPEs. The authors cut the CPSHPE into two pieces and found the presence of dynamic hydrogen bonds at the cuts, which gave the material self-healing properties due to the tendency of such non-conjugated hydrogen bonds to combine to form supramolecular networks. In addition to hydrogen bonding, there is an abundance of dynamic bonds such as dynamic disulfide bonds and ionic coordination bonds (Figure 22b) [63]. It can cause damage to generate free dynamic groups that can reassemble into a stable cross-linked network when the damaged area comes into contact again, thus restoring the material to its original properties.

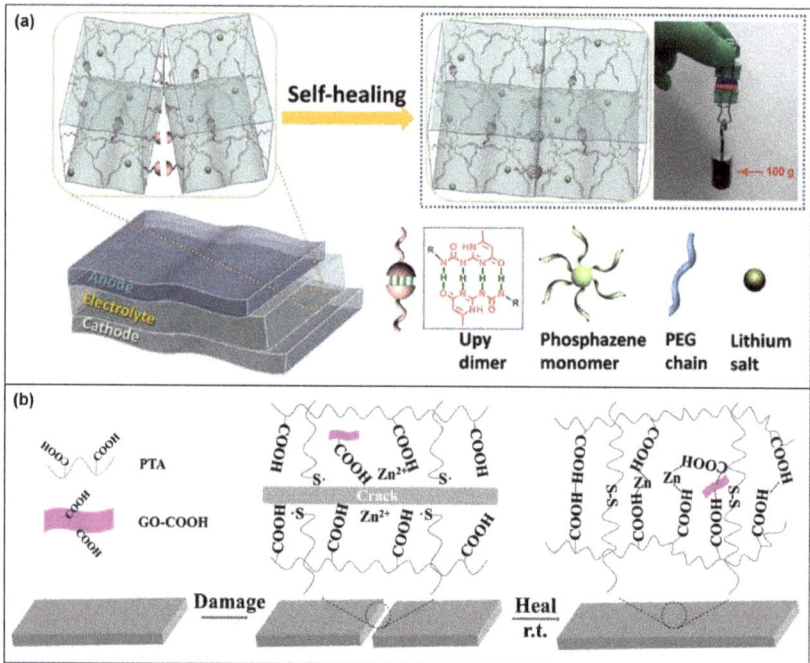

Figure 22. (a) Schematic diagram of CPSHPEs self-healing mechanism [118]. (b) Diagram of self-healing mechanism of other dynamic keys [63].

3.3.3. UV-Blocking Performance

Ultraviolet (UV) radiation has attracted extensive attention because of its negative effects on mechanical properties, discoloration, and decomposition of materials [119,120]. Compared with the flame-retardant polymer with intumescent flame retardant (IFR), SFRs can avoid accelerating the aging of the substrate under ultraviolet radiation or heating, thus reducing the flame-retardant performance of the polymer. At the same time, the degradation of mechanical properties is reduced, with excellent ultraviolet resistance [107].

MEL has a special triazine ring structure, so it can absorb ultraviolet rays and show anti-aging performance. Yuchun Li et al. [46] developed MEL-PA-MWCNTs (Figure 3c), which were applied to flame retardant PA6 and PA6 composites, showing good ultraviolet resistance. The LOI value of PA6/7% MEL-PA-MWCNTs does not decrease but increased after photoaging (Figure 23a). Mainly because MEL-PA-MWCNTs protect the substrate and migrated to the surface of the substrate under the irradiation of ultraviolet light, thus avoiding the aging and burning of PA6. The tensile strength and elongation at break of PA6 composites are much lower than those of pure PA6 (Figure 23b,c), due to the absorption of ultraviolet radiation by triazine ring groups in MEL and the capture of free radicals produced by PA in the aging process of the matrix.

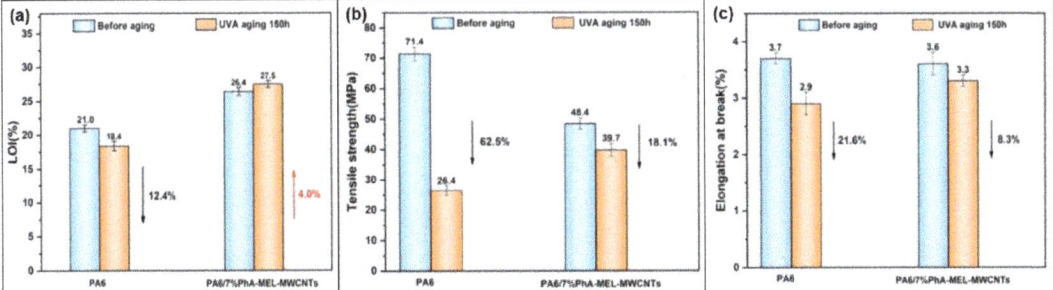

Figure 23. (a) Comparison of LOI values of PA6 and PA6/7% MEL-PA-MWCNTs after UV aging. (b) Comparison of tensile strength and (c) elongation at break after UV aging 150h [46].

4. Conclusions and Prospect

SFRs of different dimensions (including 1D, 2D, and 3D SFRs) show excellent flame-retardant efficiency in improving the fire safety of typical polymers. Similar to the flame-retardant mechanism of traditional FRs, SFRs also mainly plays a role in the gas phase and condensed phase. On the one hand, SFRs containing typical flame-retardant elements such as phosphorus and nitrogen can interfere with the free radical reaction during combustion and slow down the combustion rate. On the other hand, they can help to char yield, which acts as a heat transfer barrier. Meanwhile, the 2D and 3D structural characteristics of SFRs are conducive to blocking the transfer of O_2, flammable volatile substances, and heat.

SFRs have a good effect on improving the properties of polymers. In contrast to traditional FRs synthesis methods, SFRs are mostly self-assembled by using different primitives in non-toxic and pollution-free solvents (water or ethanol). It mainly utilizes the multiple synergies of hydrogen bonding, ion attraction, π-π stacking, and other non-covalent bonds among the units. These units usually contain groups rich in active sites, such as amino, phosphate, and hydroxyl groups. However, the selection of units is relatively limited, such as PA and MEL, which have been widely studied. More units containing flame-retardant elements with multiple active sites should be dug out for design and assembly. The application prospects of novel SFRs should be explored. At the same time, compared with some other FRs applied to the matrix in the form of simple physical dispersion, SFRs have rich active groups, can produce some crosslinking (such as multiple hydrogen bonds). They have interface interactions with the matrix and have good compatibility. Therefore, the mechanical properties of the polymer are improved.

In addition, on the basis of the improvement in fire safety performance, the performance of SFRs on other aspects of the matrix has been improved synchronously, such as ultraviolet protection, self-healing, electrochemical performance, and other properties. This is a good research idea for the high performance requirements of materials now. At present, there are few such studies, and continuous research work is needed. In particular, the self-healing performance of SFRs is worth studying. SFRs, based on the dynamic reversible non-covalent bond synthesis of structural characteristics, should be made full use of. Under the current ecological background, the recyclability, repairability, degradability and recycling of polymers will be vital topics.

Author Contributions: Writing-original draft preparation, S.X. and J.F.; Supervision, H.Y. and X.F. All authors have read and agreed to the published version of the manuscript.

Funding: The work was funded by the Fundamental Research Funds for the Central Universities (2020CDJQY-A006) and the State Key Laboratory of High Performance Civil Engineering Materials (2022CEM004).

Institutional Review Board Statement: Not applicable.

Informed Consent Statement: Not applicable.

Data Availability Statement: Not applicable.

Conflicts of Interest: The authors declare no conflict of interest.

Sample Availability: Not applicable.

References

1. Babu, R.P.; O'Connor, K.; Seeram, R. Current progress on bio-based polymers and their future trends. *Prog. Biomater.* **2013**, *2*, 8. [CrossRef] [PubMed]
2. El-Ghoul, Y.; Alminderej, F.M.; Alsubaie, F.M.; Alrasheed, R.; Almousa, N.H. Recent Advances in Functional Polymer Materials for Energy, Water, and Biomedical Applications: A Review. *Polymers* **2021**, *13*, 4327. [CrossRef]
3. Vahabi, H.; Laoutid, F.; Formela, K.; Saeb, M.R.; Dubois, P. Flame-Retardant Polymer Materials Developed by Reactive Extrusion: Present Status and Future Perspectives. *Polym. Rev.* **2022**, *62*, 919–949. [CrossRef]
4. Vahabi, H.; Laoutid, F.; Mehrpouya, M.; Saeb, M.R.; Dubois, P. Flame retardant polymer materials: An update and the future for 3D printing developments. *Mater. Sci. Eng. R-Rep.* **2021**, *144*, 100604. [CrossRef]
5. Gupta, P.; Toksha, B.; Patel, B.; Rushiya, Y.; Das, P.; Rahaman, M. Recent Developments and Research Avenues for Polymers in Electric Vehicles. *Chem. Rec.* **2022**, *22*, e202200186. [CrossRef] [PubMed]
6. Deng, K.; Zeng, Q.; Wang, D.; Liu, Z.; Wang, G.; Qiu, Z.; Zhang, Y.; Xiao, M.; Meng, Y. Nonflammable organic electrolytes for high-safety lithium-ion batteries. *Energy Storage Mater.* **2020**, *32*, 425–447. [CrossRef]
7. He, L.; Shi, Y.; Wang, Q.; Chen, D.; Shen, J.; Guo, S. Strategy for constructing electromagnetic interference shielding and flame retarding synergistic network in poly (butylene succinate) and thermoplastic polyurethane multilayered composites. *Compos. Sci. Technol.* **2020**, *199*, 108324. [CrossRef]
8. Li, J.; Zhu, C.; Zhao, Z.; Khalili, P.; Clement, M.; Tong, J.; Liu, X.; Yi, X. Fire properties of carbon fiber reinforced polymer improved by coating nonwoven flame retardant mat for aerospace application. *J. Appl. Polym. Sci.* **2019**, *136*, 47801. [CrossRef]
9. Liu, B.-W.; Zhao, H.-B.; Wang, Y.-Z. Advanced Flame-Retardant Methods for Polymeric Materials. *Adv. Mater.* **2022**, *34*, 2107905. [CrossRef]
10. Yasin, S.; Behary, N.; Curti, M.; Rovero, G. Global Consumption of Flame Retardants and Related Environmental Concerns: A Study on Possible Mechanical Recycling of Flame Retardant Textiles. *Fibers* **2016**, *4*, 16. [CrossRef]
11. Inthavong, C.; Hommet, F.; Bordet, F.; Rigourd, V.; Guerin, T.; Dragacci, S. Simultaneous liquid chromatography-tandem mass spectrometry analysis of brominated flame retardants (tetrabromobisphenol A and hexabromocyclododecane diastereoisomers) in French breast milk. *Chemosphere* **2017**, *186*, 762–769. [CrossRef] [PubMed]
12. Sharkey, M.; Harrad, S.; Abdallah, M.A.-E.; Drage, D.S.; Berresheim, H. Phasing-out of legacy brominated flame retardants: The UNEP Stockholm Convention and other legislative action worldwide. *Environ. Int.* **2020**, *144*, 106041. [CrossRef] [PubMed]
13. The New York State Senate, Senate Bill S7737. Available online: https://www.nysenate.gov/legislation/bills/2021/S773 (accessed on 5 July 2023).
14. Hobbs, C.E. Recent Advances in Bio-Based Flame Retardant Additives for Synthetic Polymeric Materials. *Polymers* **2019**, *11*, 224. [CrossRef] [PubMed]
15. Liu, Y.; Gao, Y.; Wang, Q.; Lin, W. The synergistic effect of layered double hydroxides with other flame retardant additives for polymer nanocomposites: A critical review. *Dalton Trans.* **2018**, *47*, 14827–14840. [CrossRef] [PubMed]
16. Wang, X.; Li, Y.; Meng, D.; Gu, X.; Sun, J.; Hu, Y.; Bourbigot, S.; Zhang, S. A Review on Flame-Retardant Polyvinyl Alcohol: Additives and Technologies. *Polym. Rev.* **2022**, *63*, 324–364. [CrossRef]
17. Grand View Research, Flame Retardant Market Size, Share & Trends Analysis Report by Product (Halogenated, Non-halogenated), by Application (Polyolefin, Epoxy Resins), by End Use (Electrical & Electronics, Construction), and Segment Forecasts, 2021–2028. Available online: https://www.grandviewresearch.com/industry-analysis/flame-retardant-market (accessed on 5 July 2023).
18. Huo, S.; Song, P.; Yu, B.; Ran, S.; Chevali, V.S.; Liu, L.; Fang, Z.; Wang, H. Phosphorus-containing flame retardant epoxy thermosets: Recent advances and future perspectives. *Prog. Polym. Sci.* **2021**, *114*, 101366. [CrossRef]
19. Naiker, V.E.; Mestry, S.; Nirgude, T.; Gadgeel, A.; Mhaske, S.T. Recent developments in phosphorous-containing bio-based flame-retardant (FR) materials for coatings: An attentive review. *J. Coat. Technol. Res.* **2022**, *20*, 113–139. [CrossRef]
20. Liu, X.-D.; Zheng, X.-T.; Dong, Y.-Q.; He, L.-X.; Chen, F.; Bai, W.-B.; Lin, Y.-C.; Jian, R.-K. A novel nitrogen-rich phosphinic amide towards flame-retardant, smoke suppression and mechanically strengthened epoxy resins. *Polym. Degrad. Stab.* **2022**, *196*, 109840. [CrossRef]
21. Liang, S.; Wang, F.; Liang, J.; Chen, S.; Jiang, M. Synergistic effect between flame retardant viscose and nitrogen-containing intrinsic flame-retardant fibers. *Cellulose* **2020**, *27*, 6083–6092. [CrossRef]

22. Liu, N.; Wang, H.; Xu, B.; Qu, L.; Fang, D. Cross-linkable phosphorus/nitrogen-containing aromatic ethylenediamine endowing epoxy resin with excellent flame retardancy and mechanical properties. *Compos. Part A-Appl. Sci. Manuf.* **2022**, *162*, 107145. [CrossRef]
23. Wang, W.; Wang, F.; Li, H.; Liu, Y. Synthesis of phosphorus-nitrogen hybrid flame retardant and investigation of its efficient flame-retardant behavior in PA6/PA66. *J. Appl. Polym. Sci.* **2023**, *140*, e53536. [CrossRef]
24. Yi, C.; Xu, C.; Sun, N.; Xu, J.; Ma, M.; Shi, Y.; He, H.; Chen, S.; Wang, X. Flame-Retardant and Transparent Poly(methyl methacrylate) Composites Based on Phosphorus-Nitrogen Flame Retardants. *Acs Appl. Polym. Mater.* **2022**, *5*, 846–855. [CrossRef]
25. Zhang, W.; Zhou, M.; Kan, Y.; Chen, J.; Hu, Y.; Xing, W. Synthesis and flame retardant efficiency study of two phosphorus-nitrogen type flame retardants containing triazole units. *Polym. Degrad. Stab.* **2023**, *208*, 110236. [CrossRef]
26. Dalal, A.; Bagotia, N.; Sharma, K.K.; Chatterjee, K.N.; Bansal, P.; Kumar, S. One Pot Facile Synthesis of Self-extinguishable Metal Based Flame Retardant for Cotton Fabric. *J. Nat. Fibers* **2022**, *19*, 10475–10489. [CrossRef]
27. Yang, Z.; Guo, W.; Yang, P.; Hu, J.; Duan, G.; Liu, X.; Gu, Z.; Li, Y. Metal-phenolic network green flame retardants. *Polymer* **2021**, *221*, 123627. [CrossRef]
28. Zhang, G.; Wu, W.; Yao, M.; Wu, Z.; Jiao, Y.; Qu, H. Novel triazine-based metal-organic frameworks: Synthesis and mulifunctional application of flame retardant, smoke suppression and toxic attenuation on EP. *Mater. Des.* **2023**, *226*, 111664. [CrossRef]
29. Liang, B.; Hong, X.; Zhu, M.; Gao, C.; Wang, C.; Tsubaki, N. Synthesis of novel intumescent flame retardant containing phosphorus, nitrogen and boron and its application in polyethylene. *Polym. Bull.* **2015**, *72*, 2967–2978. [CrossRef]
30. Qu, H.; Fan, R.; Yuan, J.; Liu, B.; Sun, L.; Tian, R. Preparation and Performance of a P-N Containing Intumescent Flame Retardant Based on Hydrolyzed Starch. *Polym.-Plast. Technol. Eng.* **2017**, *56*, 1760–1771. [CrossRef]
31. Wang, S.; Du, Z.; Cheng, X.; Liu, Y.; Wang, H. Synthesis of a phosphorus- and nitrogen-containing flame retardant and evaluation of its application in waterborne polyurethane. *J. Appl. Polym. Sci.* **2018**, *135*, 46093. [CrossRef]
32. Liu, L.; Xu, Y.; Di, Y.; Xu, M.; Pan, Y.; Li, B. Simultaneously enhancing the fire retardancy and crystallization rate of biodegradable polylactic acid with piperazine-1,4-diylbis (diphenylphosphine oxide). *Compos. Part B-Eng.* **2020**, *202*, 108407. [CrossRef]
33. Ollerton, K.; Greenaway, R.L.; Slater, A.G. Enabling Technology for Supramolecular Chemistry. *Front. Chem.* **2021**, *9*, 774987. [CrossRef] [PubMed]
34. Uhlenheuer, D.A.; Petkau, K.; Brunsveld, L. Combining supramolecular chemistry with biology. *Chem. Soc. Rev.* **2010**, *39*, 2817–2826. [CrossRef] [PubMed]
35. Cieslak, A.M.; Janecek, E.-R.; Sokolowski, K.; Ratajczyk, T.; Leszczynski, M.K.; Scherman, O.A.; Lewinski, J. Photo-induced interfacial electron transfer of ZnO nanocrystals to control supramolecular assembly in waterd. *Nanoscale* **2017**, *9*, 16128–16132. [CrossRef] [PubMed]
36. Hou, R.; Li, G.; Zhang, Y.; Li, M.; Zhou, G.; Chai, X. Self-Healing Polymers Materials Based on Dynamic Supramolecular Motifs. *Prog. Chem.* **2019**, *31*, 690–698. [CrossRef]
37. Wang, H.; Yang, Y.; Zheng, P.; Wang, J.; Ng, S.-W.; Chen, Y.; Deng, Y.; Zheng, Z.; Wang, C. Water-based phytic acid-crosslinked supramolecular binders for lithium-sulfur batteries. *Chem. Eng. J.* **2020**, *395*, 124981. [CrossRef]
38. Kumar, S.; Shukla, S.K. Synergistic evolution of flame-retardant hybrid structure of poly vinyl alcohol, starch and kaolin for coating on wooden substrate. *J. Polym. Res.* **2023**, *30*, 71. [CrossRef]
39. Ou, H.; Xu, J.; Liu, B.; Xue, H.; Weng, Y.; Jiang, J.; Xu, G. Study on synergistic expansion and flame retardancy of modified kaolin to low density polyethylene. *Polymer* **2021**, *221*, 123586. [CrossRef]
40. Zhang, Y.; Yang, M.; Yuan, D.; Li, C.; Ma, Y.; Wang, S.; Wang, S. Fire hazards of PMMA-based composites combined with expandable graphite and multi-walled carbon nanotubes: A comprehensive study. *Fire Saf. J.* **2023**, *135*, 103727. [CrossRef]
41. Li, N.; Li, Z.; Liu, Z.; Yang, Y.; Jia, Y.; Li, J.; Wei, M.; Li, L.; Wang, D.-Y. Magnesium hydroxide micro-whiskers as super-reinforcer to improve fire retardancy and mechanical property of epoxy resin. *Polym. Compos.* **2022**, *43*, 1996–2009. [CrossRef]
42. Yang, H.-C.; Tsai, T.-P.; Hsieh, C.-T. Enhancement on fireproof performance of construction coatings using calcium sulfate whiskers prepared from wastewater. *Chem. Pap.* **2017**, *71*, 1343–1350. [CrossRef]
43. Wang, Z.-Y.; Zhu, Y.-J.; Chen, Y.-Q.; Yu, H.-P.; Xiong, Z.-C. Flexible nanocomposite paper with superior fire retardance, mechanical properties and electrical insulation by engineering ultralong hydroxyapatite nanowires and aramid nanofibers. *Chem. Eng. J.* **2022**, *444*, 136470. [CrossRef]
44. Chen, T.; Wang, X.; Peng, C.; Chen, G.; Yuan, C.; Xu, Y.; Zeng, B.; Luo, W.; Balaji, K.; Petri, D.F.S.; et al. Efficient Flame Retardancy, Smoke Suppression, and Mechanical Enhancement of beta-FeOOH@Metallo-Supramolecular Polymer Core-Shell Nanorod Modified Epoxy Resin. *Macromol. Mater. Eng.* **2020**, *305*, 202000137. [CrossRef]
45. Shang, S.; Ma, X.; Yuan, B.; Chen, G.; Sun, Y.; Huang, C.; He, S.; Dai, H.; Chen, X. Modification of halloysite nanotubes with supramolecular self-assembly aggregates for reducing smoke release and fire hazard of polypropylene. *Compos. Part B-Eng.* **2019**, *177*, 107371. [CrossRef]
46. Li, Y.; Wang, J.; Xue, B.; Wang, S.; Qi, P.; Sun, J.; Li, H.; Gu, X.; Zhang, S. Enhancing the flame retardancy and UV resistance of polyamide 6 by introducing ternary supramolecular aggregates. *Chemosphere* **2022**, *287*, 132100. [CrossRef] [PubMed]
47. Fu, X.-L.; Ding, H.-L.; Wang, X.; Song, L.; Hu, Y. Fabrication of zirconium phenylphosphonate/epoxy composites with simultaneously enhanced mechanical strength, anti-flammability and smoke suppression. *Compos. Part A-Appl. Sci. Manuf.* **2022**, *155*, 106837. [CrossRef]

48. Gong, K.; Yin, L.; Zhou, K.; Qian, X.; Shi, C.; Gui, Z.; Yu, B.; Qian, L. Construction of interface-engineered two-dimensional nanohybrids towards superb fire resistance of epoxy composites. *Compos. Part A-Appl. Sci. Manuf.* **2022**, *152*, 106707. [CrossRef]
49. Wang, D.; Peng, H.; Yu, B.; Zhou, K.; Pan, H.; Zhang, L.; Li, M.; Liu, M.; Tian, A.; Fu, S. Biomimetic structural cellulose nanofiber aerogels with exceptional mechanical, flame-retardant and thermal-insulating properties. *Chem. Eng. J.* **2020**, *389*, 124449. [CrossRef]
50. Xu, L.; Tan, X.; Xu, R.; Xie, J.; Lei, C. Influence of functionalized molybdenum disulfide (MoS2) with triazine derivatives on the thermal stability and flame retardancy of intumescent Poly(lactic acid) system. *Polym. Compos.* **2019**, *40*, 2244–2257. [CrossRef]
51. Mokhena, T.C.; Sadiku, E.R.; Ray, S.S.; Mochane, M.J.; Matabola, K.P.; Motloung, M. Flame retardancy efficacy of phytic acid: An overview. *J. Appl. Polym. Sci.* **2022**, *139*, e52495. [CrossRef]
52. Liu, S.-H.; Xu, Z.-L.; Zhang, L. Effect of cyano ionic liquid on flame retardancy of melamine. *J. Therm. Anal. Calorim.* **2021**, *144*, 305–314. [CrossRef]
53. Pan, M.; Chen, W.; Dai, J.; Shu, Z.; Xue, H.; Xu, J.; Ou, H. Research on preparation and flame retardancy of melamine-modified flame retardant loofah composites. *J. Ind. Text.* **2022**, *52*, 15280837221113361. [CrossRef]
54. Shang, S.; Yuan, B.; Sun, Y.; Chen, G.; Huang, C.; Yu, B.; He, S.; Dai, H.; Chen, X. Facile preparation of layered melamine-phytate flame retardant via supramolecular self-assembly technology. *J. Colloid Interface Sci.* **2019**, *553*, 364–371. [CrossRef] [PubMed]
55. Wang, P.-J.; Liao, D.-J.; Hu, X.-P.; Pan, N.; Li, W.-X.; Wang, D.-Y.; Yao, Y. Facile fabrication of biobased P-N-C-containing nano-layered hybrid: Preparation, growth mechanism and its efficient fire retardancy in epoxy. *Polym. Degrad. Stab.* **2019**, *159*, 153–162. [CrossRef]
56. Li, W.-X.; Zhang, H.-J.; Hu, X.-P.; Yang, W.-X.; Cheng, Z.; Xie, C.-Q. Highly efficient replacement of traditional intumescent flame retardants in polypropylene by manganese ions doped melamine phytate nanosheets. *J. Hazard. Mater.* **2020**, *398*, 123001. [CrossRef]
57. Qian, X.; Shi, C.; Wan, M.; Jing, J.; Che, H.; Ren, F.; Li, J.; Yu, B. Novel transition metal modified layered phosphate for reducing the fire hazards of PA6. *Compos. Commun.* **2023**, *37*, 101442. [CrossRef]
58. Liu, Q.; Chen, X.; Zhu, G.; Gu, X.; Li, H.; Zhang, S.; Sun, J.; Jin, X. Preparation of a novel supramolecular intumescent flame retardants containing P/N/S/Fe/Zn and its application in polylactic acid. *Fire Saf. J.* **2022**, *128*, 103536. [CrossRef]
59. Sun, Y.; Yuan, B.; Shang, S.; Zhang, H.; Shi, Y.; Yu, B.; Qi, C.; Dong, H.; Chen, X.; Yang, X. Surface modification of ammonium polyphosphate by supramolecular assembly for enhancing fire safety properties of polypropylene. *Compos. Part B-Eng.* **2020**, *181*, 107588. [CrossRef]
60. Zhang, M.; Wang, D.; Li, T.; Jiang, J.; Bai, H.; Wang, S.; Wang, Y.; Dong, W. Multifunctional Flame-Retardant, Thermal Insulation, and Antimicrobial Wood-Based Composites. *Biomacromolecules* **2023**, *24*, 957–966. [CrossRef]
61. Li, Y.; Chen, X.; Yuan, B.; Zhao, Q.; Huang, C.; Liu, L. Synthesis of a novel prolonged action inhibitor with lotus leaf-like appearance and its suppression on methane/hydrogen/air explosion. *Fuel* **2022**, *329*, 125401. [CrossRef]
62. Wang, D.; Wang, Y.; Zhang, X.; Li, T.; Du, M.; Chen, M.; Dong, W. Preferred zinc-modified melamine phytate for the flame retardant polylactide with limited smoke release. *New J. Chem.* **2021**, *45*, 13329–13339. [CrossRef]
63. Wang, J.; Zheng, Y.; Qiu, S.; Song, L. Ethanol inducing self-assembly of poly-(thioctic acid)/graphene supramolecular ionomers for healable, flame-retardant, shape-memory electronic devices. *J. Colloid Interface Sci.* **2023**, *629*, 908–915. [CrossRef] [PubMed]
64. Malkappa, K.; Ray, S.S. Thermal Stability, Pyrolysis Behavior, and Fire-Retardant Performance of Melamine Cyanurate@Poly (cyclotriphosphazene-co-4,4′-sulfonyl diphenol) Hybrid Nanosheet-Containing Polyamide 6 Composites. *Acs Omega* **2019**, *4*, 9615–9628. [CrossRef] [PubMed]
65. Qin, P.; Yi, D.; Hao, J.; Ye, X.; Gao, M.; Song, T. Fabrication of melamine trimetaphosphate 2D supermolecule and its superior performance on flame retardancy, mechanical and dielectric properties of epoxy resin. *Compos. Part B-Eng.* **2021**, *225*, 109269. [CrossRef]
66. Cao, C.-F.; Yu, B.; Huang, J.; Feng, X.-L.; Lv, L.-Y.; Sun, F.-N.; Tang, L.-C.; Feng, J.; Song, P.; Wang, H. Biomimetic, Mechanically Strong Supramolecular Nanosystem Enabling Solvent Resistance, Reliable Fire Protection and Ultralong Fire Warning. *Acs Nano* **2022**, *16*, 20865–20876. [CrossRef] [PubMed]
67. Hou, Y.; Xu, Z.; An, R.; Zheng, H.; Hu, W.; Zhou, K. Recent progress in black phosphorus nanosheets for improving the fire safety of polymer nanocomposites. *Compos. Part B-Eng.* **2023**, *249*, 110404. [CrossRef]
68. Deng, C.; Liu, Y.; Jian, H.; Liang, Y.; Wen, M.; Shi, J.; Park, H. Study on the preparation of flame retardant plywood by intercalation of phosphorus and nitrogen flame retardants modified with Mg/Al-LDH. *Constr. Build. Mater.* **2023**, *374*, 130939. [CrossRef]
69. Sang, B.; Li, Z.-W.; Li, X.-H.; Yu, L.-G.; Zhang, Z.-J. Graphene-based flame retardants: A review. *J. Mater. Sci.* **2016**, *51*, 8271–8295. [CrossRef]
70. Zhou, K.; Gao, R.; Qian, X. Self-assembly of exfoliated molybdenum disulfide (MoS2) nanosheets and layered double hydroxide (LDH): Towards reducing fire hazards of epoxy. *J. Hazard. Mater.* **2017**, *338*, 343–355. [CrossRef]
71. Cheng, L.; Wang, J.; Qiu, S.; Wang, J.; Zhou, Y.; Han, L.; Zou, B.; Xu, Z.; Hu, Y.; Ma, C. Supramolecular wrapped sandwich like SW-Si3N4 hybrid sheets as advanced filler toward reducing fire risks and enhancing thermal conductivity of thermoplastic polyurethanes. *J. Colloid Interface Sci.* **2021**, *603*, 844–855. [CrossRef]
72. Sun, W.; Sun, Y. Growth of biobased flakes on the surface and within interlayer of metakaolinite to enhance the fire safety and mechanical properties of intumescent flame-retardant polyurea composites. *Chem. Eng. J.* **2022**, *450*, 138350. [CrossRef]

73. Feng, X.; Wang, X.; Cai, W.; Hong, N.; Hu, Y.; Liew, K.M. Integrated effect of supramolecular self-assembled sandwich-like melamine cyanurate/MoS2 hybrid sheets on reducing fire hazards of polyamide 6 composites. *J. Hazard. Mater.* **2016**, *320*, 252–264. [CrossRef] [PubMed]
74. Qiu, S.; Zhou, Y.; Xing, W.; Ren, X.; Zou, B.; Hu, Y. Conceptually Novel Few-Layer Black Phosphorus/Supramolecular Coalition: Noncovalent Functionalization Toward Fire Safety Enhancement. *Ind. Eng. Chem. Res.* **2021**, *60*, 12579–12591. [CrossRef]
75. Sui, Y.; Qu, L.; Dai, X.; Li, P.; Zhang, J.; Luo, S.; Zhang, C. A green self-assembled organic supermolecule as an effective flame retardant for epoxy resin. *Rsc Adv.* **2020**, *10*, 12492–12503. [CrossRef] [PubMed]
76. Gao, S.; Qi, J.; Qi, P.; Xu, R.; Wu, T.; Zhang, B.; Huang, J.; Yan, Y. Unprecedented Nonflammable Organic Adhesives Leading to Fireproof Wood Products. *Acs Appl. Mater. Interfaces* **2023**, *15*, 8609–8616. [CrossRef] [PubMed]
77. Chen, S.-P.; Pan, L.-L.; Yuan, Y.-X.; Shi, X.-X.; Yuan, L.-J. A Novel Supramolecular Resin Based On an Organic Acid-Base Compound. *Cryst. Growth Des.* **2009**, *9*, 2668–2673. [CrossRef]
78. Yu, S.-L.; Xiang, H.-X.; Zhou, J.-L.; Qiu, T.; Hu, Z.-X.; Zhu, M.-F. Typical Polymer Fiber Materials: An Overview and Outlook. *Acta Polym. Sin.* **2020**, *51*, 39–54. [CrossRef]
79. Wu, Q.; Xiao, L.; Chen, J.; Peng, Z. Facile fabrication of high-performance epoxy systems with superior mechanical properties, flame retardancy, and smoke suppression. *J. Appl. Polym. Sci.* **2023**, *140*, e53480. [CrossRef]
80. Zhi, M.; Yang, X.; Fan, R.; Yue, S.; Zheng, J.; Liu, Q.; He, Y. A comprehensive review of reactive flame-retardant epoxy resin: Fundamentals, recent developments, and perspectives. *Polym. Degrad. Stab.* **2022**, *201*, 109976. [CrossRef]
81. Wan, M.; Shi, C.; Qian, X.; Qin, Y.; Jing, J.; Che, H. Metal-organic Framework ZIF-67 Functionalized MXene for Enhancing the Fire Safety of Thermoplastic Polyurethanes. *Nanomaterials* **2022**, *12*, 1142. [CrossRef]
82. Huang, Y.; Zhou, J.; Sun, P.; Zhang, L.; Qian, X.; Jiang, S.; Shi, C. Green, tough and highly efficient flame-retardant rigid polyurethane foam enabled by double network hydrogel coatings. *Soft Matter* **2021**, *17*, 10555–10565. [CrossRef]
83. Wang, N.; Chen, S.; Li, L.; Bai, Z.; Guo, J.; Qin, J.; Chen, X.; Zhao, R.; Zhang, K.; Wu, H. An Environmentally Friendly Nanohybrid Flame Retardant with Outstanding Flame-Retardant Efficiency for Polypropylene. *J. Phys. Chem. C* **2021**, *125*, 5185–5196. [CrossRef]
84. Yan, X.; Fang, J.; Gu, J.; Zhu, C.; Qi, D. Flame Retardancy, Thermal and Mechanical Properties of Novel Intumescent Flame Retardant/MXene/Poly(Vinyl Alcohol) Nanocomposites. *Nanomaterials* **2022**, *12*, 477. [CrossRef] [PubMed]
85. Cai, C.; Sun, Q.; Zhang, K.; Bai, X.; Liu, P.; Li, A.; LYu, Z.; Li, Q. Flame-retardant thermoplastic polyurethane based on reactive phosphonate polyol. *Fire Mater.* **2022**, *46*, 130–137. [CrossRef]
86. Sun, Y.; Wang, Z.; Wu, D.; Wang, X.; Yu, J.; Yuan, R.; Li, F. A phosphorus-containing flame retardant with thermal feature suitable for polyamide 6 and its filaments with enhanced anti-dripping performance. *Polym. Degrad. Stab.* **2022**, *200*, 109936. [CrossRef]
87. Zuluaga-Parra, J.D.; Ramos-deValle, L.F.; Sánchez-Valdes, S.; Torres-Lubian, R.; Pérez-Mora, R.; Ramírez-Vargas, E.; Martínez-Colunga, J.G.; da Silva, L.; Vazquez-Rodriguez, S.; Lozano-Ramírez, T.; et al. Grafting of ammonium polyphosphate onto poly(lactic acid) and its effect on flame retardancy and mechanical properties. *Iran. Polym. J.* **2023**, *32*, 225–238. [CrossRef]
88. Xu, J.; Niu, Y.; Xie, Z.; Liang, F.; Guo, F.; Wu, J. Synergistic flame retardant effect of carbon nanohorns and ammonium polyphosphate as a novel flame retardant system for cotton fabrics. *Chem. Eng. J.* **2023**, *451*, 138566. [CrossRef]
89. Mochane, M.J.; Mokhothu, T.H.; Mokhena, T.C. Synthesis, mechanical, and flammability properties of metal hydroxide reinforced polymer composites: A review. *Polym. Eng. Sci.* **2022**, *62*, 44–65. [CrossRef]
90. Attia, N.F.; Elashery, S.E.A.; Zakria, A.M.; Eltaweil, A.S.; Oh, H. Recent advances in graphene sheets as new generation of flame retardant materials. *Mater. Sci. Eng. B* **2021**, *274*, 115460. [CrossRef]
91. Yasin, S.; Curti, M.; Behary, N.; Perwuelz, A.; Giraud, S.; Rovero, G.; Guan, J.; Chen, G. Process optimization of eco-friendly flame retardant finish for cotton fabric: A response surface methodology approach. *Surf. Rev. Lett.* **2017**, *24*, 1750114. [CrossRef]
92. Burgaz, E.; Kendirlioglu, C. Thermomechanical behavior and thermal stability of polyurethane rigid nanocomposite foams containing binary nanoparticle mixtures. *Polym. Test.* **2019**, *77*, 105930. [CrossRef]
93. Qi, C.; Yuan, B.; Dong, H.; Li, K.; Shang, S.; Sun, Y.; Chen, G.; Zhan, Y. Supramolecular self-assembly modification of ammonium polyphosphate and its flame retardant application in polypropylene. *Polym. Adv. Technol.* **2020**, *31*, 1099–1109. [CrossRef]
94. Liu, L.; Zhu, M.; Ma, Z.; Xu, X.; Dai, J.; Yu, Y.; Seraji, S.M.; Wang, H.; Song, P. Small multiamine molecule enabled fire-retardant polymeric materials with enhanced strength, toughness, and self-healing properties. *Chem. Eng. J.* **2022**, *440*, 135645. [CrossRef]
95. Zhang, Y.; Tian, W.; Liu, L.; Cheng, W.; Wang, W.; Liew, K.M.; Wang, B.; Hu, Y. Eco-friendly flame retardant and electromagnetic interference shielding cotton fabrics with multi-layered coatings. *Chem. Eng. J.* **2019**, *372*, 1077–1090. [CrossRef]
96. Chen, H.-Q.; Xu, Y.-J.; Jiang, Z.-M.; Jin, X.; Liu, Y.; Zhang, L.; Zhang, C.-J.; Yan, C. The thermal degradation property and flame-retardant mechanism of coated knitted cotton fabric with chitosan and APP by LBL assembly. *J. Therm. Anal. Calorim.* **2020**, *140*, 591–602. [CrossRef]
97. Islam, M.S.; van de Ven, T.G.M. Cotton-Based Flame-Retardant Textiles: A Review. *Bioresources* **2021**, *16*, 4354–4381. [CrossRef]
98. Sohail, Y.; Parag, B.; Nemeshwaree, B.; Giorgio, R. Optimizing Organophosphorus Fire Resistant Finish for Cotton Fabric Using Box-Behnken Design. *Int. J. Environ. Res.* **2016**, *10*, 313–320.
99. Liu, L.; Huang, Z.; Pan, Y.; Wang, X.; Song, L.; Hu, Y. Finishing of cotton fabrics by multi-layered coatings to improve their flame retardancy and water repellency. *Cellulose* **2018**, *25*, 4791–4803. [CrossRef]
100. Li, Y.-C.; Schulz, J.; Mannen, S.; Delhom, C.; Condon, B.; Chang, S.; Zammarano, M.; Grunlan, J.C. Flame Retardant Behavior of Polyelectrolyte−Clay Thin Film Assemblies on Cotton Fabric. *ACS Nano* **2010**, *4*, 3325–3337. [CrossRef]

101. An, W.; Ma, J.; Xu, Q.; Fan, Q. Flame retardant, antistatic cotton fabrics crafted by layer-by-layer assembly. *Cellulose* **2020**, *27*, 8457–8469. [CrossRef]
102. Li, S.; Ding, F.; Lin, X.; Li, Z.; Ren, X. Layer-by-Layer Self-assembly of Organic-inorganic Hybrid Intumescent Flame Retardant on Cotton Fabrics. *Fibers Polym.* **2019**, *20*, 538–544. [CrossRef]
103. Chen, Y.; Wu, X.; Li, M.; Qian, L.; Zhou, H. Construction of crosslinking network structures by adding ZnO and ADR in intumescent flame retardant PLA composites. *Polym. Adv. Technol.* **2022**, *33*, 198–211. [CrossRef]
104. Xue, Y.; Ma, Z.; Xu, X.; Shen, M.; Huang, G.; Bourbigot, S.; Liu, X.; Song, P. Mechanically robust and flame-retardant polylactide composites based on molecularly-engineered polyphosphoramides. *Compos. Part A Appl. Sci. Manuf.* **2021**, *144*, 106317. [CrossRef]
105. Ren, X.; Song, M.; Jiang, J.; Yu, Z.; Zhang, Y.; Zhu, Y.; Liu, X.; Li, C.; Oguzlu-Baldelli, H.; Jiang, F. Fire-Retardant and Thermal-Insulating Cellulose Nanofibril Aerogel Modified by In Situ Supramolecular Assembly of Melamine and Phytic Acid. *Adv. Eng. Mater.* **2022**, *24*, 2101534. [CrossRef]
106. Han, S.; Yang, F.; Li, Q.; Sui, G.; Kalimuldina, G.; Araby, S. Synergetic Effect of α-ZrP Nanosheets and Nitrogen-Based Flame Retardants on Thermoplastic Polyurethane. *ACS Appl. Mater. Interfaces* **2023**, *15*, 17054–17069. [CrossRef]
107. Li, K.; Jin, S.; Zhou, Y.; Luo, J.; Li, J.; Li, X.; Shi, S.Q.; Li, J. Bioinspired interface design of multifunctional soy protein-based biomaterials with excellent mechanical strength and UV-blocking performance. *Compos. Part B Eng.* **2021**, *224*, 109187. [CrossRef]
108. Anna, S.; Karolina, M.; Sylwia, C. Buckwheat Hulls/Perlite as an Environmentally Friendly Flame-Retardant System for Rigid Polyurethane Foams. *Polymers* **2023**, *15*, 1913. [CrossRef]
109. Lazar, S.T.; Kolibaba, T.J.; Grunlan, J.C. Flame-retardant surface treatments. *Nat. Rev. Mater.* **2020**, *5*, 259–275. [CrossRef]
110. Han, C.; Xing, W.; Zhou, K.; Lu, Y.; Zhang, H.; Nie, Z.; Xu, F.; Sun, Z.; Du, Y.; Yu, H.; et al. Self-assembly of two-dimensional supramolecular as flame-retardant electrode for lithium-ion battery. *Chem. Eng. J.* **2022**, *430*, 132873. [CrossRef]
111. Qiu, J.; Wu, S.; Yang, Y.; Xiao, H.; Wei, X.; Zhang, B.; Hui, K.N.; Lin, Z. Aqueous Supramolecular Binder for a Lithium-Sulfur Battery with Flame-Retardant Property. *Acs Appl. Mater. Interfaces* **2021**, *13*, 55092–55101. [CrossRef]
112. Chen, X.; Yan, S.; Tan, T.; Zhou, P.; Hou, J.; Feng, X.; Dong, H.; Wang, P.; Wang, D.; Wang, B.; et al. Supramolecular "flame-retardant" electrolyte enables safe and stable cycling of lithium-ion batteries. *Energy Storage Mater.* **2022**, *45*, 182–190. [CrossRef]
113. Malkappa, K.; Salehiyan, R.; Ray, S.S. Supramolecular Poly(cyclotriphosphazene) Functionalized Graphene Oxide/Polypropylene Composites with Simultaneously Improved Thermal Stability, Flame Retardancy, and Viscoelastic Properties. *Macromol. Mater. Eng.* **2020**, *305*, 2000207. [CrossRef]
114. Xie, L.; Liu, B.; Liu, X.; Lu, Y. Preparation of flame retardant viscose fibers by supramolecular self-assembly. *J. Appl. Polym. Sci.* **2022**, *139*, e52792. [CrossRef]
115. Wang, H.; Xia, B.; Song, R.; Huang, W.; Zhang, M.; Liu, C.; Ke, Y.; Yin, J.-F.; Chen, K.; Yin, P. Metal oxide cluster-assisted assembly of anisotropic cellulose nanocrystal aerogels for balanced mechanical and thermal insulation properties. *Nanoscale* **2023**, *15*, 5469–5475. [CrossRef]
116. Aguirresarobe, R.H.; Nevejans, S.; Reck, B.; Irusta, L.; Sardon, H.; Asua, J.M.; Ballard, N. Healable and self-healing polyurethanes using dynamic chemistry. *Prog. Polym. Sci.* **2021**, *114*, 101362. [CrossRef]
117. Lugger, S.J.D.; Houben, S.J.A.; Foelen, Y.; Debije, M.G.; Schenning, A.P.H.J.; Mulder, D.J. Hydrogen-Bonded Supramolecular Liquid Crystal Polymers: Smart Materials with Stimuli-Responsive, Self-Healing, and Recyclable Properties. *Chem. Rev.* **2022**, *122*, 4946–4975. [CrossRef] [PubMed]
118. Zhou, B.; Yang, M.; Zuo, C.; Chen, G.; He, D.; Zhou, X.; Liu, C.; Xie, X.; Xue, Z. Flexible, Self-Healing, and Fire-Resistant Polymer Electrolytes Fabricated via Photopolymerization for All-Solid-State Lithium Metal Batteries. *ACS Macro Lett.* **2020**, *9*, 525–532. [CrossRef] [PubMed]
119. Rabani, I.; Lee, S.-H.; Kim, H.-S.; Yoo, J.; Hussain, S.; Maqbool, T.; Seo, Y.-S. Engineering-safer-by design ZnO nanoparticles incorporated cellulose nanofiber hybrid for high UV protection and low photocatalytic activity with mechanism. *J. Environ. Chem. Eng.* **2021**, *9*, 105845. [CrossRef]
120. Liang, X.-Y.; Wang, L.; Wang, Y.-M.; Ding, L.-S.; Li, B.-J.; Zhang, S. UV-Blocking Coating with Self-Healing Capacity. *Macromol. Chem. Phys.* **2017**, *218*, 1700213. [CrossRef]

Disclaimer/Publisher's Note: The statements, opinions and data contained in all publications are solely those of the individual author(s) and contributor(s) and not of MDPI and/or the editor(s). MDPI and/or the editor(s) disclaim responsibility for any injury to people or property resulting from any ideas, methods, instructions or products referred to in the content.

Brief Report

Functionally Active Microheterogeneous Systems for Elastomer Fire- and Heat-Protective Materials

Victor F. Kablov [1], Oksana M. Novopoltseva [1], Daria A. Kryukova [1,2], Natalia A. Keibal [1], Vladimir Burmistrov [2] and Vladimir G. Kochetkov [1,2,*]

1. Department of Chemical Technology of Polymers and Industrial Ecology, Volzhsky Polytechnic Institute (Branch) of Volgograd State Technical University, 42a Engelsa st., Volzhsky 404121, Russia
2. Department of Organic Chemistry, Volgograd State Technical University, 28 Lenina Avenue, Volgograd 400005, Russia; crus_himself@mail.ru
* Correspondence: vg.kochetkov@mail.ru; Tel.: +7-8443-41-27-69

Abstract: Elastomeric materials are utilized for the short-term protection of products and structures operating under extreme conditions in the aerospace, marine, and oil and gas industries. This research aims to study the influence of functionally active structures on the physical, mechanical, thermophysical, and fire- and heat-protective characteristics of elastomer compositions. The physical and mechanical properties of elastomer samples were determined using Shimazu AG-Xplus, while morphological research into microheterogeneous systems and coke structures was carried out on a scanning electronic microscope, Versa 3D. Differential thermal and thermogravimetric analyses of the samples were conducted on derivatograph Q-1500D. The presence of aluminosilicate microspheres, carbon microfibers, and a phosphor–nitrogen–organic modifier as part of the aforementioned structures contributes to the appearance of a synergetic effect, which results in an increase in the heat-protective properties of a material due to the enhancement in coke strength and intensification of material carbonization processes. The results indicate an 8–17% increase in the heating time of the unheated surface of a sample and a decrease in its linear burning speed by 6–17% compared to known analogues. In conclusion, microspheres compensate for the negative impact of microfibers on the density and thermal conductivity of a composition.

Keywords: microspheres; microfibers; elastomers; fire- and heat-protective material; organoelement modifiers

Citation: Kablov, V.F.; Novopoltseva, O.M.; Kryukova, D.A.; Keibal, N.A.; Burmistrov, V.; Kochetkov, V.G. Functionally Active Microheterogeneous Systems for Elastomer Fire- and Heat-Protective Materials. *Molecules* 2023, 28, 5267. https://doi.org/10.3390/molecules28135267

Academic Editors: Xin Wang, Weiyi Xing and Gang Tang

Received: 22 May 2023
Revised: 26 June 2023
Accepted: 5 July 2023
Published: 7 July 2023

Copyright: © 2023 by the authors. Licensee MDPI, Basel, Switzerland. This article is an open access article distributed under the terms and conditions of the Creative Commons Attribution (CC BY) license (https://creativecommons.org/licenses/by/4.0/).

1. Introduction

The development of elastomer materials based on ethylene–propylene–diene rubber that can withstand high-temperature heat flows in a short amount of time is crucial for protecting structures in industries such as aerospace, rocket, and oil and gas [1,2]. These materials can be used as coatings for fire and heat protection in combustion chambers, nozzles of solid-propellant rocket engines, gas generator covers, and other applications.

Advancements in creating fire-resistant polymer materials have been explored by Walter, Bhuvaneswari, and Ahmed [3–5]. The effectiveness of elastomeric fire- and heat-protective materials (FHPMs) largely depends on a complex set of endothermic physical and chemical transformations of components and their thermal destruction, as well as changes in the material's chemical structure, such as intumescence, pore formation, and carbonization [6,7].

Intensive thermal impact on FHPMs initiates structurization and carbonization processes that lead to the formation of a protective layer with low thermal conductivity [6,7]. In most heat-protective compositions, non-organic fillers act as elements on which primary carbon appears and then coke deposits as a result of thermal decomposition. Introducing phosphorus-, boron-, or nitrogen-containing components can intensify carbonization processes. Microspheres can also be introduced to create a uniform, fine porous structure of the coke layer [7–9].

One challenge faced in FHPMs is the thinning of the material due to high-speed gas flow, which leads to a decrease in efficiency under operational conditions. This challenge can be solved by using microfiber fillers such as kaolin, basalt, carbon, and other fibers. However, introducing such fillers is associated with technical difficulties such as agglomeration and an increase in density and thermal conductivity. Various coupling agents can be used to improve the distribution of microfibers and enhance the interaction at the polymer–microfiber interface. A phosphorus-nitrogen-containing modifier (DDF) has been shown to have a positive impact on physical and mechanical, fire- and heat-protective, and thermophysical properties of a composition [10].

Research has shown that phosphorus-containing additives can inhibit flames by increasing the recombination speed of H^+ and OH^- and reactions with phosphorus oxides and oxyacids [11–13]. In the condensed phase, residues of phosphoric acid are formed, which act as a dehydrating agent, enabling the formation of carbonized structures. Phosphorus compounds and their degradation products can also act as cross-linking agents, causing dehydration, cyclization, ligation, aromatization, and graphitization [14–16].

Microcapsulated phosphorus can also be used as a modifying agent, increasing fire resistance in epoxipolymers without affecting physical and mechanical or dielectric parameters [16].

Under operating conditions, fire- and heat-resistant materials are exposed not only to elevated temperatures and pressure but also to high-speed gas flow. This leads to the removal of the upper layer of the material, thinning the material itself and decreasing its efficiency. This issue can be resolved by incorporating microfiber fillers that form a "network" to increase the erosion resistance of the material. Examples of such fillers include kaolin, basalt, carbon, and other fibers. The introduction of these fillers into polymers involves various interactions at the polymer–filler interface that affect the mechanical, physicochemical properties, and thermooxidative stability of the composite material [17–21].

However, the introduction of such fillers comes with a number of technological difficulties. Fibers, especially asbestos, tend to agglomerate, which reduces the homogeneity of the material. Additionally, there is an increase in the density and thermal conductivity of the material, which is unacceptable in some cases.

To improve the distribution of microfibers and enhance the interaction at the polymer-microfiber interface, various finishes can be used. One criterion for selecting a sizing system is its positive effect on not only the physical and mechanical properties but also on the fire and heat protection and thermophysical properties of the composition. As our study has shown, the DDP modifier (dimethylcarbamyl(diaminomethyl)phosphoramide) synthesized by us can be such a system.

To compensate for the negative effect of microfibers on density and thermal conductivity, their combination with microspheres is possible [22–26]. By selecting sizing systems and varying the ratio of microspheres and microfibers, it is possible to obtain in situ organizing structures that combine the advantages of both components. The efficiency of fire- and heat-resistant elastomeric materials is primarily determined by a complex set of endothermic physical and chemical transformations of the components, as well as their thermal degradation. Additionally, the material's chemical structure undergoes changes such as swelling, pore formation, and coking. Under high-temperature conditions, these processes are initiated, leading to the formation of a protective layer with low thermal conductivity through coke formation and structuring [25–29].

2. Results

Depending on the proportions of microspheres and microfibers used, we anticipated the formation of various functionally active structures. Table 1 outlines the proportions of microspheres, microfibers, and DDF that were studied (Supplementary Materials). Scanning electronic microscopy revealed the formation of three distinct types of structures (as depicted in Figure 1), indicating the modification of the surface of the microdispersed components through the appearance of phosphorus atom peaks in the elementograms.

Table 1. Proportions studied of microspheres, microfibers, and DDF.

Ingredient	Sample Number		
	5MUV:10MSF	10MUV:5MSF	15MUV:5MSF
	Content, wt. pts. per 100 wt. pts. Rubber		
Aluminosilicate microspheres	10	5	5
Carbon microfibers	5	10	15
DDF	1	1	1

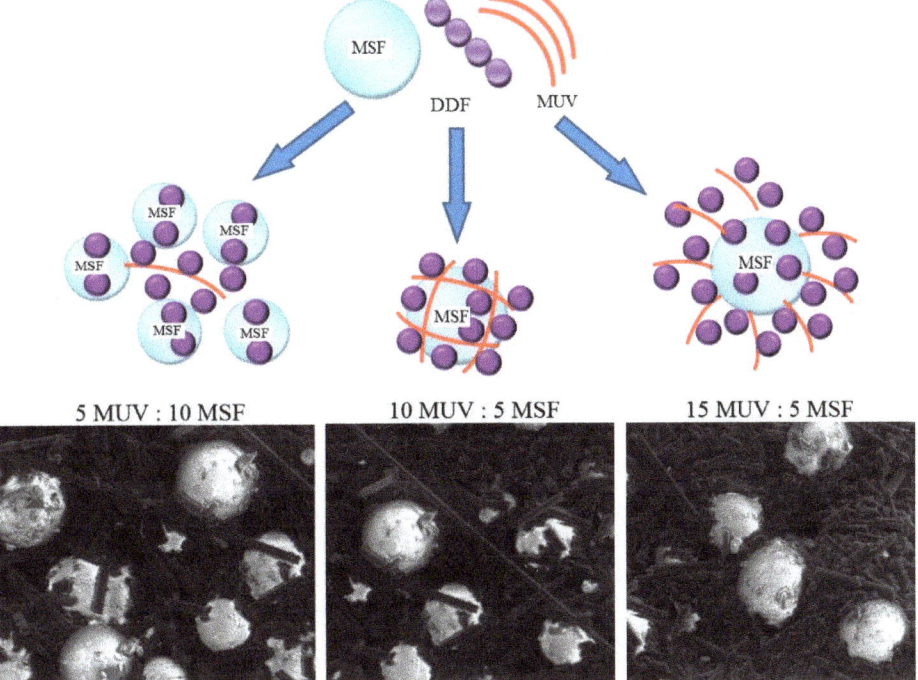

Figure 1. Diagram of the assembly of functionally active structures and SEM microphotos thereof (magnification ×10,000).

As illustrated in the microphotograph (Figure 2), the introduced functionally active structures (FASs) retain their integrity. The existing intermediate variants in the form of treated microspheres or microfibers also positively influence the fire and heat protection characteristics of the material: the heating time of the unheated surface of a sample up to 100 °C increases to 5.5–7.5%; the linear burning speed decreases to 4.3–6.7%.

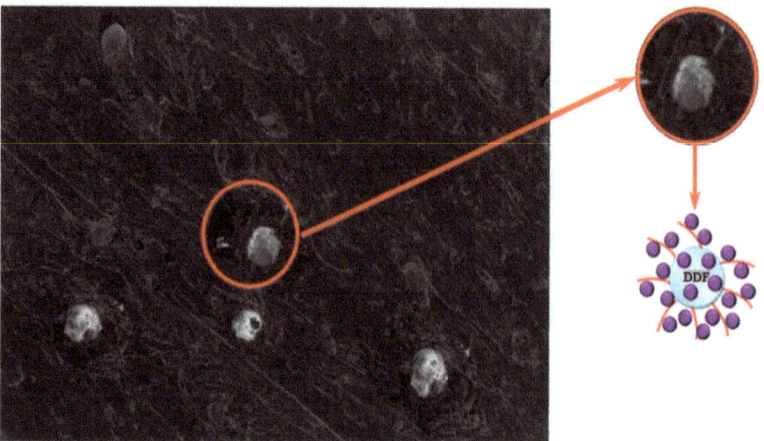

Figure 2. SEM microphotograph of vulcanized stock 10MUV: 5MSF (magnification ×10,000).

Table 2 presents the properties of vulcanized samples of the elastomeric compositions under study.

Table 2. Vulcanized Rubber Properties.

Parameter	Ref.	Control Sample	5MUV:10MSF	10MUV:5MSF	15MUV:5MSF
Tensile strength f_t, MPa	Not less than 6.0	16.5	8.5	11.8	12.4
Breaking elongation ε_{rel}, %	Not less than 300	450	400	350	380
Permanent elongation θ_{perm}, %	Not more than 30	20	25	18	18
Density ρ, kg m^{-3}	Not more than 1100	1080	1065	1082	1105
Heating time of unheated surface of a sample up to 100 °C, s	–	62	70	87	83
Coke number CCV, %	–	2.4	14.8	15.9	16.7
Linear burning speed $V_{l.b.}$, mm min^{-1}	–	32.1	30.4	26.7	25.4
Coke layer tear propagation strength σ, mPa	–	37.3	40.1	41.4	41.6

3. Discussion

The introduction of functionally active components, which form relatively large microstructures, results in a certain decrease in the strength characteristics. However, their values still remain higher than the standard ones (Table 2). More importantly, the density of the compositions decreases with an increase in the content of microspheres and approaches the control variant. A decrease in this parameter is crucial for creating protective materials for aircraft and rocketry.

Previous studies [10,30] have demonstrated that the optimal balance between fire and heat protection and the physical and mechanical properties of the material can be achieved by incorporating 5–15 mass parts of microfibers and 3–7 mass parts of microspheres.

Samples containing FASs with a ratio of microspheres/microfibers = 5:10 exhibit the most efficient fire and heat protection characteristics. When heat flow passes through the thickness of a fire- and heat-protective material, a number of adaptive processes occur: the destruction of the polymer matrix occurs in the upper layers of the material, and the pre-introduced functionally active components enable the formation of a denser fine porous coke layer (Figure 3), which is reinforced with microfibers. In the deeper levels of the material, carbonization processes are initiated owing to the presence of a phosphorus–boron–organic modifier on the surface of the microspheres. These processes enable a decrease in the heating-up speed.

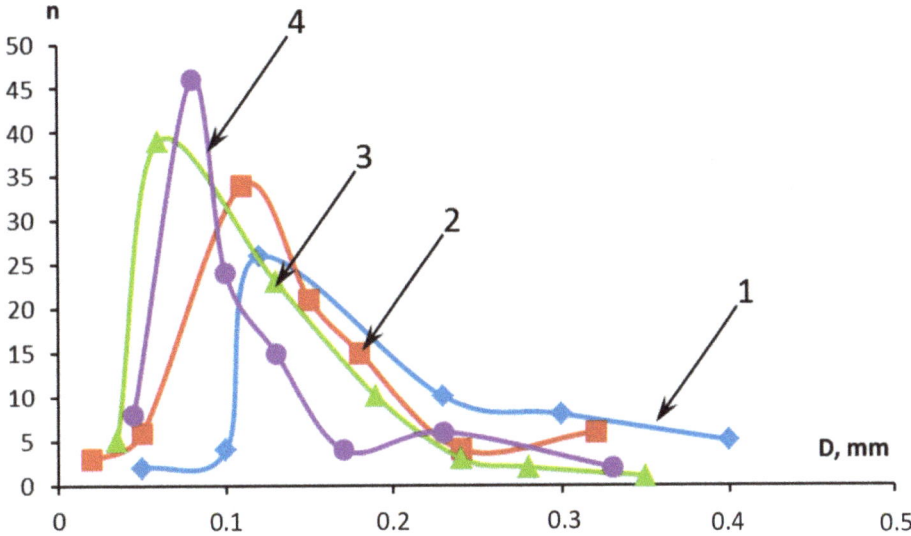

Figure 3. Pore distribution in the pre-pyrolysis layer: 1—control sample; 2—5MUV:10MSF; 3—10MUV:5MSF; 4—15MUV:5MSF.

The efficiency of the additives researched is confirmed with the DTA and TG analysis methods (Supplementary Materials). The value of the energy consumed for the structurization processes, carbonization of the material, degradation, and chemical transformation of the modifier under the impact of a heat flow can be evaluated by the square of the endothermic peak on the DTA curve. When FASs are introduced, an increase in the carbon residue occurs at 4–30% and a growth in the square of the endothermic peak takes place at 24%.

The studied samples are shown in Figure 4 after being tested for erosion strength under conditions of high-speed heat flow. The control sample (Figure 4b) is characterized by greater weight loss and little start time of combustion. The presence of FASs intensifies the carbonization processes and the microfibers enable the creation of strong coke with low thermal conductivity (Figure 4e).

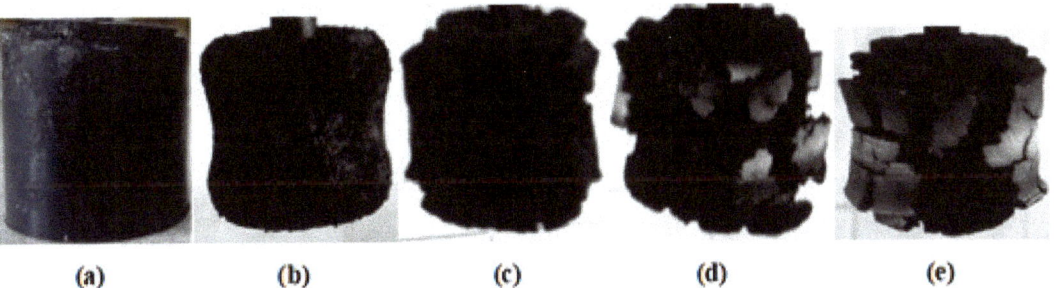

Figure 4. The appearance of the sample surface under the impact of high-speed heat flow: (**a**) the control sample before testing; (**b**) the control sample after testing; (**c**) 5MUV:10MSF; (**d**) 10MUV:5MSF; (**e**) 15MUV:5MSF.

The experimental data obtained have allowed us to propose a mechanism for the fire- and heat-protective action of materials containing functionally active structures (as depicted in Figure 5). The introduced functionally active components are transformed into a reinforced coke layer with increased strength against erosion carry-over and decreased thermal conductivity under the impact of high-temperature heat flow. In this case, the microspheres act as carbonization centers, while a layer of DDF on their surface initiates this process and retains the microfibers necessary to enhance the coke strength under the conditions of material erosion carry-over.

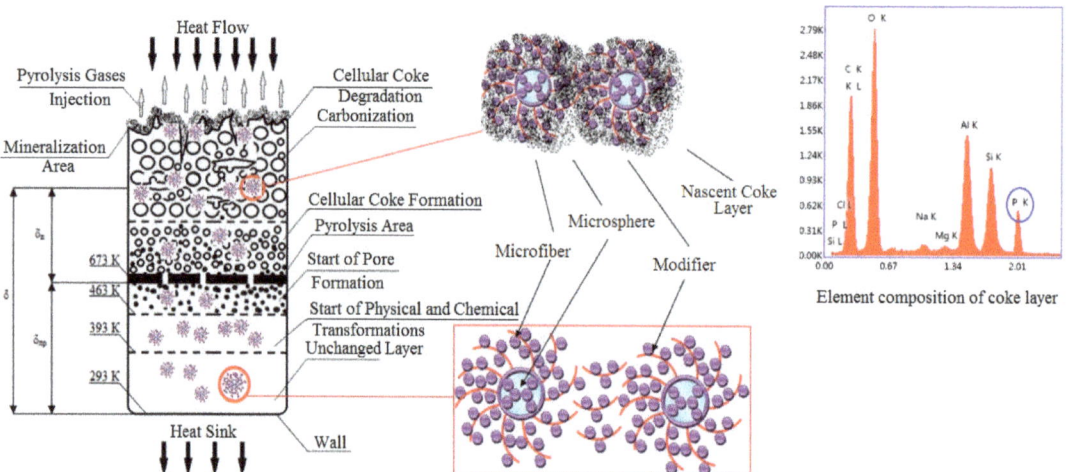

Figure 5. The presumed mechanism of fire- and heat-protective action of materials with the content of functionally active structures.

When subjected to high-temperature impact in the carbonization area, fusion and destruction of the microspheres occur. However, microfibers remain even on the surface of the shards (Figure 6), continuing to perform their function of reinforcing the coke.

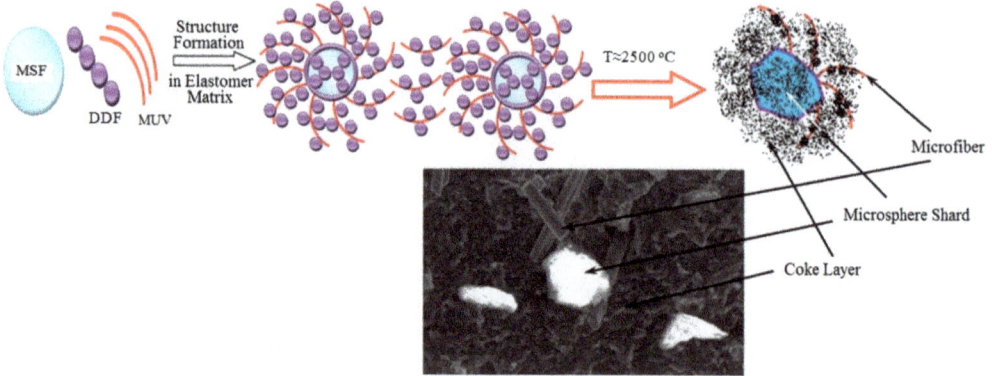

Figure 6. SEM microphotograph of the coke layer being formed after a high-temperature impact.

4. Materials and Methods

Materials

The materials under investigation were rubbers based on triple ethylene–propylene–diene rubber (EPDM-40), manufactured by the Nizhnekamsk Synthetic Rubber Factory.

This particular rubber was selected for its relatively low density and unsaturation degree, as well as its high heat resistance [31].

The compositions being studied are listed in Table 3, with previous works [7,10] determining the optimal weight content of microspheres, microfibers, and an organic element modifier at 5, 10, and 1 wt pts., respectively, per 100 wt pts of rubber.

Table 3. Rubber Formulas.

Ingredient	Sample Number			
	Control Sample	5MUV:10MSF	10MUV:5MSF	15MUV:5MSF
	Content, wt. pts. per 100 wt. pts. Rubber			
EPDM-40	100	100	100	100
BS-120	30	30	30	30
Zinc oxide	5	5	5	5
Stearine	1	1	1	1
Captax	2	2	2	2
Sulphur	2	2	2	2
FAS	0	16	16	16
Total	140	156	156	156

Elastomer compositions without functionally active additives were used as control samples. The properties of compositions containing single components of functionally active structures were previously studied by our team [30,32].

The rubber mixes were prepared in two stages, using a high-speed laboratory micromixer of the "Brabender" type (Polimermash, Saint Petersburg, Russia) for the first stage. A calculated amount of the functionally active modifier was added to the compounded masterbatch at a temperature of 105–110 °C and homogenized. In the second stage, sulfur, vulcanization accelerators, and activators were added on laboratory rollers (Polimermash, Saint Petersburg, Russia) at a temperature of 45–50 °C after a 24 h rest. Vulcanization of the samples was then performed in a PHG-2 212/4 vulcanizing press (Carver, Aldridge, Hungary) under the optimum mode determined using a flow meter MDR 3000 Professional (MonTech, Columbia City, IN, USA) (165 °C, 40 min.).

The physical and mechanical properties of standard elastomer samples were determined using a tearing machine Shimazu AG-Xplus 1.0 kN (Shimadzu, Kyoto, Japan) in accordance with ISO 37-2017 [33]. To evaluate the fire and heat resistance of the samples, the temperature dependence on the unheated surface of the sample from the time of open plasma torch fire impact, the loss of sample weight, and the linear burning speed were determined according to a developed procedure. The sample used was a disc with a diameter of 100 mm and thickness of 10 mm. High-temperature heating created a temperature around 2000 °C on the surface of the sample.

To evaluate the erosion strength of the material under high-temperature conditions, a sample was placed on a rotating shaft and tangentially heated by a plasma torch. The start time of combustion and coke detachment were recorded during the test, followed by the determination of the sample diameter after the test. The disrupted layer was removed, and the thickness of the non-disrupted layer was determined.

The strength characteristics of the coke were determined by considering the tear-off forces realized on the interface between the coke layer and non-degraded material during the tear-off of the coke layer under the impact of centrifugal forces when a cylindrical sample was rotated at a constant speed during high-temperature heating in a plasma torch flame. The tear propagation strength of the carbonized layer was determined using Formula (1):

$$\sigma = \frac{2 \cdot (\pi \cdot \omega)^2 \cdot (R_0^2 - R^2) \cdot \rho_K}{(R_0 - R)^2} \quad (1)$$

where ω—angular velocity, rps; R_0—initial radius, mm; R—radius before the boundary of the pyrolysis layer, along which the tear-off occurs, and mm; ρ_k—coke density, kg·m^{-3}.

The preassembly of a functionally active structure was carried out by treating the surface of the microspheres and microfibers with a 5% aqueous solution of DDF (Figure 7), followed by drying until constant weight.

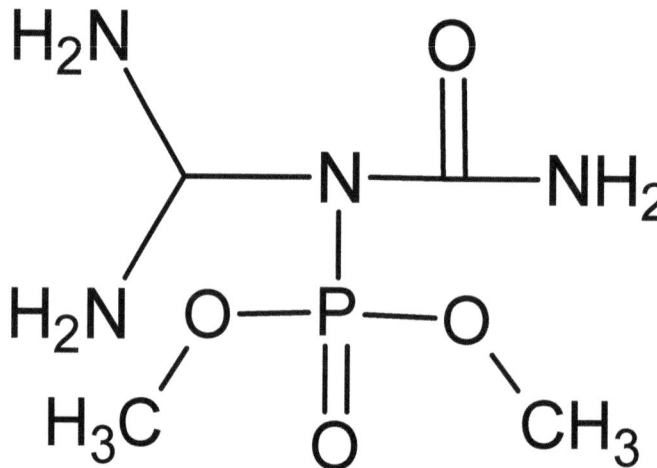

Figure 7. Structural formula of dimethylcarbamyl(diaminomethyl)phosphoramide [34].

Morphological research of the FAS, vulcanized stocks, and coke structures was carried out on the scanning electronic microscope Versa 3D (FEI Company, Hillsboro, OR, USA). Differential thermal and thermogravimetric analyses of the samples were carried out on a derivatograph Q-1500D (MOM Szerviz Kft, Budapest, Hungary).

5. Conclusions

The results indicate that incorporating pre-assembled functionally active systems into the composition of elastomeric fire- and heat-protective materials enhances their interaction with the elastomeric matrix, resulting in better distribution. This targeted delivery of the modifier into the interphase layer enhances the coke formation processes at the interface. The combined introduction of microspheres and microfibers during initiated coke formation creates structures where microfibers are grouped around the microspheres, reinforcing the coke layer and increasing its erosion resistance. Morphological analysis of the FAS, vulcanized stocks, and coke structures was conducted using the scanning electronic microscope Versa 3D. Differential thermal and thermogravimetric analyses of the samples were carried out using a derivatograph Q-1500D.

Supplementary Materials: The following supporting can be downloaded at: https://www.mdpi.com/article/10.3390/molecules28135267/s1. Raw data of DTA and TG analyzes of the studied samples.

Author Contributions: Conceptualization, V.F.K. and V.G.K.; formal analysis, O.M.N. and N.A.K.; investigation, D.A.K. and V.B.; methodology, O.M.N.; writing—original draft, V.B. and V.G.K.; writing—review and editing, V.B. and V.G.K. All authors have read and agreed to the published version of the manuscript.

Funding: Financial support was provided by the Ministry of Science and Higher Education of the Russian Federation FZUS-2021-0013 and Scholarship of the President of the Russian Federation to young scientists and Postgraduates (SP-1507.2022.1).

Institutional Review Board Statement: Not applicable.

Informed Consent Statement: Not applicable.

Data Availability Statement: The data presented in this study are available in Supplementary Materials.

Conflicts of Interest: The authors declare no conflict of interest.

Sample Availability: Not available.

References

1. Zaikov, G.E.; Kalugina, E.V.; Gumargalieva, K.Z. *Fundamental Regularities of Thermal Oxidation of Heat-Resistant Heterochain Polymers—Thermal Stability of Engineering Heterochain Thermoresistant Polymers*; Utrecht: Boston, MA, USA, 2004; pp. 15–57.
2. Mikhailin, Y.A. *Structural Polymeric Composite Materials*; Nauka: St. Petersburg, Russia, 2010; pp. 25–67.
3. Walter, M.D.; Wajer, M.T. *Overveiw of Flame Retardants Including Magnesium Hydroxine*; Martin Marietta Magnesia Specialties LLC: Nottingham, MD, USA, 2010.
4. Bhuvaneswari, C.M.; Surehkumar, M.S.; Kakade, S.D. Manoj Gupta Ethylene-propylene Diene Rubber as a Futuristic Elastomer for Insulation of Solid Rocket Motors. *Def. Sci. J.* **2006**, *56*, 309–320. [CrossRef]
5. Ahmed, A.F.; Hoa, S.V. Thermal insulation by heat resistant polymers for solid rocket motor insulation. *J. Compos. Mater.* **2012**, *46*, 1544–1599. [CrossRef]
6. Mariappan, T. Fire Retardant Coatings. In *New Technologies in Protective Coatings*; BoD—Books on Demand: Paris, France, 2017. [CrossRef]
7. Kablov, V.F.; Novopoltseva, O.M.; Kochetkov, V.G.; Lapina, A.G.; Pudovkin, V.V. Elastomer heat-shielding materials containing aluminosilicate microspheres. *Russ. Eng. Res.* **2017**, *37*, 1059–1061. [CrossRef]
8. Markowski, J. Cenospheres. Anuniversal construction material. *Przem. Chem.* **2019**, *98*, 940–943. [CrossRef]
9. Bezzaponnaya, O.V.; Golovina, E.V. Effect of mineral fillers on the heat resistance and combustibility of an intumescent fire proofing formulation on silicone base. *Russ. J. Appl. Chem.* **2018**, *91*, 96–100. [CrossRef]
10. Kablov, V.F.; Keibal, N.A.; Kochetkov, V.G.; Motchenko, A.O.; Antonov, Y.M. Research of the influence of carbon microfiber on the properties of elastomer fire-protective materials. *Russ. J. Appl. Chem.* **2018**, *91*, 1160–1164. [CrossRef]
11. Jang, B.N.; Wilkie, C.A. The effects of triphenylphosphate and recorcinolhis on the thermal degradation of polycarbonate in air. *Thermochim. Acta* **2005**, *433*, 1–12. [CrossRef]
12. Xiao, J.; Hu, Y.; Yang, L.; Cai, Y.; Song, L.; Chen, Z.; Fan, W. Fire retardant synergism between melamine and triphenyl phosphate in poly(butylene terephthalate). *Polym. Degrad. Stab.* **2006**, *91*, 2093–2100. [CrossRef]
13. Pawlowski, K.H.; Schartel, B.; Fichera, M.A.; Jager, C. Flame Retardancy Mechanisms of Bisphenol a- Bis(diphenyl phosphate) in Combination with Zinc Borate in Bisphenol a-Polycarbonate/Acrylonitrile-Butadiene-Styrene Blends. *Thermochim. Acta* **2010**, *49*, 92–99. [CrossRef]
14. Kashiwagi, T. *Flame Retardant Mechanism of the Nanotubes-Based Nanocomposites*; Final Report. NIST GCR 07-912; National Institute of Standards and Technology: Gaithersburg, MD, USA, 2007; p. 65.
15. Bourbigot, S.; Le Bras, M. Flame Retardant Plastics. In *Plastics Flammability Handbook*; Manser: Munich, Germany, 2004; Volume 5, pp. 133–172. [CrossRef]
16. Beach, M.W.; Rondan, N.G.; Froese, R.D.; Gerhart, B.B.; Green, J.G.; Stobby, B.G.; Shmakov, A.G.; Shvartsberg, V.M.; Korobeinichev, O.P. Studies of degradation enhancement of polystyrene by flame retardant additives. *Polym. Degrad. Stab.* **2008**, *9*, 1664–1673. [CrossRef]
17. Koo, J.H. *Polymer Nanocomposites: Processing, Characterization, and Applications*; McGraw-Hill Professional: New York, NY, USA, 2006; pp. 159–176.
18. Koo, J.H.; Miller, M.J.; Weispfenning, J.; Blackmon, C. Silicone polymer composites for thermal protection system: Fiber reinforcements and microstructures. *J. Compos. Mater.* **2011**, *45*, 1363–1380. [CrossRef]
19. Li, J.; Hu, B.; Hui, K.; Li, K.; Wang, L. Erosion resistance of ethylene propylene diene monomer insulations reinforced with precoated multi-walled carbon nanotubes. *Acta Astronaut.* **2022**, *198*, 251–257. [CrossRef]
20. Al-Saleh, M.H.; Sundararaj, U. Review of the mechanical properties of carbon nanofiber/polymer composites. *Compos. Part A Appl. Sci. Manuf.* **2011**, *42*, 2126–2142. [CrossRef]
21. Rybiński, P.; Syrek, B.; Marzec, A.; Szadkowski, B.; Kuśmierek, M.; Śliwka-Kaszyńska, M.; Mirkhodjaev, U.Z. Effects of Basalt and Carbon Fillers on Fire Hazard, Thermal, and Mechanical Properties of EPDM Rubber Composites. *Materials* **2021**, *14*, 5245. [CrossRef]
22. Raask, E. Cenospheres in pulverized fuel ash. *J. Inst. Fuel* **1968**, *43*, 339–344.
23. Rose, N.L. Inorganie flu-ash spheres as pollution. *Trasers Environ. Pollut.* **1996**, *91*, 245–252. [CrossRef]
24. Pandey, G.S.; Gain, V.K. Cenosphere-load in coal ash diseharge ofthermal power plant. *Res. Ind.* **1993**, *38*, 99–100.
25. Wang, S.; Wang, L.; Su, H.; Li, C.; Fan, W.; Jing, X. Enhanced thermal resistance and ablation properties of ethylene-propylene-diene monomer rubber with boron-containing phenolic resins. *React. Funct. Polym.* **2022**, *170*, 105136. [CrossRef]
26. Guo, M.; Li, J.; Wang, Y. Effects of carbon nanotubes on char structure and heat transfer in ethylene propylene diene monomer composites at high temperature. *Compos. Sci. Technol.* **2021**, *211*, 108852. [CrossRef]
27. Sun, H.; Yang, Y.; Ge, X. Effect of various fibers on the ablation resistance of poly(diaryloxyphosphazene) elastomer. *J. Appl. Polym. Sci.* **2020**, *137*, 48534. [CrossRef]
28. Iqbal, N.; Sagar, S.; Khan, M.B.; Rafique, H.M. Elastomeric ablative nanocomposites used in hyperthermal environments. *Polym. Eng. Sci.* **2014**, *54*, 255–263. [CrossRef]

29. Tate, J.S.; Gaikwad, S.; Theodoropoulou, N.; Trevino, ...; Koo, J.H. Carbo... ...hermal protection material in aerospace applications. *J. Compos.* **2013**, *2013*, 403656. [CrossRef]
30. Kablov, V.F.; Novopol'tseva, O.M.; Kochetkov, V.G.; Pudovkin, V.V. Physicomechani... ...-ret... ...operties of elastomer compounds based on ethylene–propylene–diene rubber and filled with ... ...osilicate microspheres. *Russ. J. Appl. Chem.* **2017**, *90*, 257–261. [CrossRef]
31. Ravishankar, P.S. Treatise on EPDM. *Rubber Chem. Technol.* **2012**, *85*, 327–349. [CrossRef]
32. Kablov, V.F.; Bondarenko, S.N.; Vasilkova, L.A. Properties of fire-retardant coating. *Nov. Mater.* **2013**, *5*, 61–66.
33. *ISO Standard 37-2013*; Rubber, Vulcanized or Thermoplastic; Determination of Tensile Stress-Strain Properties. Standartinform Publ.: Moscow, Russia, 2014; p. 32.
34. Kablov, V.F.; Kochetkov, V.G.; Keibal, N.A.; Novopol'tseva, O.M.; Kryukova, D.A. Modifier Based on Dicyandiamide and Dimethyl Phosphite for Fire and Heat Resistant Elastomer Materials. *Russ. J. Appl. Chem.* **2022**, *95*, 661–668. [CrossRef]

Disclaimer/Publisher's Note: The statements, opinions and data contained in all publications are solely those of the individual author(s) and contributor(s) and not of MDPI and/or the editor(s). MDPI and/or the editor(s) disclaim responsibility for any injury to people or property resulting from any ideas, methods, instructions or products referred to in the content.

Article

The Effect of Flame-Retardant Additives DDM-DOPO and Graphene on Flame Propagation over Glass-Fiber-Reinforced Epoxy Resin under the Influence of External Thermal Radiation

Oleg P. Korobeinichev [1,*], Egor A. Sosnin [1,2], Artem A. Shaklein [3], Alexander I. Karpov [3], Albert R. Sagitov [1,2], Stanislav A. Trubachev [1], Andrey G. Shmakov [1], Alexander A. Paletsky [1] and Ilya V. Kulikov [1]

1. Voevodsky Institute of Chemical Kinetics and Combustion SB RAS, 630090 Novosibirsk, Russia; e.sosnin@g.nsu.ru (E.A.S.); rasagitov@gmail.com (A.R.S.); satrubachev@gmail.com (S.A.T.); shmakov@kinetics.nsc.ru (A.G.S.); paletsky@kinetics.nsc.ru (A.A.P.); kulikovsas2001@gmail.com (I.V.K.)
2. Department of Physics, Novosibirsk State University, 630090 Novosibirsk, Russia
3. Udmurt Federal Research Center, 426067 Izhevsk, Russia; shaklein@udman.ru (A.A.S.); karpov@udman.ru (A.I.K.)
* Correspondence: okorobeinichev@gmail.com

Abstract: The flammability of various materials used in industry is an important issue in the modern world. This work is devoted to the study of the effect of flame retardants, graphene and DDM-DOPO (9,10-dihydro-9-oxa-10-phosphaphenanthrene-10-oxide-4,4′-diamino-diphenyl methane), on the flammability of glass-fiber-reinforced epoxy resin (GFRER). Samples were made without additives and with additives of fire retardants: graphene and DDM-DOPO in various proportions. To study the flammability of the samples, standard flammability tests were carried out, such as thermogravimetric analysis, the limiting oxygen index (LOI) test, and cone calorimetry. In addition, in order to test the effectiveness of fire retardants under real fire conditions, for the first time, the thermal structure of downward flame propagation over GFRER composites was measured using thin thermocouples. For the first time, the measured thermal structure of the flame was compared with the results of numerical simulations of flame propagation over GFRER.

Keywords: glass-fiber-reinforced epoxy resin; flammability; fire retardancy; phosphorus-containing flame retardant; graphene; limiting oxygen index; cone calorimetry; numerical simulation

1. Introduction

Composite materials based on epoxy resin are promising materials for various industries, such as aviation, mechanical engineering, shipbuilding, etc. [1]. These materials have gained great popularity due to their high strength, flexibility, chemical resistance and thermal-insulation properties, together with their low specific weight. However, epoxy resin is a flammable material [2,3]; therefore, increased fire-resistance requirements are applied to composites based on it. When exposed to an external heat source, the combustibility of these materials can increase even more, which makes the issue especially important for products installed near a heat source (aircraft nacelles, car radiator mounts, etc.).

A prevalent way to improve the fire resistance of epoxy resin is to add fire retardants. Previously, halogen compounds have been used as epoxy-polymer flame-retardant additives for a long time. However, halogen-containing compounds, when burned, emit acrid and toxic smoke, which is harmful to humans and the environment [4,5]. Numerous studies have been carried out to find an alternative to halogen-containing flame retardants, and as a result, a switch to phosphorus-containing flame retardants has been proposed. Phosphorus-based additives such as 9,10-dihydro-9-oxa-10-phosphafenentren-10-oxide (DOPO) and its derivatives [6,7] can serve as an example. Moreover, some studies show that polymer composites containing both phosphorus and nitrogen compounds show a significant improvement in flame-retardant performance [8–12]. Therefore, due to their special structure, DOPO-based phosphonamidates can be effective flame retardants [13–15]. So, a fire retardant such as DDM-DOPO [16] can be used in a smaller amount as part of an epoxy binder, which favorably affects the physical and mechanical properties of fiberglass-reinforced plastics made from such materials.

Nano-material additive, such as graphene, has recently become another popular halogen-free flame retardant due to its ability to increase the formation of char on the surface of the polymer and to strengthen it due to graphitization [17,18]. Graphene nanoparticles can effectively reduce the melt flow and inhibit the flammable drips of epoxy resin during combustion, which generally prevents the spread of flames [19]. In [20] it was shown that graphene has adsorbing properties, so it can capture volatile combustible compounds on the sample surface, which also increases the fire resistance of the material.

To assess the combustibility of polymeric materials, usually thermogravimetric analysis (TGA), a UL-94 test, limiting oxygen index (LOI), and cone calorimetry are used. Often, the fire resistance of materials is evaluated only on the basis of these methods [7,21]. However, these tests do not always provide reliable information about the behavior of the material under fire conditions [22]. In this case, more accurate information can be obtained from experimental studies of flame propagation over the surface of the material. However, hardly flammable materials such as reinforced composites do not support combustion under standard atmospheric conditions [23,24]. Such composites can be studied under conditions of high oxygen concentration [25], in the presence of external thermal radiation [26] or in the presence of an external flame source [22].

In 1970–1980, Williams [27], Fernandez-Pello [28,29], and Hirano [30] et al. performed experimental studies of flame propagation over a solid surface under the action of an external heat flux. Work [31] studied the horizontal flame propagation over thermoplastic polymers: polymethyl methacrylate (PMMA), polypropylene (PP) and polystyrene (PS) plates, under the action of one-sided heating, and it was shown that the effect of sample thickness on the nature of flame propagation is less pronounced than the effect heat flux of external radiation. The same effect is also shown in [32]. Moreover, the application of additional heat flux caused the flame to propagate more quickly over PMMA slabs. The qualitative influence of the heat flow was observed during the combustion of wood-based materials, which do not allow the flame to spread to the upper part of the sample, but allow it to spread in the presence of an external heat flow. In addition, preheating the particle board samples (with a 2.2 kW/m^2 radiant flux) increased the surface temperature by more than 100 °C, and therefore required less time and energy to reach the pyrolysis temperature. The preheated sample showed significantly faster flame propagation than the sample that was shielded prior to ignition. It was shown in [33] that under the influence of an increasing external heat flux, the rate of flame propagation of polyurethane foam and expanded polystyrene demonstrate different trends. The rate of flame propagation over polyurethane foam increases all the time, and fluctuates over expanded polystyrene. In [34], the thermomechanical analysis of PMMA slabs was carried out under the action of a powerful heater installed on one side of the sample. Most investigations on flame propagation under external thermal radiation source conditions are devoted to such polymers as PMMA, PP, etc. However, a relatively small number of works are devoted to flame propagation over the surface of composites based on epoxy resin (ER). For example, in work [25], the authors studied downward flame propagation over the surface of carbon-fiber-reinforced epoxy resin at various oxygen concentrations. In addition, there are no data on flame propagation over reinforced composites with the addition of fire retardants, with the exception of a small number of works [22]. To predict the behavior of polymeric materials in different fire scenarios, a number of numerical models were developed and tested for the purpose of comparison with the experimental data [35,36]. Meanwhile, little attention has been paid to development of models of flame propagation over reinforced polymers and polymers with the addition of flame retardants, and there are no widely accepted approaches to the theoretical description of the pyrolysis and combustion of inhibited polymers.

Thus, the purpose of this work was to study the effect of flame retardants DDM-DOPO and graphene on flammability and downward flame propagation over glass-fiber-reinforced epoxy resin (GFRER) under the action of external thermal radiation and to develop an effective model that would predict the flame propagation behavior for GFRER composites.

2. Results and Discussion

2.1. TGA, DTG, UL-94 HB, LOI Results

Figure 1 shows the data of thermogravimetric analysis in inert atmosphere (Ar) for GFRER composites. It can be seen in Figure 1a that the addition of flame retardants reduces the amount of residue at the temperature of 580 °C, indicating an increase in the yield of pyrolysis products into the gas phase in the case of samples inhibited by flame retardants. Addition of 1.5% DDM-DOPO leads to a decrease in the residue mass by ~3% relative to the sample without the additives, while the addition of 3% DDM-DOPO leads to a decrease in the residue mass by ~7%. A partially registered decrease in the residue mass during the decomposition of mixtures with DDM-DOPO can also be associated with decomposition of the additive itself in the mixture. However, the individual substance DDM-DOPO is degraded up to 90% [16]; therefore, the maximum contribution to the residue mass reduction for the ER + DDM-DOPO mixture from DDM-DOPO can be 1.3% in 3% and 2.7% in 7% for samples with 1.5% and 3% DDM-DOPO (Table 1), respectively. The addition of 1.5% graphene has practically no effect on the amount of the residue; however, 3% graphene leads to a decrease in the weight of the residue in TGA by ~7%.

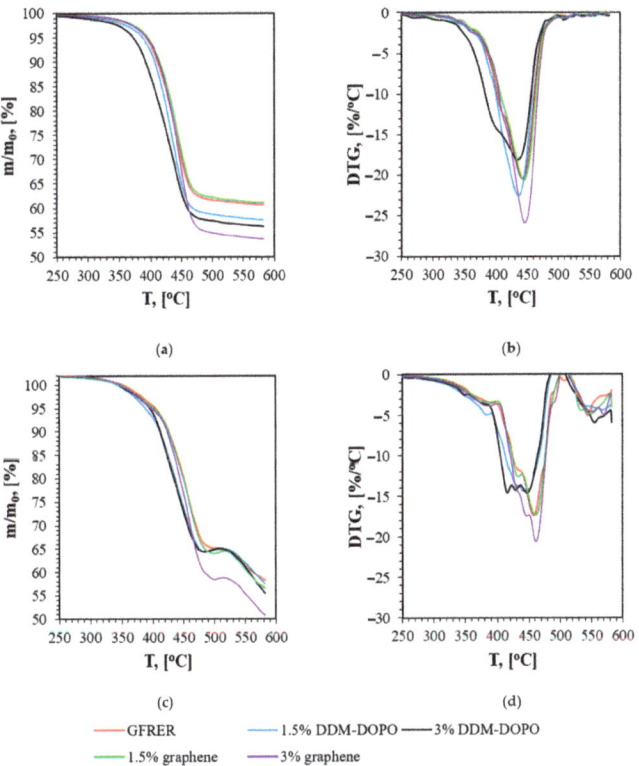

Figure 1. Thermogravimetric analysis data for GFRER with and without additives (**a**) TG data in Ar, (**b**) DTG data in Ar, (**c**) TG data in 79 v% He + 21 v% O_2 and (**d**) DTG data in in 79 v% He + 21 v% O_2.

Table 1. UL-94, LOI, mass loss (ML) and soot mass (SM) in downward flame-spread experiment, max T and residue mass in TGA.

Sample	UL-94 HB, with ROS mm/min	LOI, % Thin Samples	ML in % of the Initial Mass	SM in % of the Initial Mass	T_{max} DTG in Inert, °C	Residue Mass in TGA, %
GFRER	84 ± 2 [1]	21.0	18	12	442 ± 2	59.1 ± 1.5
1.5% DDM-DOPO	66 ± 2 [2]	21.9	14	17	438 ± 2	56.5 ± 1.5
3% DDM-DOPO	64 ± 2 [2]	21.9	13	20	444 ± 2	55.0 ± 1.5
1.5% graphene	65 ± 2 [2]	21.7	13	18	442 ± 2	59.1 ± 1.5
3% graphene	58 ± 2 [2]	21.6	11	23	448 ± 2	53.7 ± 0.5
6.5% DDM-DOPO	23 ± 1 [2]	26.1	-	-	-	-
9.8% DDM-DOPO	26 ± 1 [2]	26.4	-	-	-	-

[1] no rating; [2] HB rating.

Assuming that the pyrolysis reaction occurs in one stage and is of the first order, from the data in Figure 1, the kinetic parameters of pyrolysis were obtained using the established method [37]. The obtained kinetic parameters for GFRER are presented in Table 2. The accuracy of determining the pyrolysis rate constant by this method is factor 5. Despite small differences in the temperature of the maximum thermal decomposition rate of the samples (Table 1), the pyrolysis kinetics barely changes with the addition of flame retardants (Figure 2) for high temperatures.

Table 2. The measured flame spread rate depending on the magnitude of the heat flux and the concentration of flame retardants.

Heat Flux, kW/m²	Samples				
	GFRER	1.5% DDM-DOPO	3% DDM-DOPO	1.5% Graphene	3% Graphene
4.2	1.17 ± 0.08	0.87 ± 0.05	0.85 ± 0.1	1.07 ± 0.14	0.88 ± 0.05
3	0.75 ± 0.03	0.61 ± 0.03	0.51 ± 0.03	0.67 ± 0.04	0.58 ± 0.02
2.5	0.63 ± 0.07	0.53 ± 0.05	0.45 ± 0.04	0.54 ± 0.04	0.49 ± 0.02
2	0.57 ± 0.03	0.43 ± 0.03	Self-extinguished	0.45 ± 0.03	0.40 ± 0.02
1.5	0.52 ± 0.02	Self-extinguished	Self-extinguished	Self-extinguished	Self-extinguished

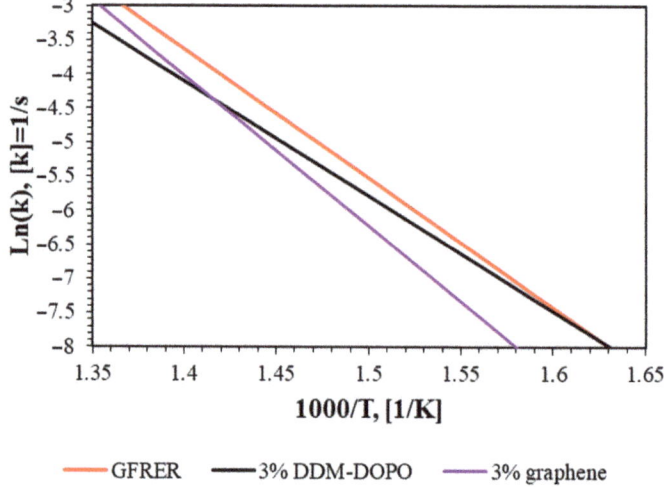

Figure 2. Pyrolysis rate constants of GFRER compositions in the Arrhenius form.

Table 1 shows UL-94 HB data with the measured average rate of spread (ROS) and the LOI test. The 50% binder GFRER failed the UL-94 test and had a burn rate of 84 mm/min (Table 1). It is to be noted that in the UL-94HB test, the speed incombustibility criterion for the HB rating for specimens up to 3 mm thick was 75 mm/min.

Table 1 also shows that both flame retardants improve the fire resistance of GFRER in the UL-94 and LOI tests. At the same time, the additions of 3% DDM-DOPO and 3% graphene result in a decrease in the pyrolysis rate constant. In the UL-94 horizontal flame spread test, all flame-retardant-inhibited samples were rated as "HB" (at 1.5% and 3% additive concentrations). The addition of 1.5% DDM-DOPO to the sample reduced the flame propagation rate in the UL-94 HB test by ~27% relative to pure GFRER, while the addition of 3% DDM-DOPO reduced the rate by ~31%. Thus, a twofold increase in the additive concentration does not lead to a significant increase in the inhibition effect (the saturation effect). In the case of 1.5% graphene, the difference in the rate relative to pure GFRER was ~29%, while for 3% graphene, the difference increased to ~45%. In fact, an increase in graphene concentration leads to a noticeable increase in the inhibition effect in this test. Figure 3 shows the dependence of the UL-94HB rate of flame spread (ROS) on the concentration of flame retardant in the composition of GFRER. For graphene, a close to linear dependence of ROS on concentration is observed, and for DDM-DOPO, a decrease in efficiency occurs as the flame retardant concentration rises in the UL94 test. At the same time, at 1.5% of the additive, the effectiveness of both fire retardants is practically the same. In order to show the effectiveness of the flame retardants used, the LOI of 2 mm thick samples with a higher mass fraction of flame retardants in the composition was additionally measured. Increasing the thickness of the samples and the mass fraction of flame retardants leads to an increase in LOI, which indicates the high fire resistance of GFRER under standard atmospheric conditions. Further experiments with these samples were not carried out in this work, since it turned out to be impossible to support their combustion with an external heat flux.

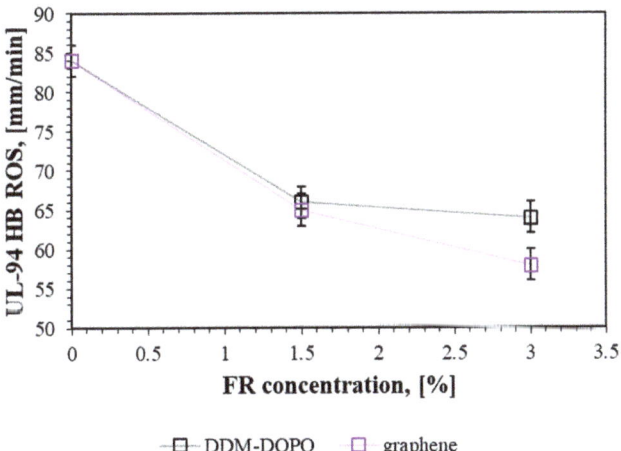

Figure 3. The effect of flame retardant concentration on the burning rate in the UL-94HB test.

2.2. The Flame Spread Rate over the Samples Surface

The dependence of the rate of flame spread (ROS) from a flame front position under incident heat flux (IHF) 3 kW/m² is shown in Figure 4. The flame spreads over the sample surface uniformly, without acceleration. ROS fluctuations in a particular part of the sample may be connected with the inhomogeneity of the manufactured samples. The addition of the flame retardant led to a decrease in ROS over GFRER surface. It can also be noted that an increase in the flame-retardant concentration to 3% led to a greater decrease in ROS of

1.5%. The sample incorporated with 3% DDM-DOPO showed the lowest flame spread rate, although the sample incorporated with 3% graphene showed the lowest flame spread rate in the UL-94 HB test.

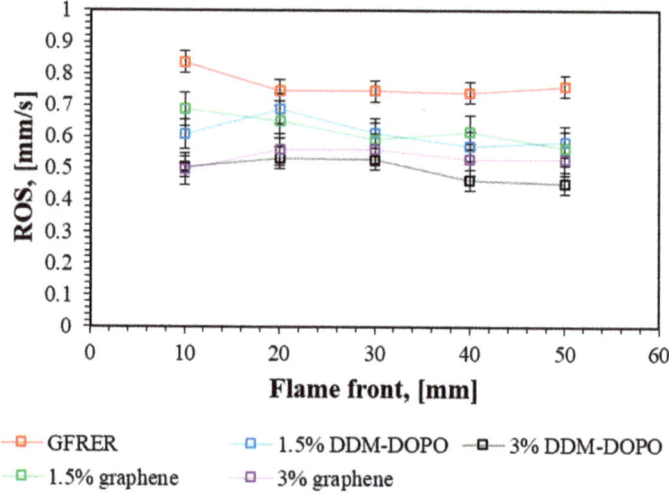

Figure 4. Dependence of the flame spread rate over the sample surface from the flame front position under the incident heat flux 3 kW/m².

The dependence of the flame spread rate on different heat fluxes and flame-retardant concentrations is shown in Table 2 and Figure 5. The slope of the curves shows that the linear increase in the heat flux magnitude leads to the linear growth in ROS. The samples incorporated with 3% additive have a lower ROS compared to the samples with 1.5% additive under the heat fluxes lower than 3 kW/m². The same tendency is shown in Figure 5. The sample without the additives does not demonstrate self-sustaining combustion at the external heat flux 1 kW/m², the samples with 1.5% DDM-DOPO and 1.5% and 3% graphene extinguish at the external heat flux 1.5 kW/m², and the sample with 3% DDM-DOPO extinguishes at the heat flux 2 kW/m². When the heat flux was increased to 4.3 kW/m², the ROS for GFRER sample began to deviate from the linear dependence. It can be connected with the early pyrolysis of the sample surface ahead of the flame front. Whereas the efficiency of the 1.5% graphene additive does not depend on the heat flux, the efficiency of the other additives increases when the heat flux increases. However, at the external heat flux 4.3 kW/m², the efficiency of 1.5% DDM-DOPO, 3% DDM-DOPO, and 3% graphene are almost the same. In the case of DDM-DOPO, this is probably due to H and OH radicals' equilibrium concentration in the flame. In the case of graphene, an increase in the concentration of the additive from 1.5% to 3% leads to a further decrease in ROS. The same results were obtained in UL-94 HB (Table 1), where the flame spread rate of UL-94HB for DDM-DOPO was not strongly affected by the increasing concentration, while an increase in the graphene concentration continued to reduce ROS.

The calculated ROS for GFRER and GFRER with flame retardant additives are satisfactorily predicted by the model (Figure 5b,c) described in Section 3.6. However, the flame extinguishes at a low heat flux that is not predicted by the model. Additionally, in the case of 1.5% DDM-DOPO, the model overpredicts ROS, while with 3% DDM-DOPO it slightly underpredicts ROS. This is due to the simplicity of the flame retardant action assumption in the model in the gas phase, based on the decrease in the reaction rate constant of a one-stage global reaction. The DDM-DOPO effectiveness coefficient Ψ in the model depends on the additive concentration, while in the experiments (Table 1, Figure 5), a decrease in the additive effectiveness with increasing concentration was observed. The results of elemental analysis obtained by scanning electron microscopy (Quantitative Energy-dispersive X-ray spectroscopy microanalysis) after the flame propagation experiments under IHF 3 kW/m^2 for the samples incorporated with DDM-DOPO, showed that from 30 to 50% of the initial phosphorus went into the gas phase. The SEM microphotographs (Figures S1 and S2) are presented in the Supplementary Materials. All things considered, the results of elemental analysis confirm the assumption about the gas-phase mechanism inhibition of the DDM-DOPO additive. The influence of flame retardants containing phosphorus in the composition of polymeric materials was previously studied [38–41]. The main mechanism of action of these additives is believed to be participation of this species and its decomposition products in chain-termination reactions.

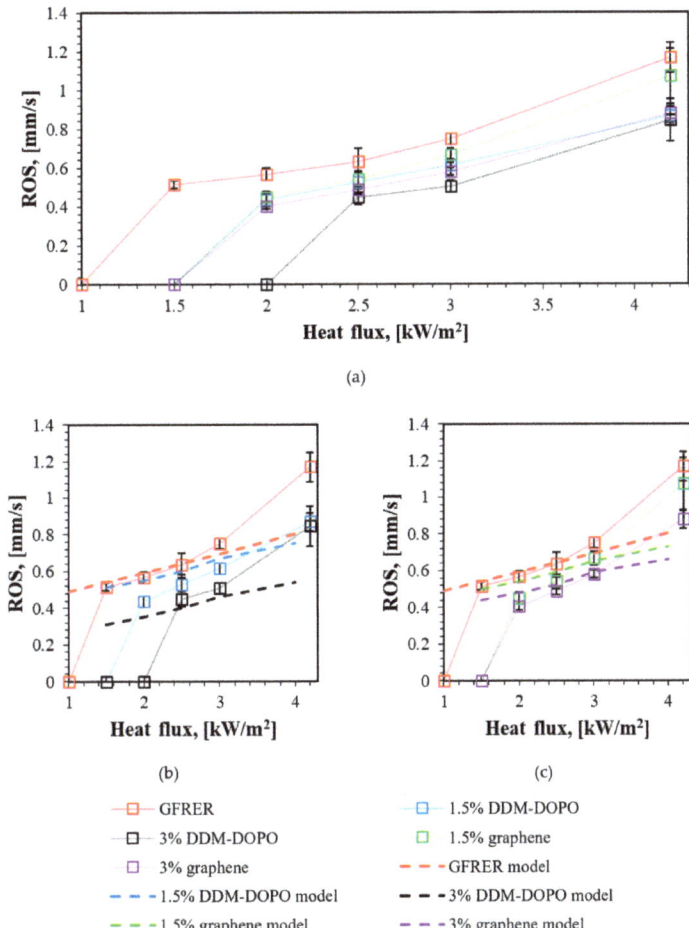

Figure 5. An average flame spread rate over the samples surface depending on the incident heat flux. (**a**) The experimental results. (**b**) The model and the experimental results for DDM-DOPO. (**c**) The model and the experimental results for graphene.

Mass loss (ML) and soot mass (SM) dependence versus flame retardant concentration and the incident heat flux obtained after the downward flame propagation experiments are shown in Figure 6. Figure 6a shows the total mass loss in % of the initial sample mass. Figure 6b shows soot mass formed during flame propagation in % of the initial sample mass.

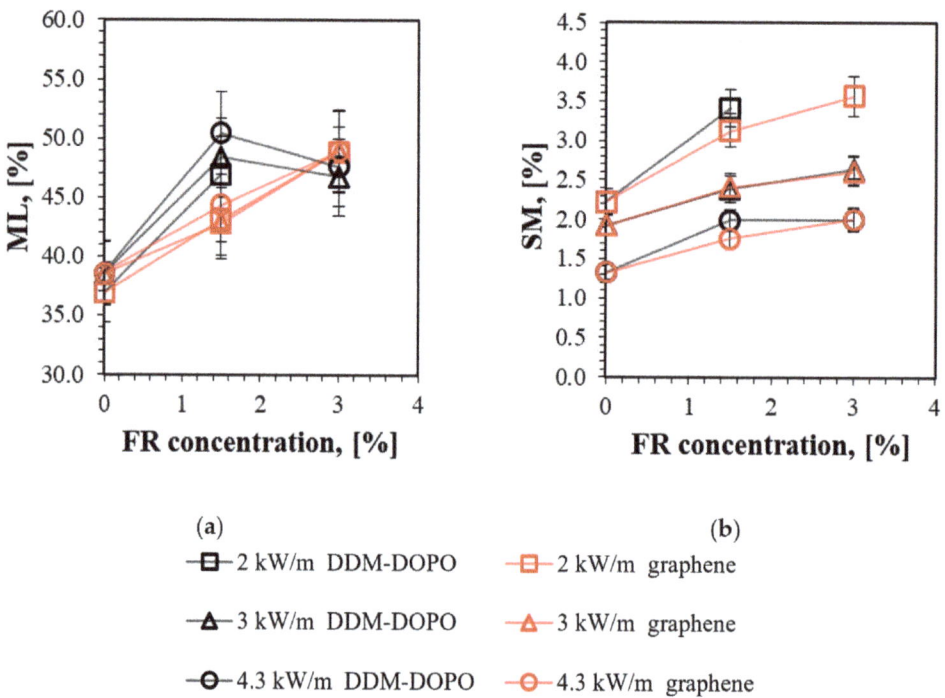

Figure 6. Comparison (**a**) mass loss (ML) in % of the initial sample mass depending on the additive concentration and the incident heat flux (**b**) soot mass (SM) on the sample surface in % of initial sample mass, depending on the additive concentration and the incident heat flux.

The samples incorporated with 1.5% DDM-DOPO showed an increase in ML by ~10% compared to the sample without the additives under all the tested heat flux values. However, increasing the DDM-DOPO concentration to 3% did not affect ML within the experimental accuracy, since almost all of the binder left the sample in this case. The samples incorporated with graphene showed a linear increase in ML (5% for 1.5% graphene and 10% for 3% graphene) with increasing additive concentration for all the tested heat flux values. This is due to an increase in the gas yield for samples with additives, which was observed in the TGA (Figure 1c). It can also be seen that the heat flux magnitude has little effect on ML in the experiment.

Figure 6b shows that SM decreases for all samples with increasing heat flux. This is probably due to the more intense combustion of the sample under a higher heat flux, which leads to soot oxidation on the sample surface. An increase in the concentration of DDM-DOPO and graphene additives led to the same SM for all the cases. The addition of graphene probably leads to an increase in the proportion of non-combustible volatile pyrolysis products, which dilutes combustible gases and extinguishes the flame. DDM-DOPO decomposes under the action of temperature and its phosphorus-containing decomposition products act in the gas phase. In this case, the effect of reducing the efficiency of DDM-DOPO with an increase in its initial concentration in the polymer was observed. An increase in the soot yield for the samples incorporated with graphene compared to the samples without the additives and samples incorporated with DDM-DOPO under elevated oxygen concentration conditions was observed in work [22].

2.3. Thermal Flame Structure

The temperature profiles in the gas and condensed phases obtained in the experiment and the model are compared in Figure 7. The flame structures obtained by scanning with a thermocouple and the flame structure obtained from the model are shown in Figure 8. The temperature profile gradients for the condensed phase in the preheating zone show good agreement between the experiment and the model. Yet, the major contribution to the rate of flame front shifting (ROS) is made by the temperature gradient from the flame to the sample surface near the flame front, where, as the observations have shown, there is no soot accumulation; therefore, the ROS is well predicted by the model. The maximum surface temperature in the model has a lower value regarding the experiment due to the underestimation of soot formation on the surface. In the region after the maximum, the model better predicts the temperature profile for samples with 3% DDM-DOPO and 3% graphene additives, while for the GFRER sample, the temperature in the model is overestimated. Thus, the model satisfactorily predicts the width of the combustion zone for samples with additives.

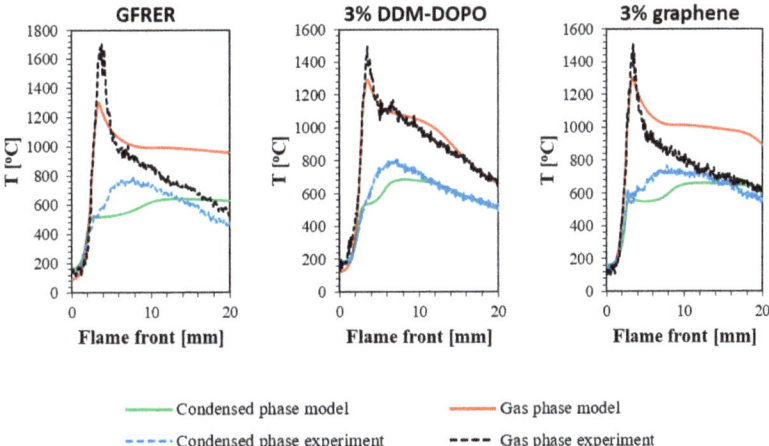

Figure 7. Comparison of the temperature profiles in the condensed phase and in the gas phase at the 1 mm distance from the sample surface under the incident heat flux 3 kW/m^2, obtained from the experiment and from the model.

Figure 7 shows the comparison of the temperature profiles in the gas phase at a distance of 1 mm from the sample surface obtained in the model and in the experiment. In the gas phase, the model also predicts the temperature profile gradient well, but underestimates the maximum temperature. The maximum temperature in the gas phase for all the samples was approximately 1600–1650 °C (Figure 8) within the thermocouple measurement error (±50 °C). However, for GFRER without the additives, the flame was pressed against the sample surface strongly; therefore, the maximum temperature recorded for GFRER at 1 mm distance from the surface (Figure 7) was higher than for the samples with 3% DDM-DOPO and 3% graphene. In the model, the maximum temperature in the gas phase at a distance of 1 mm from the surface was underestimated. Figure 8 shows that this is due to the fact that in the model the flame is not so strongly pressed against the surface and the zone of the maximum temperature is located slightly higher. This deviation is connected with a single-stage global macro reaction of combustion, because in the experiment near the combustion surface, many thermal decomposition reactions and the combustion of the volatile pyrolysis products of epoxy resin take place [42], which are not considered in the model. At the same time, volatile combustible products only appear on the sample surface in the model.

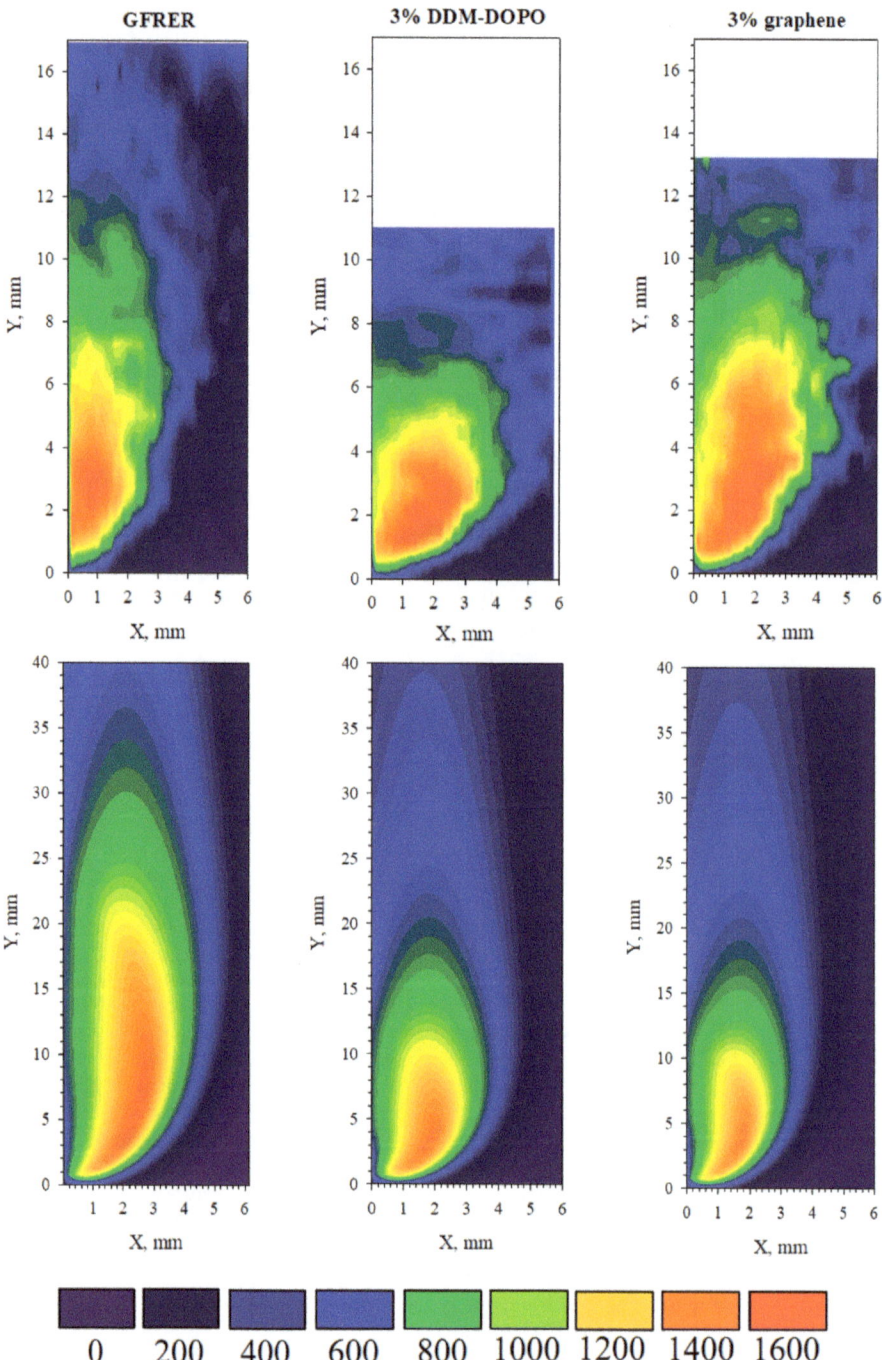

Figure 8. Comparison of the thermal flame structure in the experiment and in the model under the incident heat flux 3 kW/m^2.

The flame height was determined as its projection onto the sample plane from the beginning of the flame front to the region of 800 °C. The flame height for GFRER without the additives was 12 mm. The addition of the flame retardants reduced the length of the luminous flame zone by 4 mm in the case of DDM-DOPO and had almost no effect in the case of graphene. In the case of 3% graphene, the model reduced the length of the luminous flame zone and made it comparable in length with the experiment; however, the model incorrectly reproduced the trend that was observed in the experiment, where the addition of 3% graphene had almost no effect on the length of the luminous flame zone. It is to be emphasized that in the DDM-DOPO case, the model also reduced the length of the luminous flame zone and made it closer to the experiment, and, more importantly, the model decreased the luminous flame zone compared to pure GFRER observed in the experiment. This can explain the underprediction of the temperature values by the model in the gaseous and condensed phases (Figure 7) after passing the maximum on the temperature profiles GFRER without the additives and GFRER incorporated with 3% graphene and, consequently, it can explain the good temperature value agreement for 3% DDM-DOPO. Finally, it can be concluded that the use of a one-step inhibition mechanism for DDM-DOPO in the gas phase, connected with a decrease in the pre-exponential coefficient for the burning rate in the gas phase k_g, allows one to predict the DDM-DOPO fire-retardant effect on the flame behavior for GFRER with satisfactory accuracy. On the other hand, the mechanism of the graphene action, which is considered to force some of the volatile pyrolysis products to become non-combustible and not to participate in the combustion, requires more detailed study and the refinement of the model.

3. Materials and Methods

3.1. Sample Preparation

3.1.1. Epoxy Matrix

For glass fiber-reinforced epoxy-resin preparation, a polymer matrix was prepared with the composition presented in Table 3. ED-22 is a commercially available epoxy resin. Graphene and DDM-DOPO were used as flame retardants (9,10-dihydro-9-hydroxy-10-phosphaphenantrene-10-oxide-diaminodiphenylmethane).

Table 3. Epoxy matrix composition.

Component	Composition, Mass Fraction						
	1	2	3	4	5	6	7
Epoxy resin ED-22	100	100	100	100	100	100	100
Curing agent #9 [24]	5	5	5	5	5	13	13
DDM-DOPO	-	1.5	3	-	-	6.5	9.8
Graphene	-	-	-	1.5	3	-	-

Before being introduced, graphene was dispersed in an ultrasonic bath in acetone in the ratio of 1 g graphene to 100 mL acetone for 1 h at room temperature. After that, epoxy resin was added to the graphene dispersion in acetone and was stirred for 2 h. Then, the mixture was heated in an oil bath at the temperature of 130 °C during 0.5 h to remove acetone [43]. After that, the mixture was cooled to 50 °C, finely ground DDM powder was added to it, and the mixture was stirred for 2 h. Before DDM-DOPO and the curing agent (DDM) were introduced into epoxy resin, they were ground to fine powder and then mixed with the resin during 2 h at 50 °C. The obtained mixtures of resin, graphene and DDM, as well as of resin, DDM-DOPO and DDM, were used for making prepregs. The glass-fiber matrix was oriented in one direction for all the layers of the prepreg based on T-15(P)-76(92) glass fabric, indicating the one-directional structure of the fiber reinforcement. The curing mode was as follows: 100 °C for 2 h, 150 °C for 2 h. There were 2 layers in the glass fiber fabric. The binder content in the prepreg was 50%mass. Densities of the samples with compositions 1–5 were 1.44–1.55 g/cm^3 and with compositions 6–7 were ~1.1 g/cm^3.

For samples 6 and 7 (thickness ~2 mm), flame-propagation experiments were not carried out due to the low combustibility of these materials (they do not support flame spread under the action of a heat flux created by the radiation panels used in the work).

3.1.2. Additives

DDM-DOPO was provided for the tests by Prof. Yuan Hu from the USTC China. The method of synthesizing DDM-DOPO was previously described in [16].

Graphene was produced by the RUSGRAPHENE company (https://rusgraphene.com/, accessed on 1 June 2023).

3.2. Thermal Degradation Analysis

The thermal decomposition of the samples was studied using TGA. Pieces of GFRER slabs weighing 3–4 mg were placed into an aluminum or platinum crucible using a synchronous TG/DSC analyzer STA 409 PC (Netzsch, Selb, Germany) in a 100 v% argon and 79 v% He + 21 v% O_2 flow with a volumetric velocity of 27 cm^3/min (NTP). The samples were heated from 30 to 580 °C at the heating rate of 30 K/min. All the experiments were repeated at least 3 times. The accuracy of determining the decomposition residue was ~3%.

3.3. Elemental Analysis

Elemental analysis of the samples with the addition of DDM-DOPO to test the presence of phosphorus on the surface before and after combustion in the experiment, under the action of external radiation, was carried out using a JSM-6460LV SEM (JEOL, Tokyo, Japan) microscope (using Quantitative Energy-dispersive X-ray spectroscopy microanalysis).

3.4. Flammability Tests

The accuracy of determining the burning velocity in UL-94HB test was ±3–5%. The limiting oxygen index test was carried out in accordance with ISO 4589-2. The accuracy of the LOI was ±0.1%.

3.5. Description of the Experimental Setup for Downward Flame Propagation under the Action of Bilateral External Heating

The experiment on downward flame propagation under the action of two heaters was carried out on the setup shown in Figure 9.

The sample dimensions were 75 × 25 × 0.4 mm^3. The sample was installed in an aluminum frame 1 mm thick to prevent the flame spread along the side surfaces and to limit the width of the combustion zone to 20 mm. Using a tripod, the sample was fixed between two heaters (Almac IK5 Infrared Electric Heater, 500 Watt) powered by 220 V AC. The value of the incident heat flux (IHF) to the sample surface, depending on the distance from the heater from 25 to 100 mm, varied from 1 to 3 kW/m^2. The heat flux of 4.2 kW/m^2 was obtained by increasing the power supply voltage to 260 V. The sample was placed between two heaters at an equal distance from its surface. Calibration of the heating panels was previously carried out using the heat flux sensor described in [44]. The heat flux setting accuracy was ±0.3 kW/m^2. After installing the sample, the heating panels were switched on and heated for 15 min until a constant heat flux was established. To form a uniform flame front, the sample was ignited using a torch soaked in alcohol, which covered the entire length of the upper face of the sample. The sample was marked by horizontal lines (every 10 mm). The rate of flame spread over the surface of the sample was determined by the position of the luminous flame front, determined by video recording with a FujiFilm x-A20 (Fujifilm Holdings Corporation, Tokyo, Japan) camcorder (the shooting frequency was 30 frames per second). The temperature on the sample surface and in the gas phase was measured with a Pt-Pt + 10% Rh thermocouple (S-type thermocouple, wire diameter was 50 μm thick) [24]. The radiation correction was taken into account using the formula [45]. To determine the thermal flame structure, a Pt-Pt + 10% Rh thermocouple was used, coated with a layer of SiO_2 to prevent catalytic reactions on the surface of the thermocouple. The thermocouple was mounted on a biaxial positioning system with two stepper motors. In the initial position, the thermocouple junction was placed at the center of the sample with respect to the X axis at a distance of 6 mm from the sample plane. The thermocouple moved to the surface of the sample to the surface with a horizontal speed of 1 mm/s and a vertical speed equal to the speed of the flame. Flame scanning was carried out with a step of 1 mm along the Y axis. All the thermocouples were connected to a multi-channel 14 bit analog to digital converter (ADC) E14-140M, which was connected to a PC. To determine the weight loss of the samples, as well as the amount of soot on the surface of the burnt sample, additional experiments were carried out without connecting thermocouples. After the burning of the sample, soot was collected from its surface, and its weight was measured. The complete yield of the combustion products into the gas phase, including gaseous volatile products and soot deposits, was determined as the difference in mass between the original sample and the burnt sample, which was normalized to the mass of the original sample.

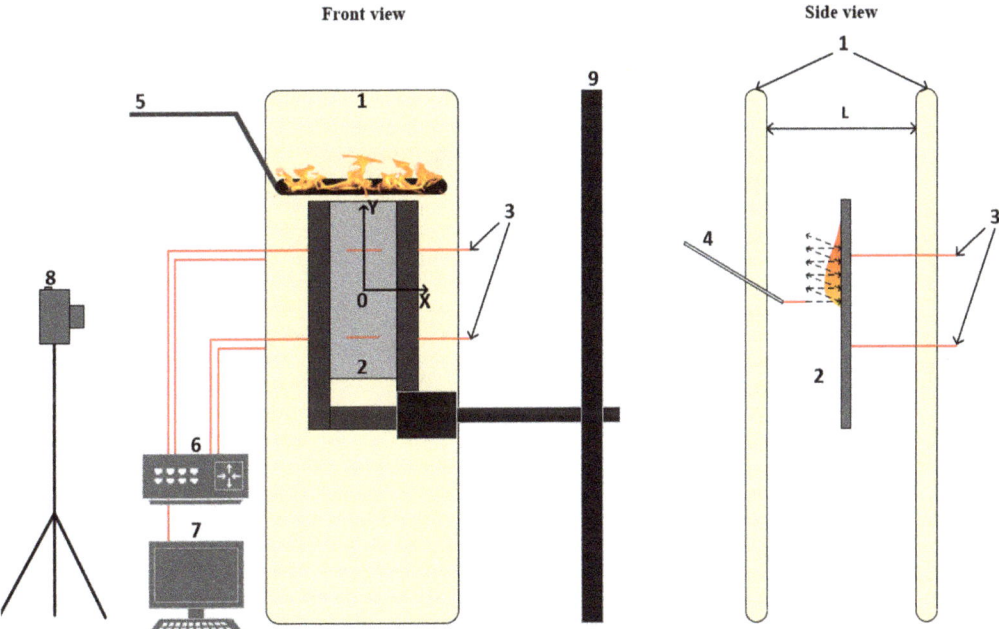

Figure 9. The configuration of the experimental setup for downward flame propagation under the action of a heater. 1—heaters, 2—sample in an aluminum frame, 3—thermocouples on the sample surface and in the gas phase, 4—scanning thermocouple, 5—ignition torch, 6—ADC, 7—PC, 8—camera, 9—sample holder.

3.6. Numerical Approach

The following mathematical model was proposed to predict a behavior of flame spread over polymers combined with flame retardants. The mathematical model takes into account coupled heat and mass transfer between gas phase flame and solid fuel, multicomponent reacting gas flow, gas phase combustion, heat transfer and pyrolysis in a solid material [22,24,36]:

$$\frac{\partial \rho}{\partial t} + \frac{\partial \rho u_j}{\partial x_j} = 0, \qquad (1)$$

$$\rho \frac{\partial u_i}{\partial t} + \rho u_j \frac{\partial u_i}{\partial x_j} = -\frac{\partial p}{\partial x_i} + \frac{\partial}{\partial x_j} \mu \frac{\partial u_i}{\partial x_j} + (\rho_a - \rho) g_i, \qquad (2)$$

$$\rho C \frac{\partial T}{\partial t} + \rho u_j C \frac{\partial T}{\partial x_j} = \frac{\partial}{\partial x_j} \lambda \frac{\partial T}{\partial x_j} + \rho W Q - \frac{\partial q_j^r}{\partial x_j}, \quad (3)$$

$$\rho \frac{\partial Y_k}{\partial t} + \rho u_j \frac{\partial Y_k}{\partial x_j} = \frac{\partial}{\partial x_j} \rho D \frac{\partial Y_k}{\partial x_j} + \nu_k \rho W, \quad (4)$$

$$\rho = p/RT. \quad (5)$$

Here $x_i = \{x, y\}$, $u_i = \{u, v\}$, $k = \{F, O, P\}$, $\nu_k = \{-1, -\nu_O, 1 + \nu_O\}$, $g_i = \{g, 0\}$.

A single-step mechanism is employed here for a gas phase combustion reaction as follows

$$F + \nu_O O + I \to (1 + \nu_O) P + I, \quad (6)$$

in which reaction rate is expressed in an Arrhenius form as

$$W = k Y_F Y_O \exp(-E/R_0 T). \quad (7)$$

The model of a solid material is extended to resolve the thermal degradation of the composite material. Such a composite consists of a combustible part (binder) reinforced with a noncombustible (e.g., glass fibers). The model was formulated to take into account these two components. Also, the model takes into account the anisotropy of thermal conductivity of a solid material. The energy conservation equation of solid material was expressed as:

$$\rho_s C_s \frac{\partial T_s}{\partial t} = \frac{\partial}{\partial x_j} \lambda_s^j \frac{\partial T_s}{\partial x_j} + \eta_b^0 \rho_b Q_b W_b, \quad (8)$$

The reaction rate of the pyrolysis reaction is given by

$$W_b = (1 - \alpha)^n k_b \exp(-E_b/R_0 T_s), \quad (9)$$

where the conversion degree varies from 0 to 1 and is defined as

$$\frac{d\alpha}{dt} = W_b. \quad (10)$$

The density of the solid material was defined taking into account binder burnout as follows:

$$\rho_s = \eta_b^0 (1 - \alpha) \rho_b + \left(1 - \eta_b^0\right) \rho_f, \quad (11)$$

The local mass burning rate (combustible volatiles gasification rate) at a burning surface was defined as

$$\dot{m}_b = \eta_b^0 \rho_b \int_{-L_s}^{0} W_b dy. \quad (12)$$

According to the previous studies [22,36], a DOPO-based flame retardant was considered to have an effect in the gas phase. Thus, the pre-exponential factor of the gas phase combustion reaction is reduced in the following way:

$$k_{g,DOPO} = (1 - \psi_{DOPO} Y_{DOPO}) k_g, \quad (13)$$

where Y_{DOPO} is the initial mass fraction of DDM-DOPO in the solid material, and ψ_{DOPO} is the DOPO inhibition effect coefficient.

A graphene-based flame retardant was proposed [22] to have an effect by a reduction in the amount of gaseous pyrolysates available for combustion in the following way. Only the $\psi_{gr} Y_{gr} \dot{m}_b$ part of the local mass burning rate given by Equation (12) was set to go to the gaseous fuel (F), while $\left(1 - \psi_{gr} Y_{gr}\right) \dot{m}_b$ part supplied a non-combustible gas, which was set to be the products (P). Here, ψ_{gr} represents the inhibition effect coefficient of graphene, and Y_{gr} is the graphene mass fraction.

The boundary conditions were set according to the computational domain scheme (Figure 10) as follows:

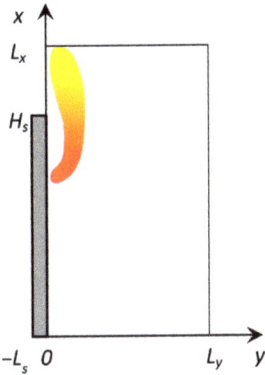

Figure 10. Arrangement of the computation domain.

$x = 0$: $T = T_a, Y_O = Y_{O,a}, Y_F = 0, Y_P = 0, \partial u/\partial x = 0, v = 0;$
$x = L_x$: $\partial \phi/\partial x = 0, \phi = \{u, v, T, Y_k\}, k = \{F, O, P\};$
$y = L_y$: $T = T_a, Y_O = Y_{O,a}, Y_F = 0, Y_P = 0, u = 0, \partial v/\partial y = 0;$
$y = 0, H_s < x < L_x$: $v = 0, \partial \phi/\partial y = 0, \phi = \{u, T, Y_k\}, k = \{F, O, P\};$
$y = -L_s, 0 < x < H_s$
$x = 0, -L_s < y < 0$: $\partial T_s/\partial n = 0;$
$x = H_s, -L_s < y < 0$
$y = 0, 0 < x < H_s$: $u = 0, \rho v = \rho_s u_s, T_s = T,$
$-\lambda_s \frac{\partial T_s}{\partial y} + (\rho u CT)_s = -\lambda \frac{\partial T}{\partial y} + \rho v CT + q_w^r + q_{ext},$
$-\rho D \frac{\partial Y_F}{\partial y} + \rho v Y_F = \rho_s u_s,$
$-\rho D \frac{\partial Y_k}{\partial y} + \rho v Y_k = 0, k = \{O, P\};$

Table 4 presents the inhibition effect values in the gas phase used in the model and the kinetic parameters of sample pyrolysis determined from TGA data. In the model for samples with flame retardants, the pyrolysis kinetics determined for the maximum concentration of the additive (3%) was used, since the effect of the additives on the thermal decomposition rate constant was relatively small.

Table 4. Inhibitor efficiency parameters in the gas phase (determined in the model) and kinetic parameters of the pyrolysis rate constant in a one-stage approximation (determined experimentally).

	Ψ	k_b, 1/s	F_b, kJ/mole
GFRER	0	$10^{9.93}$	157.3
GFRER + DDM-DOPO	29	$10^{8.49}$	140.4
GFRER + graphene	2/3	$10^{11.68}$	183.5

4. Conclusions

An experimental and numerical study of the DDM-DOPO-and-graphene flame retardant's impact on the flammability of the epoxy resin-based composites was carried out using both traditional flammability tests (UL-94, LOI), thermogravimetry and a downward flame propagation study under the action of bilateral external heating. For the 2 mm thick samples incorporated with 9.8% DDM-DOPO, the LOI was 26.4%; therefore, this composition can be applied in practice. For the first time, using thin thermocouples, the thermal structure of the downward spreading flame of fiberglass-reinforced (~50 wt.%) epoxy resin thin samples (~0.4 mm) with and without the flame retardants DDM-DOPO and graphene (1.5 and 3 wt.%) were measured in a quiescent air atmosphere under the action of an additional heat flux (up to 4.2 kW/m^2). For the first time, the measured thermal flame structure was compared with the numerical simulation results of the flame propagation over the GFRER surface. The flame retardants were shown to be effective (a decrease in the flame spread rate) in improving the fire resistance of the epoxy resin composites. The flame spread over the surface of the sample almost uniformly, without acceleration. The flame spread rate had an almost linear dependence on the heat flux, which was well predicted by the model (up to 3 kW/m^2). As the external heat flux increased, an increase in the flame spread rate was observed, which was associated with the approach of the sample surface temperature to the epoxy resin pyrolysis temperature. Under the heat flux 4.2 kW/m^2, an increase in the additive concentration from 1.5% to 3% for DDM-DOPO had almost no effect on the flame spread rate. A similar result was observed in UL-94 HB, where the flame propagation rate did not change with increasing DDM-DOPO concentration. Also, an increase in the DDM-DOPO concentration had almost no effect on the mass consumption during downward flame propagation.

The numerical model of the downward flame propagation under the action of bilateral external heating over the fiberglass-reinforced epoxy resins with and without the flame retardants DDM-DOPO and graphene was developed. The numerical model well predicted the temperature gradient in the condensed and gaseous phases, whereby the model well predicted the flame spread rate over the samples surface. It should be emphasized, the effectiveness of DDM-DOPO in the experiment decreases with increasing flame-retardant concentration, while in the model, the effectiveness of DDM-DOPO was linear. This is due to the linear dependence of the burning rate in the gas phase from the flame-retardant concentration in the model. The comparison of the flame structure from the model and from the experiment showed that the model for DDM-DOPO predicted the trend towards a decrease in the length of the luminous flame zone well. In the case of graphene, the model incorrectly predicted the trend, reducing the length of the luminous zone; however, it correctly described the decrease in the flame spread rate and the surface temperature. In the model, the DDM-DOPO inhibition mechanism was limited by a one-step reaction, and graphene worked only at the solid material boundary, reducing the proportion of combustible pyrolysis products. Such simplicity of the mechanism inevitably entails limitations and certain disagreements with the experiment, and, of course, using a more detailed inhibition mechanism for DDM-DOPO or graphene will result in a better agreement with the experiment. The obtained data may be used for designing effective reinforced non-combustible composites applied in the aircraft industry and for determining a detailed mechanism of the effect of flame-retardant additives.

Supplementary Materials: The following supporting information can be downloaded at: https://www.mdpi.com/article/10.3390/molecules28135162/s1, Figure S1: SEM micrographs and X-ray energy dispersive spectra for glass fiber reinforced epoxy resin samples with 1.5% DOPO-DDM additives, Figure S2: SEM micrographs and X-ray energy dispersive spectra for glass fiber reinforced epoxy resin samples with 3% DOPO-DDM additives.

Author Contributions: Conceptualization, O.P.K. and A.I.K.; methodology, E.A.S., A.I.K., A.R.S., S.A.T., A.G.S. and A.A.P.; software, A.A.S.; validation, E.A.S., A.I.K., A.A.S., A.R.S. and S.A.T.; investigation, E.A.S., A.R.S.,A.G.S., A.A.P. and I.V.K.; writing—original draft preparation, O.P.K., E.A.S. and S.A.T.; writing—review and editing, O.P.K., A.I.K., E.A.S., S.A.T. and A.G.S.; supervision, O.P.K., A.KI., A.R.S., S.A.T. and A.A.S.; project administration, O.P.K. All authors have read and agreed to the published version of the manuscript.

Funding: This research was funded by Russian Science Foundation, grant number 20-19-00295.

Institutional Review Board Statement: Not applicable.

Informed Consent Statement: Not applicable.

Data Availability Statement: Not applicable.

Acknowledgments: This research was funded by the Russian Science Foundation, grant number 20-19-00295. The authors acknowledge the utility of the Multi-Access Chemical Service Center SB RAS in the part of the spectral and analytical measurements made. The authors are thankful to I. Shundrina for the TGA measurements. The authors are thankful to Yuan Hu for synthesis of DDM-DOPO.

Conflicts of Interest: The authors declare no conflict of interest.

References

1. Visakh, P.M.; Yoshihiko, A. *Flame Retardants: Polymer Blends, Composites and Nanocomposites*; Springer: Berlin/Heidelberg, Germany, 2015. ISBN 978-3-319-03466-9.
2. Shahari, S.; Fathullah, M.; Abdullah, M.M.A.B.; Shayfull, Z.; Mia, M.; Budi Darmawan, V.E. Recent Developments in Fire Retardant Glass Fibre Reinforced Epoxy Composite and Geopolymer as a Potential Fire-Retardant Material: A Review. *Constr. Build. Mater.* **2021**, *277*, 122246. [CrossRef]
3. Mngomezulu, M.E.; John, M.J.; Jacobs, V.; Luyt, A.S. Review on Flammability of Biofibres and Biocomposites. *Carbohydr. Polym.* **2014**, *111*, 149–182. [CrossRef]
4. Liu, H.; Xu, K.; Ai, H.; Zhang, L.; Chen, M. Preparation and Characterization of Phosphorus-Containing Mannich-Type Bases as Curing Agents for Epoxy Resin. *Polym. Adv. Technol.* **2009**, *20*, 753–758. [CrossRef]
5. Zang, L.; Wagner, S.; Ciesielski, M.; Müller, P.; Döring, M. Novel Star-Shaped and Hyperbranched Phosphorus-Containing Flame Retardants in Epoxy Resins. *Polym. Adv. Technol.* **2011**, *22*, 1182–1191. [CrossRef]
6. Wang, J.; Ma, C.; Wang, P.; Qiu, S.; Cai, W.; Hu, Y. Ultra-Low Phosphorus Loading to Achieve the Superior Flame Retardancy of Epoxy Resin. *Polym. Degrad. Stab.* **2018**, *149*, 119–128. [CrossRef]
7. Perret, B.; Schartel, B.; Stöß, K.; Ciesielski, M.; Diederichs, J.; Döring, M.; Krämer, J.; Altstädt, V. Novel DOPO-Based Flame Retardants in High-Performance Carbon Fibre Epoxy Composites for Aviation. *Eur. Polym. J.* **2011**, *47*, 1081–1089. [CrossRef]
8. Gu, L.; Chen, G.; Yao, Y. Two Novel Phosphorus–Nitrogen-Containing Halogen-Free Flame Retardants of High Performance for Epoxy Resin. *Polym. Degrad. Stab.* **2014**, *108*, 68–75. [CrossRef]
9. Yuan, Y.; Yang, H.; Yu, B.; Shi, Y.; Wang, W.; Song, L.; Hu, Y.; Zhang, Y. Phosphorus and Nitrogen-Containing Polyols: Synergistic Effect on the Thermal Property and Flame Retardancy of Rigid Polyurethane Foam Composites. *Ind. Eng. Chem. Res.* **2016**, *55*, 10813–10822. [CrossRef]
10. Qian, L.; Qiu, Y.; Liu, J.; Xin, F.; Chen, Y. The Flame Retardant Group-Synergistic-Effect of a Phosphaphenanthrene and Triazine Double-Group Compound in Epoxy Resin. *J. Appl. Polym. Sci.* **2014**, *131*. [CrossRef]
11. Yang, S.; Wang, J.; Huo, S.; Cheng, L.; Wang, M. The Synergistic Effect of Maleimide and Phosphaphenanthrene Groups on a Reactive Flame-Retarded Epoxy Resin System. *Polym. Degrad. Stab.* **2015**, *115*, 63–69. [CrossRef]
12. Qin, Y.; Xu, G.; Wang, Y.; Hu, J. Preparation of Phosphorus-Containing Epoxy Emulsion and Flame Retardancy of Its Thermoset. *High Perform. Polym.* **2014**, *26*, 526–531. [CrossRef]
13. Gaan, S.; Liang, S.; Mispreuve, H.; Perler, H.; Naescher, R.; Neisius, M. Flame Retardant Flexible Polyurethane Foams from Novel DOPO-Phosphonamidate Additives. *Polym. Degrad. Stab.* **2015**, *113*, 180–188. [CrossRef]
14. Jian, R.; Wang, P.; Duan, W.; Wang, J.; Zheng, X.; Weng, J. Synthesis of a Novel P/N/S-Containing Flame Retardant and Its Application in Epoxy Resin: Thermal Property, Flame Retardance, and Pyrolysis Behavior. *Ind. Eng. Chem. Res.* **2016**, *55*, 11520–11527. [CrossRef]
15. Jiang, P.; Gu, X.; Zhang, S.; Wu, S.; Zhao, Q.; Hu, Z. Synthesis, Characterization, and Utilization of a Novel Phosphorus/Nitrogen-Containing Flame Retardant. *Ind. Eng. Chem. Res.* **2015**, *54*, 2974–2982. [CrossRef]
16. Wang, P.; Fu, X.; Kan, Y.; Xin, W.; Hu, Y. Two High-Efficient DOPO-Based Phosphonamidate Flame Retardants for Transparent Epoxy Resin. *High Perform. Polym.* **2018**, *31*, 095400831876203. [CrossRef]
17. Wang, X.; Song, L.; Pornwannchai, W.; Hu, Y.; Kandola, B. The Effect of Graphene Presence in Flame Retarded Epoxy Resin Matrix on the Mechanical and Flammability Properties of Glass Fiber-Reinforced Composites. *Compos. Part A Appl. Sci. Manuf.* **2013**, *53*, 88–96. [CrossRef]
18. Huang, G.; Gao, J.; Wang, X.; Liang, H.; Ge, C. How Can Graphene Reduce the Flammability of Polymer Nanocomposites? *Mater. Lett.* **2012**, *66*, 187–189. [CrossRef]
19. Liu, S.; Yan, H.; Fang, Z.; Wang, H. Effect of Graphene Nanosheets on Morphology, Thermal Stability and Flame Retardancy of Epoxy Resin. *Compos. Sci. Technol.* **2014**, *90*, 40–47. [CrossRef]
20. Zhou, D.; Cheng, Q.-Y.; Cui, Y.; Wang, T.; Li, X.; Han, B.-H. Graphene–Terpyridine Complex Hybrid Porous Material for Carbon Dioxide Adsorption. *Carbon* **2014**, *66*, 592–598. [CrossRef]
21. Wang, Z.-Y.; Liu, Y.; Wang, Q. Flame Retardant Polyoxymethylene with Aluminium Hydroxide/Melamine/Novolac Resin Synergistic System. *Polym. Degrad. Stab.* **2010**, *95*, 945–954. [CrossRef]
22. Korobeinichev, O.; Shaklein, A.; Trubachev, S.; Karpov, A.; Paletsky, A.; Chernov, A.; Sosnin, E.; Shmakov, A. The Influence of Flame Retardants on Combustion of Glass Fiber-Reinforced Epoxy Resin. *Polymers* **2022**, *14*, 3379. [CrossRef]

23. Kobayashi, Y.; Terashima, K.; Oiwa, R.; Tokoro, M.; Takahashi, S. Opposed-Flow Flame Spread over Carbon Fiber Reinforced Plastic under Variable Flow Velocity and Oxygen Concentration: The Effect of in-Plane Thermal Isotropy and Anisotropy. *Proc. Combust. Inst.* **2021**, *38*, 4857–4866. [CrossRef]
24. Korobeinichev, O.; Karpov, A.; Shaklein, A.; Paletsky, A.; Chernov, A.; Trubachev, S.; Glaznev, R.; Shmakov, A.; Barbot'ko, S. Experimental and Numerical Study of Downward Flame Spread over Glass-Fiber-Reinforced Epoxy Resin. *Polymers* **2022**, *14*, 911. [CrossRef]
25. Kobayashi, Y.; Oiwa, R.; Tokoro, M.; Takahashi, S. Buoyant-Flow Downward Flame Spread over Carbon Fiber Reinforced Plastic in Variable Oxygen Atmospheres. *Combust. Flame* **2021**, *232*, 111528. [CrossRef]
26. Snegirev, A.; Kuznetsov, E.; Korobeinichev, O.; Shmakov, A.; Paletsky, A.; Shvartsberg, V.; Trubachev, S. Fully Coupled Three-Dimensional Simulation of Downward Flame Spread over Combustible Material. *Polymers* **2022**, *14*, 4136. [CrossRef] [PubMed]
27. Williams, F.A. Mechanisms of Fire Spread. *Symp. (Int.) Combust.* **1977**, *16*, 1281–1294. [CrossRef]
28. Fernandez-pello, A.C. Upward Laminar Flame Spread under the Influence of Externally Applied Thermal Radiation. *Combust. Sci. Technol.* **1977**, *17*, 87–98. [CrossRef]
29. Fernandez-Pello, A.C. Downward Flame Spread under the Influence of Externally Applied Thermal Radiation. *Combust. Sci. Technol.* **1977**, *17*, 1–9. [CrossRef]
30. Hirano, T.; Sato, K. Effects of Radiation and Convection on Gas Velocity and Temperature Profiles of Flames Spreading over Paper. *Symp. (Int.) Combust.* **1975**, *15*, 233–241. [CrossRef]
31. Zhou, Y.; Xu, B.; Zhang, X.; Yang, Y. A Comparative Study on Horizontal Flame Spread Behaviors of Thermoplastic Polymers with Different Melt Flow Indexes under External Radiation. *Therm. Sci. Eng. Prog.* **2022**, *35*, 101463. [CrossRef]
32. Brehob, E.G.; Kulkarni, A.K. Experimental Measurements of Upward Flame Spread on a Vertical Wall with External Radiation. *Fire Saf. J.* **1998**, *31*, 181–200. [CrossRef]
33. Zhou, Y.; Xiao, H.; Yan, W.; An, W.; Jiang, L.; Sun, J. Horizontal Flame Spread Characteristics of Rigid Polyurethane and Molded Polystyrene Foams Under Externally Applied Radiation at Two Different Altitudes. *Fire Technol* **2015**, *51*, 1195–1216. [CrossRef]
34. Vincent, C.; Corn, S.; Longuet, C.; Aprin, L.; Rambaud, G.; Ferry, L. Experimental and Numerical Thermo-Mechanical Analysis of the Influence of Thermoplastic Slabs Installation on the Assessment of Their Fire Hazard. *Fire Saf. J.* **2019**, *108*, 102850. [CrossRef]
35. Bhattachariee, S.; King, M.D.; Takahashi, S.; Nagumo, T.; Wakai, K. Downward Flame Spread over Poly(Methyl)Methacrylate. *Proc. Combust. Inst.* **2000**, *28*, 2891–2897. [CrossRef]
36. Trubachev, S.A.; Korobeinichev, O.P.; Karpov, A.I.; Shaklein, A.A.; Glaznev, R.K.; Gonchikzhapov, M.B.; Paletsky, A.A.; Tereshchenko, A.G.; Shmakov, A.G.; Bespalova, A.S.; et al. The Effect of Triphenyl Phosphate Inhibition on Flame Propagation over Cast PMMA Slabs. *Proc. Combust. Inst.* **2021**, *38*, 4635–4644. [CrossRef]
37. Korobeinichev, O.P.; Paletsky, A.A.; Gonchikzhapov, M.B.; Glaznev, R.K.; Gerasimov, I.E.; Naganovsky, Y.K.; Shundrina, I.K.; Snegirev, A.Y.; Vinu, R. Kinetics of Thermal Decomposition of PMMA at Different Heating Rates and in a Wide Temperature Range. *Thermochim. Acta* **2019**, *671*, 17–25. [CrossRef]
38. Schartel, B. Phosphorus-Based Flame Retardancy Mechanisms—Old Hat or a Starting Point for Future Development? *Materials* **2010**, *3*, 4710–4745. [CrossRef]
39. Korobeinichev, O.P.; Gonchikzhapov, M.B.; Paletsky, A.A.; Tereshchenko, A.G.; Shundrina, I.K.; Kuibida, L.V.; Shmakov, A.G.; Hu, Y. Counterflow Flames of Ultrahigh-Molecular-Weight Polyethylene with and without Triphenylphosphate. *Combust. Flame* **2016**, *169*, 261–271. [CrossRef]
40. Trubachev, S.A.; Korobeinichev, O.P.; Kostritsa, S.A.; Kobtsev, V.D.; Paletsky, A.A.; Kumar, A.; Smirnov, V.V. An Insight into the Gas-Phase Inhibition Mechanism of Polymers by Addition of Triphenyl Phosphate Flame Retardant. *AIP Conf. Proc.* **2020**, *2304*, 020019. [CrossRef]
41. Korobeinichev, O.P.; Trubachev, S.A.; Joshi, A.K.; Kumar, A.; Paletsky, A.A.; Tereshchenko, A.G.; Shmakov, A.G.; Glaznev, R.K.; Raghavan, V.; Mebel, A.M. Experimental and Numerical Studies of Downward Flame Spread over PMMA with and without Addition of Tri Phenyl Phosphate. *Proc. Combust. Inst.* **2021**, *38*, 4867–4875. [CrossRef]
42. Ma, C.; Sánchez-Rodríguez, D.; Kamo, T. A Comprehensive Study on the Oxidative Pyrolysis of Epoxy Resin from Fiber/Epoxy Composites: Product Characteristics and Kinetics. *J. Hazard. Mater.* **2021**, *412*, 125329. [CrossRef] [PubMed]
43. Yan, W.; Zhang, M.-Q.; Yu, J.; Nie, S.-Q.; Zhang, D.-Q.; Qin, S.-H. Synergistic Flame-Retardant Effect of Epoxy Resin Combined with Phenethyl-Bridged DOPO Derivative and Graphene Nanosheets. *Chin. J. Polym. Sci.* **2019**, *37*, 79–88. [CrossRef]
44. Korobeinichev, O.P.; Kumaran, S.M.; Raghavan, V.; Trubachev, S.A.; Paletsky, A.A.; Shmakov, A.G.; Glaznev, R.K.; Chernov, A.A.; Tereshchenko, A.G.; Loboda, E.L.; et al. Investigation of the Impact of Pinus Silvestris Pine Needles Bed Parameters on the Spread of Ground Fire in Still Air. *Combust. Sci. Technol.* **2022**, 1–23. [CrossRef]
45. Singh, A.V.; Gollner, M.J. A Methodology for Estimation of Local Heat Fluxes in Steady Laminar Boundary Layer Diffusion Flames. *Combust. Flame* **2015**, *162*, 2214–2230. [CrossRef]

Disclaimer/Publisher's Note: The statements, opinions and data contained in all publications are solely those of the individual author(s) and contributor(s) and not of MDPI and/or the editor(s). MDPI and/or the editor(s) disclaim responsibility for any injury to people or property resulting from any ideas, methods, instructions or products referred to in the content.

Review

Facile Ball Milling Preparation of Flame-Retardant Polymer Materials: An Overview

Xiaming Feng [1,*], Xiang Lin [1], Kaiwen Deng [1], Hongyu Yang [1,*] and Cheng Yan [2]

1. College of Materials Science and Engineering, Chongqing University, 174 Shazhengjie, Shapingba, Chongqing 400044, China
2. Department of Mechanical Engineering, Southern University and A&M College, Baton Rouge, LA 70813, USA
* Correspondence: fengxm@cqu.edu.cn (X.F.); yhongyu@cqu.edu.cn (H.Y.)

Abstract: To meet the growing needs of public safety and sustainable development, it is highly desirable to develop flame-retardant polymer materials using a facile and low-cost method. Although conventional solution chemical synthesis has proven to be an efficient way of developing flame retardants, it often requires organic solvents and a complicated separation process. In this review, we summarize the progress made in utilizing simple ball milling (an important type of mechanochemical approach) to fabricate flame retardants and flame-retardant polymer composites. To elaborate, we first present a basic introduction to ball milling, and its crushing, exfoliating, modifying, and reacting actions, as used in the development of high-performance flame retardants. Then, we report the mixing action of ball milling, as used in the preparation of flame-retardant polymer composites, especially in the formation of multifunctional segregated structures. Hopefully, this review will provide a reference for the study of developing flame-retardant polymer materials in a facile and feasible way.

Keywords: ball milling; flame retardants; polymer composites; exfoliation; fire retardancy

Citation: Feng, X.; Lin, X.; Deng, K.; Yang, H.; Yan, C. Facile Ball Milling Preparation of Flame-Retardant Polymer Materials: An Overview. *Molecules* 2023, 28, 5090. https://doi.org/10.3390/molecules28135090

Academic Editor: Gaëlle Fontaine

Received: 31 May 2023
Revised: 23 June 2023
Accepted: 26 June 2023
Published: 29 June 2023

Copyright: © 2023 by the authors. Licensee MDPI, Basel, Switzerland. This article is an open access article distributed under the terms and conditions of the Creative Commons Attribution (CC BY) license (https://creativecommons.org/licenses/by/4.0/).

1. Introduction

Polymer materials have contributed significantly to the development of modern society, due to their excellent properties, including high workability, low price, and good chemical resistance [1–6]. However, the flammability of most polymer materials greatly limits their wide range of applications [7–10]. One proposed solution to this problem is the development of flame retardants [11–14], which could restrain the ignition and fire-spreading of polymer materials, specifically decreasing heat release and smoke production. To date, various flame retardants have been developed to achieve satisfactory fire resistance in different polymer materials, such as halogenated flame retardants [15–19], inorganic layered compounds [20–23], and phosphorous–nitrogen intumescent flame retardants [24–27]. Flame-retardant polymers and their composites have been used widely in construction, electronics, transportation, and so on [28–30]. Regarding the preparation of flame retardants, the conventional method is liquid-phase synthesis [31,32], in which organic solvents and complicated purification are always required. Moreover, the environmental toxicity and environmental accumulation of various flame retardants have become increasingly important [33–35]. Therefore, for high efficiency and environmental protection, a more efficient and greener approach toward the facile preparation of flame retardants is highly desirable.

As a typical sub-factor in mechanochemistry [36], ball milling has been developed for crushing, mixing, and reacting, due to the impact and shear forces generated by high-speed rotation and high-temperature surroundings [37–40]. It is widely used in fabricating flame retardants, for its superior properties of easy processing, low cost, and large-scale production [39]. The most conventional application of ball milling in the field of flame retardants is grinding to reduce particle size [41,42]. Notably, ball milling is effective in preparing inorganic compounds on the nanoscale [43–45], including flame-retardant synergists. As is well known, the size of additives for polymer composites is closely related to the dispersion state and the final performance. Recently, with the rise of two-dimensional nanomaterials (e.g., graphene), the shear force generated by ball milling is utilized to achieve the facile exfoliation of layered compounds, such as graphite [46–48], boron nitride [49–52], and black phosphorus [53,54]. Another important usage of ball milling is to conduct chemical reactions, including the simple surface modification of particles, and complicated chemical synthesis [14,55]. As for flame retardants, surface modification by ball milling is prominent in improving properties, such as hydrolysis resistance [56,57], interfacial compatibility [56,57], and flame-retardant efficiency [58,59]. Moreover, ball milling plays an important role in fabricating flame-retardant polymer composites, including simple mixing, and customizing specific structures (e.g., a segregated structure for electromagnetic wave shielding) [60,61].

This review aims to summarize the technological progress made in utilizing ball milling for facilely preparing flame retardants, followed by a study and discussion of the fabrication of flame-retardant polymer composites. Firstly, the basic concept and the category of ball milling are briefly discussed. Next, the crushing, exfoliating, modifying, and reacting actions of ball milling in developing flame retardants are primarily examined. Then, ball milling for the mixing of flame retardants and the polymer matrix, which concern segregated structures and flame retardancy, is overviewed. Finally, conclusions are proposed, and insights are given.

2. Ball Milling Methods

Ball milling is a technique that is widely utilized to crush powders into small particles. According to the operational mode, the types of ball milling primarily include planetary ball milling, tumbler ball milling, vibration ball milling, and attrition ball milling. Each type of mill is developed to achieve a specific purpose, and each undoubtedly has relative weaknesses. For example, attrition ball milling generates higher surface contact, while vibration ball milling could produce higher milling force [62,63]. Comparatively, planetary ball milling is widely used owing to its compact size and low cost. According to the generated energy, ball milling can be mainly classified as low-energy ball milling and high-energy ball milling. The rotation speed of most planetary ball mills is 0~500 rpm, while high-energy ball mills can reach up to 1800 rpm. During ball milling, shear force, impact force, and friction force can be generated. Shear force is in favor of exfoliating layered compounds, and impact force can efficiently grind the powders into fine particles. As for high-energy ball milling, the heat generated by friction can be used to induce the chemical reaction and phase transition, such as the conversion of red phosphorus to black phosphorus. Regarding the materials of the ball milling tank and bead, the most commonly used is made from steel, due to the high hardness and processability. Other widely used materials include agate, zirconia, and nylon, which are appropriate for those raw materials that can react with steel. That is to say, the ball milling method satisfies almost all requirements when developing flame retardants, including crushing, exfoliating, surface modifying, and reacting, as well as mixing flame-retardant additives and the polymer matrix without the limitation of containers.

3. Ball Milling-Assisted Fabrication of Flame Retardants

3.1. Ball Milling for Crushing

The basic application of ball milling is crushing powders, including flame-retardant additives. As is known, the particle size plays an important role in influencing the properties. For example, the specific surface area increases with the reduction in particle size, which is crucial for flame retardants with catalytic effects. As for mechanical properties, large flame-retardant particles lead to a stress concentration within polymer composites. As a result, the mechanical strength is always decreased. Therefore, crushing flame retardants into fine powders is necessary. Bocz and coworkers studied the influence of flame retardant size on the fire retardancy and mechanical properties of polypropylene (PP) composites [64]. The flame-retardant system consists of pentaerythritol (PER) and ammonium polyphosphate (APP) in a weight ratio of 1 to 2. Ball milling was utilized to reduce the particle size of the APP/PER mixture to a large extent, as shown in Figure 1. Upon undergoing the ball milling process, the average particle size of APP was decreased from 15 μm to 8 μm, while the PER was crushed from microparticles (~200 μm) to submicronic particles. An outperformed flame retardancy was observed in the PP composite with smaller APP/PER particles. It is believed that the better distribution of additives and the modified degradation mechanism contribute significantly to the formation of a protective char layer. Moreover, the smaller APP/PER particles are in favor of enhancing the mechanical performance (10% higher tensile strength) of flame-retarded PP composites.

Figure 1. SEM micrographs of APP and PER additives before and after milling. Reproduced from ref. [64] with permission from John Wiley and Sons.

In Bao's research, the attapulgite clay (ATP) was treated by ball milling to reduce the particle size. It was then grafted onto cotton fabric to improve flame retardancy [65]. To comprehensively evaluate the effect of the milling process, the ball milling was conducted for 4, 5, 6, and 8 h, under rotational speeds of 200, 300, and 400 r/min, in the mass ratio of zirconia ball to ATP of 3:1, 4:1, 5:1, and 6:1, respectively. It was observed that the ball milling parameters (speed, time, and mass ratio) strongly influenced the particle size of the milled ATP, as well as the final fire retardancy of the resultant cotton fabrics. The uniformly distributed stable oxide layer decomposed by the ATP was believed to be the key contributor. Following the same idea, Üreyen and partners crushed zinc borate (ZnB) from 9 μm to a submicron scale by wet milling, and subsequent high-shear-fluid processing, to reduce the flammability of polyethylene terephthalate (PET) woven fabrics [66]. The crushed ZnB was dispersed in alkyl phosphonate and organophosphorus flame retardants, to obtain a homogenous dispersion, which was then applied to the fabrics by the pad-dry-cure method. A synergistic effect between ZnB and organophosphorus flame retardants was proposed, specifically decreasing the mean CO, total smoke release, and total smoke production.

In recent years, the concept of sustainable development has been defined as essential to combating climate change. The combination of facile preparation and biomass raw materials attracts much attention in developing flame retardants. Among these, the ball milling of various biomass materials is of great importance. For example, Jawaid and coworkers reported the preparation of nano flame retardants from date palm biomass [67]. They performed the agricultural waste, date palm trunk fiber as biobased raw material upon the chemical process, and high-energy ball milling to fabricate new flame retardants on the nanoscale. The particle size analysis results indicated that the as-prepared products were in the mean size range of 274.5–289.7 nm. The multielement composition (carbon, oxygen, silicon, sulfur, calcium, and potassium) and high decomposition temperature of nano-sized fillers suggest their potential application in the field of flame retardants [68,69]. Besides the biomass raw materials, another important issue in consideration of critical environmental pollution is how to reuse the existing petroleum-based material [70–72]. Wang and coauthors proposed a novel recycling strategy for fabricating fire retardants from polyphenylene sulfide waste textiles [73]. A sequence of thermal aging, ball milling, and screening was conducted to upcycle waste polyphenylene sulfide (PPS) filter bags into PPS powders (75–100 μm), which were first used to decrease the flammability of epoxy resin (EP). The high thermal stability and good charring ability of PPS are believed to constitute the main reason for the decreased heat and smoke/toxic gases released from EP composites. This work presented a promising pattern in the upcycling of solid wastes into flame retardants.

3.2. Ball Milling for Exfoliation

As a typical type of inorganic flame-retardant additive with a physical barrier effect, nano-sized layered compounds have been a hotspot for flame-retardant research in recent years, from clay to graphene to MXene. The preparation of nano-sized layered flame retardants generally includes the bottom-up approach and top-down processing. Comparatively, the top-down approach features the superiority of easy processing, large-scale production, and low cost. Owing to the shear forces, among other aspects, ball milling is considered one of the most promising techniques for exfoliating layered compounds and producing two-dimensional flame retardants.

3.2.1. Graphene-Based Flame Retardants

The discovery of graphene has opened up a whole new field of material research. Graphene has been regarded as a promising material in various fields, including electronics [74,75], catalysts [76,77], and semiconductors [78,79], as well as flame retardants [80,81]. However, the high cost of preparing graphene is always a critical problem, especially in the flame-retardant field, which sees such high consumption. Researchers have developed various methods for fabricating graphene on an acceptable production scale. Ball milling is always a good choice. Kim and coworkers initially reported the fabrication of graphene phosphonic acid (GPA) flame retardants via the ball milling of graphite and red phosphorus [46]. As shown in Figure 2, the graphite is firstly crushed and exfoliated into thin layer graphene. Upon exposure to the high energy generated by ball milling, the graphitic C-C bonds cleave, and react with phosphorus to form C-P bonds. Subsequently, the unstable P converts into a phospho-oxygen compound, and then phosphate compounds (highest oxidation state), in the presence of oxygen and moisture. The obtained GPA can be easily dispersed in various solvents to form a stable solution, which could be used to treat papers and fabrics, to allow fire retardancy.

Inspired by Kim's work, Jeon and coworkers prepared a heavily aluminated graphene (AlGnP) flame retardant through the ball milling of graphite and solid aluminum (Al) beads. Approximately 30.9 wt% of element Al in AlGnP was detected [47]. Subsequently, the poly(vinyl alcohol)/AlGnP composite films were fabricated via a simple solution blending and casting method, thanks to the excellent dispersibility of AlGnP. As expected, an improved flame retardancy was observed. Additionally, Chen and coworkers prepared Sn-doped graphene (GnPSn) as an efficient flame retardant, via ball milling expandable graphite and Sn powder in a wet condition [48]. The synergistic flame-retardant effect of GnPSn and hexaphenoxy cyclotriphosphazene (HPCTP) for EP resin was proposed. An

LOI value of 33.6%, and UL-94 of V0 grade, were observed for the EP composite with 6.3 wt% HPCTP and 2.7 wt% GnPSn. The compactness of the residual char in the condensed phase was highlighted in discussing the specific flame-retardant mechanism.

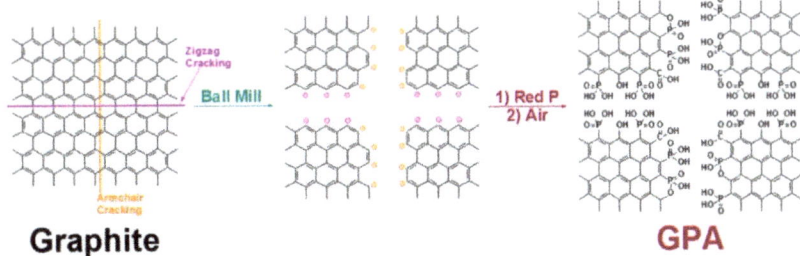

Figure 2. Schematic representation of the mechanochemical cracking of a graphite flake in a ball-mill crusher containing stainless steel balls (diameter 5 mm), in the presence of red phosphorus, and subsequent exposure to air moisture to produce GPA. Reproduced from ref. [46] with permission from the American Chemical Society.

For the efficient exfoliation of graphite, a combination of ball milling and other techniques has been proposed, such as ball milling coupled with ultrasonication, reported by Wang et al. [82], thermal shock combined with the ball milling method [83], and a combination of the ball milling and microwave-assisted methods [84]. Moreover, to meet the goal of sustainable development, various green and biobased materials have been applied to assist the ball milling-induced exfoliation of graphite, such as waste fish deoxyribonucleic acid and Acacia mangium tannin (AMT) [80,85]. As expected, the green exfoliating agent can not only enhance the exfoliation efficiency of graphite, but can also improve the dispersibility and flame-retardant capability.

3.2.2. Boron-Nitride-Based Flame Retardants

Boron nitride is another important layered compound, which is highly thermally conductive, but electrically insulated [51]. Different from graphene, boron nitride nanosheets are widely used to fabricate polymer nanocomposites with high thermal conductivity, while maintaining electric insulation properties, which are promising in the field of thermal management with high fire risk [50,52]. Certainly, boron nitride nanosheets can be an efficient type of flame retardant, due to their high thermal stability and specific nano-size effect. Based on the same exfoliation mechanism, ball milling is often used to achieve the exfoliation of bulk boron nitride. Qiu and coworkers realized the scalable production of hydroxyl-functionalized BN (OBN) by simple ball milling and annealing under air conditions, in the presence of sodium hydroxide [49]. The shear-force-induced exfoliating and chemical peeling were believed to be the key points. The resultant OBN could be a platform to load and graft various flame-retardant units, to improve the fire safety of EP resin. As a result, the EP composites exhibited not only enhanced fire retardancy, but also an improved friction performance.

Inspiringly, Han et al. included the conventional flame-retardant ammonium phosphate in the above-mentioned ball milling exfoliation process, as an assistant agent [51]. A synergetic action between the shear and chemical peeling of ammonium phosphate and sodium hydroxide was observed. Density functional theory (DFT) calculations were performed, to reveal the possible mechanochemical reaction mechanisms. As expected, the resultant BN nanosheets endowed EP resin with exceptional fire retardancy, including 60.9%, 35.7%, 44.3%, and 38.8% reductions in PHRR, THR, SPR, and TSP, respectively. The catalytic charring effect and physical barrier action of BN were highlighted in the improvement of flame retardancy. Subsequently, the same group reported the fabrication of ionic liquid-wrapped boron nitride nanosheets (BNNS@IL) using the ball milling process [52]. The exfoliation and functionalization were achieved via a similar mechanochemical action.

Based on the BNNS@IL, a fire resistant EP-based thermally conductive layered film with aligned BN nanoflakes was developed, which showed high anisotropic thermal conductivity ($K_\parallel$ of 8.3 and $K_\perp$ of 0.8 W m^{-1} K^{-1}) and excellent flame retardancy, suggesting new possibilities in electrical device and thermal management. Additionally, the nitrogen-phosphorus-doped boron nitride (BN@APP) was prepared by Xu and coworkers via the ball milling method [86]. The effect of BN@APP on the flame retardancy and thermal conductivity of polybutylene succinate (PBS) was comprehensively studied. A 62.8% increase in thermal conductivity, and a 44.8% decrease in TSP, were reported.

To improve the exfoliation efficiency of boron nitride, a sugar-assisted mechanochemical exfoliation (SAMCE) method was developed by Chen's group [50]. As shown in Figure 3, the sugar (sucrose) molecules can be covalently grafted to BN nanosheets during ball milling, which efficiently prevents restacking, and leads to the high exfoliation yield of 87.3%. The obtained BN can be uniformly dispersed in water and organic solvents, due to the grafted sucrose molecules. As a result, the exfoliated BN can greatly reinforce the flexible and transparent poly(vinyl alcohol) (PVA) film, in terms of improved tensile strength, thermal dissipation capability, and fire retardancy. It is believed that this SAMCE method can be extended to the exfoliation of other layered materials, such as black phosphorus.

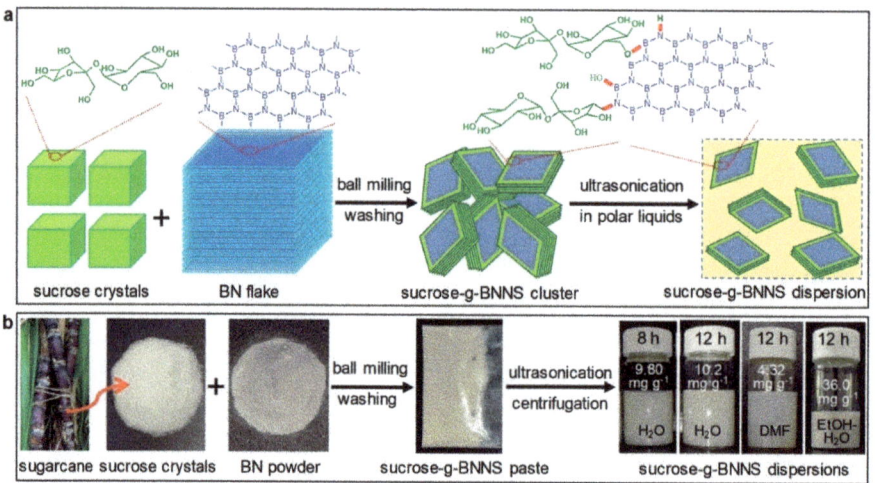

Figure 3. (a) Scheme of the exfoliation procedure. (b) Photos of the raw materials, the sucrose-g-BNNS paste, and sucrose-g-BNNS dispersions in H$_2$O, DMF, and ethanol (EtOH)-H$_2$O mixture [50]. Reproduced with permission from John Wiley and Sons.

3.2.3. Black-Phosphorus-Based Flame Retardants

Elementary phosphorus is an efficient flame retardant, including red phosphorus [87,88] and black phosphorus [89,90]. Compared to amorphous red phosphorus, layered black phosphorus integrates not only the gas phase and condensed phase flame-retardant mechanism, but also the physical barrier effect, similar to graphene, which has drawn much attention in recent decades. Similarly, ball milling can be used to exfoliate black phosphorus into few-layer nanosheets called phosphorene [53]. The difference is that black phosphorus nanosheets are not stable in open air, and can rapidly degrade into phosphate compounds upon oxidation and hydrolysis. Therefore, for black phosphorus, exfoliation and subsequent protection are of equal importance. Qu and coworkers reported the preparation of aminated black phosphorene (BP-NH$_2$) via the ball milling method [91], as shown in Figure 4. The graphene oxide (GO) was covalently bonded to black phosphorus through the reaction of -NH$_2$ and -COOH in the presence of a catalyst. The obtained product was assembled into a flexible film (RPNG) with ultrahigh thermal conductivity and remarkable flame retardancy, which exhibited a fantastic application in fire alarm sensors. Along the same lines, they bonded the multi-walled carbon nanotubes (MWCNTs) to black phosphorus nanosheets via the -NH-CO- linkage [92]. The resultant nanofiller (BP-MWCNTs) could endow cellulose nanofiber (CNF) with satisfactory thermal conductivity and fire retardancy, specifically an in-plane thermal conductivity of 22.38 ± 0.39 W m^{-1} K^{-1}, and a cross-plane thermal conductivity of 0.36 ± 0.03 W m^{-1} K^{-1}, UL-94 V-0 grade, and a LOI value of 29.9%.

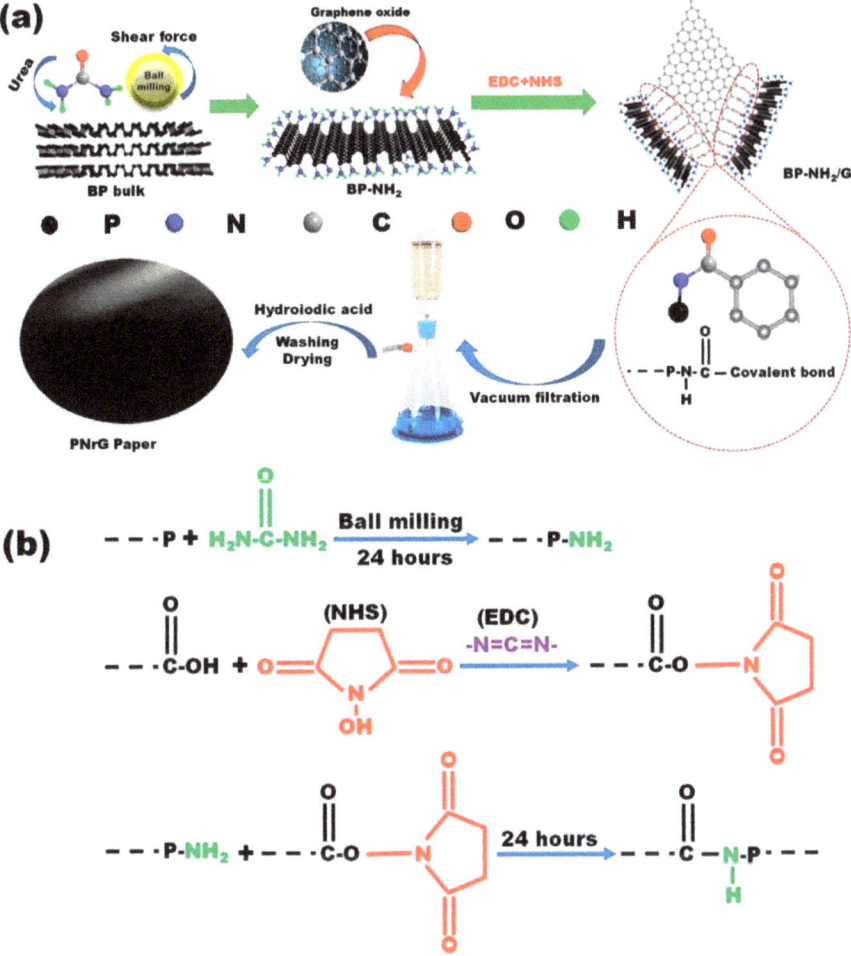

Figure 4. (a) Fabrication process of the RPNG film; (b) the mechanism of the covalent bond connection between BP-NH$_2$ and GO [91]. Reproduced with permission from the American Chemical Society.

Guo and coworkers reported the simultaneous exfoliation and functionalization of black phosphorus via sucrose-assisted ball milling, with N-methyl pyrrolidone (NMP) intercalating for high efficiency [53]. They found that the sucrose molecules could protect black phosphorus from oxidating, and promote the dispersion of black phosphorus nanosheets in solvents. The sucrose-grafted BP dramatically enhanced the mechanical performance and flame-retardant property of PVA, in terms of a 131.2% increase in tensile strength, and a 52.5% reduction in PHRR. The encapsulation effect of sucrose was highlighted in the exceptional air stability of PVA nanocomposite films. In Duan's work, ball milling, liquid exfoliation, and electrochemical exfoliation were applied to prepare black phosphorus nanosheets with different sizes, to clarify the size-dependent flame retardancy of black phosphorus nanosheets [54]. EP resin was selected as the polymer matrix. They found that the liquid ball milled BP (lb-BP) was the best at dispersing in the EP matrix, and endowed the EP with the highest flame retardancy. The barrier and carbonization catalyst action of lb-BP was believed to be the primary cause of the delayed combustion.

3.2.4. MoS$_2$-Based Flame Retardants

Molybdenum disulfide (MoS$_2$) consists of the elements Mo and S. Mo is a transition metal and demonstrates catalytic ability and smoke suppression performance [93,94]; S is also a flame-retardant element. MoS$_2$ has a layered structure, and can be exfoliated into nanosheets, the high specific area of which is in favor of improving catalytic performance. Therefore, the exfoliation of MoS$_2$ for developing high-performance flame retardant is highly desirable. Qiu and coworkers exfoliated MoS$_2$ into nanolayers via a ball milling method. Subsequently, a high-temperature polymerization was conducted to obtain the polyphosphazene nanoparticle (PPN) functionalized MoS$_2$ nanosheets (MoS$_2$@PPN) [95]. It was revealed that the loaded PPN could prevent the restacking of MoS$_2$ nanolayers, and improve the flame-retardant capability. Upon the incorporation of MoS$_2$@PPN, the flame retardancy and friction properties of the EP composite were improved. Based on the ball milling-exfoliated MoS$_2$ nanosheets, Zou's group constructed a phosphorus/nitrogen-co-doped MoS$_2$/cobalt borate nanostructure as a flame-retardant and anti-wear additive [93], as displayed in Figure 5. The GNDC and HCCP were used to assist in the exfoliation and modification of the MoS$_2$. After annealing, a two-dimensional cobalt borate (Co−Bi) nanosheet could be generated onto the MoS$_2$ nanosheets, resulting in a novel MoS$_2$-based flame retardant (PNMoS$_2$@Co−Bi). With only 2 wt% addition of PNMoS$_2$@Co−Bi, the EP composite exhibited a much-decreased flammability, and the detailed mechanism was clarified.

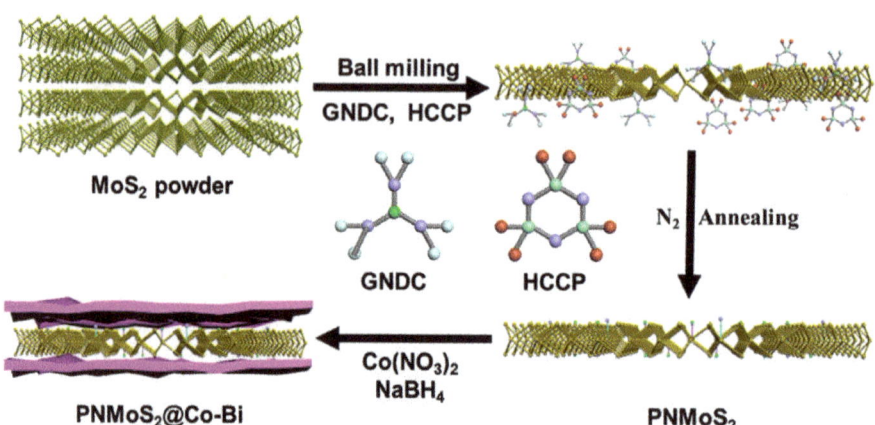

Figure 5. Synthetic route of the nanostructure PNMoS$_2$@Co−Bi [93]. Reproduced with permission from the American Chemical Society.

3.2.5. Covalent-Organic-Framework-Based Flame Retardants

Covalent organic frameworks (COFs) have a layered structure and flame-retardant action, due to their unique elementary composition [96,97]. Mu and coworkers explored the exfoliation of COFs by ball milling, and demonstrated their flame-retardant performance on thermoplastic polyurethanes (TPU) and polypropylene (PP) [98]. Firstly, they proposed the preparation of novel melamine/o-phthalaldehyde COF nanolayers. Details of the preparation, exfoliation, dispersion state, and flame-retardant performance of melamine/o-phthalaldehyde COFs were discussed. Then, to improve flame-retardant ability and smoke suppression, the Co$_3$O$_4$/COF nanohybrids were prepared based on the ball milling-induced exfoliation of the COFs [99]. The fire retardancy, smoke and carbon monoxide (CO) suppression, and thermal stability of the PP composites were characterized. A synergistic effect between Co$_3$O$_4$ and COFs was concluded.

3.2.6. Layered Oyster Waste

Most recently, Chen and his group have focused their attention to recycling layered oyster wastes as flame retardants. The layered oyster consisted of 95% layered CaCO$_3$, and 5% organic adhesives. However, the flame-retardant action of the bare oyster waste powders was unsatisfactory. Therefore, proper processing was highly desired in order to achieve the deconstruction and modification of layered oyster waste [100]. Firstly, they applied a simple ball milling of oyster powders, to obtain a phosphorus-free hybrid flame-retardant (TOSP), as shown in Figure 6. The successful exfoliation of layered oyster waste was confirmed. Moreover, the flame retardancy of the EP composite with the addition of milled layered oysters was investigated. This work opened a concept-new way to upcycle oyster waste, as high-value flame retardant, with the assistance of ball milling. To improve the flame-retardant ability of oyster wastes, the same group successively used ammonia phytate (PAA) [101] and chitosan-modified ammonium polyphosphate (CS@APP) [102] in assisting the ball milling-induced exfoliation of layered oyster wastes. This revealed that the resultant TOSP@PAA and TOSP@CS@APP were of a specific layer-crosslinking structure, by the –NH^{3+}–O$^-$P– bonds. As a result, the flame-retardant actions of TOSP@PAA for EP, and TOSP@CS@APP for cotton fabric, were higher than the common TOSP.

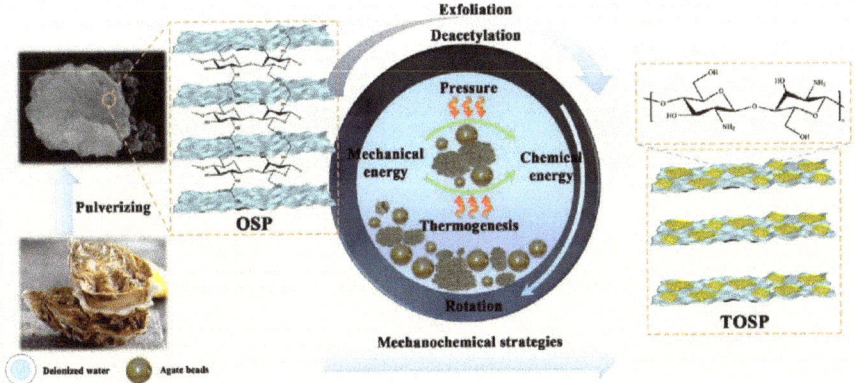

Figure 6. The preparation procedure for the TOSP [100]. Reproduced with permission from Elsevier.

3.2.7. Others

Besides the above-mentioned layered materials, some other layered flame-retardant compounds can be exfoliated by the ball milling method, such as layered double hydroxide (LDH) [103], Mxene [55], flake-NiNH$_4$PO$_4$·H$_2$O (IL-ANP) [104], and kaolin [105]. The primary thinking is the same for the processing of these compounds via the ball milling method, including shearing for exfoliation, and in situ modification for improving dispersibility and flame-retardant performance. For example, Huang's group ball-milled the LDH and red phosphorus to prepare P-LDH flame retardants for TPU [103]. It is believed that the anion substitution and high exfoliation of the LDH nanosheets greatly contributed to the high performance of the TPU composites. He and his coauthors conducted the exfoliation and functionalization of MXenes in the presence of poly(diallyldimethylammonium chloride) (PDDA) by ball milling, to improve the flame retardancy of polyurethane [55]. Notably, the 3 wt% PDDA-modified MXene could efficiently reduce heat release and smoke production.

3.3. Ball Milling for Modification

Another important application of ball milling is achieving the surface modification and functionalization of flame retardants for various purposes. For example, to enhance the hydrophobic property of aluminum hypophosphite, a rare earth-based coupling agent (REA) was utilized to modify the aluminum hypophosphite (AHP) through one-step ball milling [106], as shown in Figure 7. This revealed that the AHP modified by 4 wt% REA (RaAHP-4) had an outstanding hydrophobic performance, with a water contact angle of 94.6. Additionally, the EP/6RaAHP-4 composites behaved the best at reducing fire risk, including a decreased heat release and CO production. Guo and coworkers applied the titanate coupling agent NDZ-201, to modify the conventional IFRs composed of melamine (MEL), APP, and pentaerythritol (PER), via ball milling, to enhance thermal stability and dispersity. A cooperative effect on the fire retardancy of the PP composites was clarified. To address the same problem, Yan and coworkers conducted the surface modification using the silane coupling agent KH-550, via wet ball milling [38]. The resultant modified IFRs were used as flame retardants for polyphenylene oxide (PPO). A synergistic effect was observed between PPO and IFR in improving thermal stability and fire retardancy.

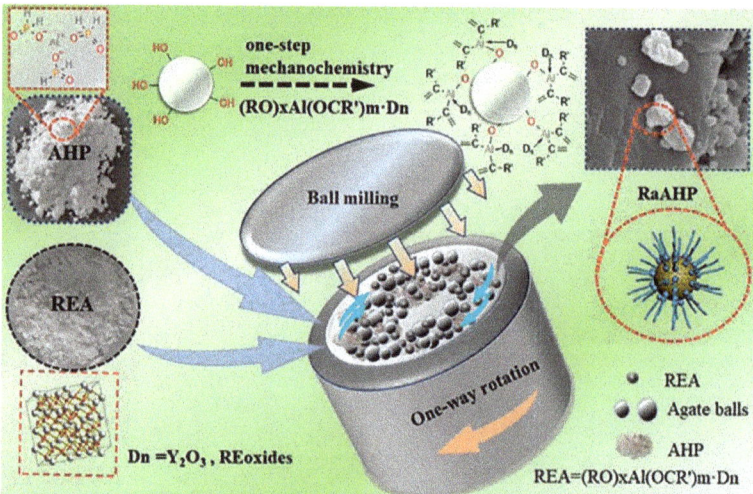

Figure 7. The synthesis process of RaAHP [106]. Reproduced with permission from John Wiley and Sons.

Filler additives have drawn much attention in the development of polymer composites, due to the high length-diameter ratio. However, the interfacial compatibility always limits the properties and applications of fiber-reinforced polymer composites. Członka and coworkers modified the walnut shell filler with selected mineral compounds: perlite, montmorillonite, and halloysite, via ball milling [107]. The rheological properties, mechanical properties, thermal properties, and fire retardancy of the fiber-reinforced polyurethane (PUR) composites were comprehensively investigated. A considerable reduction in heat release and smoke production was observed. Similarly, the vermiculite fillers were modified with casein, chitosan, and potato protein, with the assistance of ball milling, to reinforce the flame retardancy of polyurethane foams [108]. Approximately 2 wt% of vermiculite fillers were added. The rheological, thermal, and mechanical properties, and fire resistance, were explored in detail. Additionally, to the same ball milling method, Członka et al. reported the surface modification of lavender fillers with kaolinite and hydroxyapatite, for developing flame-retardant PU composites [109]. Notably, the ball milling-assisted surface modification could be extended to metal oxide flame retardants, such as Sb_2O_3 [110] and ZnO [111].

3.4. Ball Milling for Reaction

The heat and force generated by ball milling are sufficient for achieving some chemical reactions, including the solvent-free solid–solid reaction. Although it is not a dominant trend, there are indeed some reports on ball milling in the synthesis of novel flame retardants. Chen and coworkers reported a solvent-free ball milling method of fabricating phosphorus-containing hyper-crosslinked aromatic polymer (HCAP) from triphenylphosphine [112], as shown in Figure 8. Subsequently, nitrogen-rich graphitized carbon nitride was added, to synthesize a series of phosphorus and nitrogen-containing heterojunction flame retardants known as HCN. The molecular structures of HCAP and HCN were well characterized through experimental and computational approaches. The flame-retardant effect of HCN on EP resin was systematically investigated. It was found that the addition of 5 wt% HCN could endow EP resin with a UL-94 V0 rating, and dramatically reduce heat release and smoke production. Moreover, the machine learning method was performed, to evaluate the combined scores of multiple flame-retardant properties. It was believed that charring ability dominated the exceptional flame-retardant effect. This report displayed a brand-new pathway for developing high-performance flame retardants.

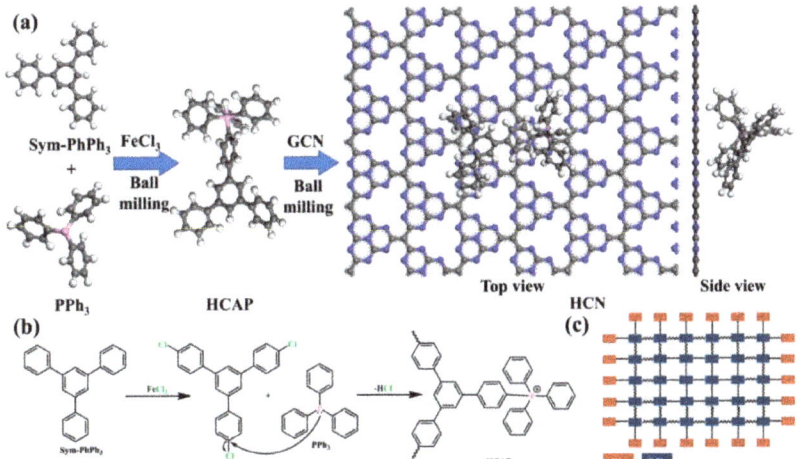

Figure 8. (a) Preparation scheme of HCN heterostructure. (b,c) A possible mechanism of HCAP synthesis [112]. Reproduced with permission from Elsevier.

How to endow biobased materials (e.g., cellulose crystals) with a flame-retardant effect is a critical problem when considering their high-value application. Apart from the conventional solution method based on corrosive concentrated phosphoric acid, Fiss and coworkers reported a phosphorylation process in cellulose nanocrystals, as well as some polymers, via ball milling [113]. It provided a feasible method to develop various flame retardants based on natural products, such as chitin nanofibril [39] and cellulose nanofibril [114]. Zhang and coworkers successively prepared chitin nanofibril-based flame retardants and cellulose nanofibril-based flame retardants, to improve the fire resistance of papers [39]. The ball milling was conducted in the presence of chitin/cellulose and P_2O_5. The degree of phosphorylation was examined in detail using X-ray photoelectron spectroscopy (XPS). As for chitin nanofibril-based flame-retardant-treated papers, an LOI of 30%, and a 62% reduction in PHRR, were obtained, compared to the control paper. Therefore, ball milling has been proven to be an efficient technique in developing novel flame retardants, which is likely to generate further interest.

4. Ball Milling for Mixing Flame Retardants and Polymer Matrices

As well as the preparation of flame-retardant additives, ball milling is often used to uniformly mix flame retardants and the polymer matrix, which is, namely, simply blending different particles in solid form. For example, Xu and coworkers conducted the mechanical ball mixing of montmorillonite (MMT), nano-Sb_2O_3, BEO, and PP by a high-energy ball milling machine, to fabricate flame-retardant PP composites [115]. To achieve the processing and improve compatibility, Liu and coworkers performed solid-state shear milling for magnesium hydroxide (MH) flame-retardant PP [116]. Upon milling, the pulverization of the PP, high-degree blending, uniform dispersion of MH, and chemical interaction between PP and MH could be obtained at the same time. Compared to conventional melt mixing, the milled sample exhibited better melt flowability, flame retardancy, and mechanical strength. Following the same idea, Prabhakar and coworkers reported the fabrication of flame-retardant thermoplastic starch/flax fabric green composites [117]. In this work, the starch was first plasticized into thermoplastic starch, via a ball milling process. The starch, flax fabric (FF), and APP were then mixed, to develop the flame-retardant composites. Pi and coworkers studied the effect of high-energy ball milling on the poly(vinyl chloride) (PVC)/zinc borate (ZB)/aluminum trihydrate (ATH) systems [118]. The first sample was PVC with ZB, and the second was PVC with ZN-ATH. The third sample was PVC with a mixture of ZB and ATH. It was found that high-energy ball milling induced the chemical bonding between PVC and ZB or ZB–ATH. As a result, an enhancement in the LOI and mechanical properties was observed. Moreover, the PVC/ZB and PVC/ZB–ATH composites exhibited better fire retardancy in terms of the suppressed release of aromatic compounds.

In addition to the ball milling-induced mixing, a subsequent hot pressing is always performed to construct a flame-retardant polymer composite with a segregated structure, which is favorable for achieving high electrical conductivity and electromagnetic-wave-shielding performance. Gao and coworkers developed segregated polystyrene (PS) composites with exceptional flame retardancy and electromagnetic wave shielding (EMI) properties, with the assistance of ball milling [119]. As shown in Figure 9, the PS particles, silicon-wrapped ammonium polyphosphate (SiAPP), and MWCNT were ball milled first, to obtain PS/SiAPP/MWCNT granules. It demonstrated that the SiAPP and MWCNT were uniformly distributed onto PS spheres. After hot pressing, a segregated structure s-PAC composite could be obtained, as displayed in the SEM image in Figure 9. Only 7 wt% MWCNT endowed the composites with promising thermal stability and fire retardancy. Specifically, a 60.5% and 33.9% reduction in PHRR and THR, respectively, was reported. Additionally, the EMI shielding property could reach 11 dB. The synergistic effect between MWCNT and SiAPP was believed to be the main contributor to exceptional fire safety, by forming a compact protective char layer on the bottom PS materials. The detailed EMI shielding mechanism was also clarified. Luo and coworkers followed the same strategy, in developing electrically conductive and flame-retardant low-density polyethylene composites with phosphorus-nitrogen-based flame retardant and MWCNTs, using a ball milling and hot-pressing method [120]. Notably, a dramatic decrease in PHRR (49.8%) and THR (51.9%) was observed. Therefore, the combination of ball milling and hot pressing is feasible to construct polymer composites with segregated structures, to achieve multifunctionality, including fire retardancy, thermal and electrical conductivity, and EMI performance. It is an up-and-coming technique for achieving the facile preparation of flame-retardant polymer materials.

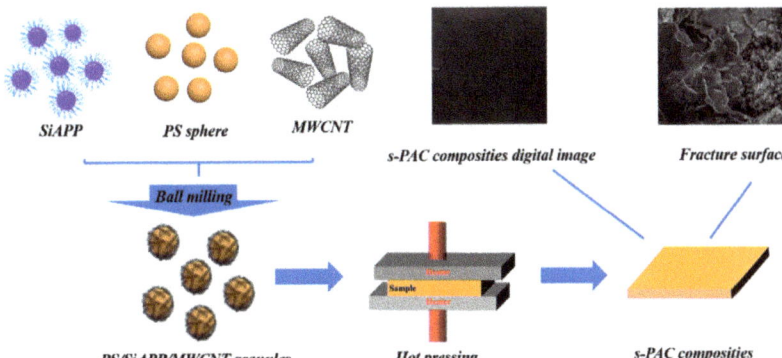

Figure 9. Schematic illustration of the fabrication route of s-PAC composites via ball milling [119]. Reproduced with permission from Elsevier.

5. Challenges and Prospects

Although progress has been made in the ball milling-promoted facile preparation of flame-retardant polymer materials, some key points need to be addressed when considering its extensive application. Firstly, limited by the basic operation principle, the products obtained by ball milling are not always homogeneous. Compared to conventional wet-chemistry synthesis, the scalability and consistency of ball milling are always limited [121,122]. To avoid this problem, the ball milling procedures can be varied, including changing the rotation direction appropriately. Sometimes, the impurities from the ball mill tank and beads are nonnegligible, due to the violent collision. Selecting the milling containers and balls with higher hardness is preferable, to reduce the chance of contamination. Furthermore, the ball milling process always lasts for a long time, which is time-consuming, and produces flame retardants in batch mode [123]. To improve efficiency, a processing additive is highly recommended. For example, the intercalators are efficient in assisting the exfoliation of layered compounds using the ball milling method. The configuration of the ball milling machine to achieve the preparation of flame retardants in flow mode is challenging, and of high importance. In addition, ball milling for the solid-state synthesizing of new flame retardants is in its infancy at present, and requires extensive research work to achieve its full potential and industrial value. Moreover, the mixing action of ball milling for a flame-retardant additive and a polymer matrix can be multipurpose, rather than the simple blending of various particles. Using the proper design, some promising polymer composites could be fabricated, such as composites with segregated structures. More customized and specific structures are highly desirable, with the assistance of ball milling.

6. Conclusions

In this paper, we review the progress in the development of flame retardants and flame-retardant polymer composites using the ball milling method. Starting from the basic concept and category of ball milling, two types of ball milling machine—planetary ball milling, and high-energy ball milling—are introduced. Due to the generated impact and shear forces, and the high-temperature surroundings, ball milling can achieve the crushing, exfoliating, modification, and chemical reaction required in the preparation of flame retardants. The simultaneous exfoliation and functionalization of layer-compound-based flame retardants are clearly introduced. In addition, the mixing action of the flame retardants and the polymer matrix using the ball milling approach is described, especially the part in which segregated structures are constructed for multifunctional purposes. Flame-retardant polymer composites with segregated structure have exhibited a promising application in electric products, which always require electrical conductivity and an electromagnetic shielding function, while facing high fire risk. Despite the rapid development of ball-milled flame retardants, some challenges and prospective developments are proposed, to promote their practical applications, including heterogeneity issues, impurity problems, and the process being time-consuming. Nevertheless, the authors believe that in the next 10 years, many more flame retardants and flame-retardant polymer composite products prepared by the ball milling method will be coming out of research labs, and will be commercialized across industries.

Author Contributions: Conceptualization, X.F.; writing—original draft preparation, X.F. and X.L.; writing—review and editing, C.Y. and K.D.; supervision, X.F. and H.Y. All authors have read and agreed to the published version of the manuscript.

Funding: The work was financially supported by the Chongqing University Start-up Funding (Grant#: 02110011080003), the State Key Laboratory of High Performance Civil Engineering Materials (No.2022CEM004), and the National Natural Science Foundation of China (No. 21374111).

Institutional Review Board Statement: Not applicable.

Informed Consent Statement: Not applicable.

Data Availability Statement: Not applicable.

Conflicts of Interest: The authors declare no conflict of interest.

Sample Availability: Not applicable.

References

1. Alexandre, M.; Dubois, P. Polymer-layered silicate nanocomposites: Preparation, properties and uses of a new class of materials. *Mater. Sci. Eng. R Rep.* **2000**, *28*, 1–63. [CrossRef]
2. Stuart, M.A.C.; Huck, W.T.S.; Genzer, J.; Muller, M.; Ober, C.; Stamm, M.; Sukhorukov, G.B.; Szleifer, I.; Tsukruk, V.V.; Urban, M.; et al. Emerging applications of stimuli-responsive polymer materials. *Nat. Mater.* **2010**, *9*, 101–113. [CrossRef] [PubMed]
3. Qiu, H.Y.; Feng, K.; Gapeeva, A.; Meurisch, K.; Kaps, S.; Li, X.; Yu, L.M.; Mishra, Y.K.; Adelung, R.; Baum, M. Functional polymer materials for modern marine biofouling control. *Prog. Polym. Sci.* **2022**, *127*, 101516. [CrossRef]
4. Fico, D.; Rizzo, D.; Casciaro, R.; Corcione, C.E. A Review of Polymer-Based Materials for Fused Filament Fabrication (FFF): Focus on Sustainability and Recycled Materials. *Polymers* **2022**, *14*, 465. [CrossRef] [PubMed]
5. Huang, Q.D.; Chen, J.A.; Shao, X.C.; Zhang, L.; Dong, Y.J.; Li, W.J.; Zhang, C.; Ma, Y.G. New electropolymerized triphenylamine polymer films and excellent multifunctional electrochromic energy storage system materials with real-time monitoring of energy storage status. *Chem. Eng. J.* **2023**, *461*, 141974. [CrossRef]
6. Oladapo, B.I.; Kayode, J.F.; Akinyoola, J.O.; Ikumapayi, O.M. Shape memory polymer review for flexible artificial intelligence materials of biomedical. *Mater. Chem. Phys.* **2023**, *293*, 126930. [CrossRef]
7. Chen, R.S.; Ahmad, S.; Gan, S.; Salleh, M.N.; Ab Ghani, M.H.; Tarawneh, M.A. Effect of polymer blend matrix compatibility and fibre reinforcement content on thermal stability and flammability of ecocomposites made from waste materials. *Thermochim. Acta* **2016**, *640*, 52–61. [CrossRef]
8. Kalali, E.N.; Zhang, L.; Shabestari, M.E.; Croyal, J.; Wang, D.Y. Flame-retardant wood polymer composites (WPCs) as potential fire safe bio-based materials for building products: Preparation, flammability and mechanical properties. *Fire Saf. J.* **2019**, *107*, 210–216. [CrossRef]
9. Mohanty, D.; Chen, S.Y.; Hung, I.M. Effect of Lithium Salt Concentration on Materials Characteristics and Electrochemical Performance of Hybrid Inorganic/Polymer Solid Electrolyte for Solid-State Lithium-Ion Batteries. *Batteries* **2022**, *8*, 173. [CrossRef]
10. Sahinoz, M.; Aruntas, H.Y.; Guru, M. Processing of polymer wood composite material from pine cone and the binder of phenol formaldehyde/PVAc/molasses and improvement of its properties. *Case Stud. Constr. Mater.* **2022**, *16*, e01013.
11. Vahabi, H.; Jouyandeh, M.; Parpaite, T.; Saeb, M.R.; Ramakrishna, S. Coffee Wastes as Sustainable Flame Retardants for Polymer Materials. *Coatings* **2021**, *11*, 1021. [CrossRef]
12. Zhao, P.; Tian, L.; Guo, Y.; Lv, B.; Mao, X.; Li, T.; Cui, J.; Guo, J.; Yang, B. A facile method to prepare high-performance thermal insulation and flame retardant materials from amine-linked porous organic polymers. *Eur. Polym. J.* **2022**, *162*, 110918. [CrossRef]
13. Vahabi, H.; Laoutid, F.; Formela, K.; Saeb, M.R.; Dubois, P. Flame-Retardant Polymer Materials Developed by Reactive Extrusion: Present Status and Future Perspectives. *Polym. Rev.* **2022**, *62*, 919–949. [CrossRef]
14. Chen, W.; Liu, P.; Liu, Y.; Liu, Z. Recent advances in Two-dimensional $Ti_3C_2T_x$ MXene for flame retardant polymer materials. *Chem. Eng. J.* **2022**, *446*, 137239. [CrossRef]
15. Salamova, A.; Hermanson, M.H.; Hites, R.A. Organophosphate and Halogenated Flame Retardants in Atmospheric Particles from a European Arctic Site. *Environ. Sci. Technol.* **2014**, *48*, 6133–6140. [CrossRef]

16. Harrad, S.; Drage, D.; Sharkey, M.; Stubbings, W.; Alghamdi, M.; Berresheim, H.; Coggins, M.; Rosa, A.H. Elevated concentrations of halogenated flame retardants in waste childcare articles from Ireland. *Environ. Pollut.* **2023**, *317*, 120732. [CrossRef]
17. Ren, H.L.; Ge, X.; Qi, Z.H.; Lin, Q.H.; Shen, G.F.; Yu, Y.X.; An, T.C. Emission and gas-particle partitioning characteristics of atmospheric halogenated and organophosphorus flame retardants in decabromodiphenyl ethane-manufacturing functional areas. *Environ. Pollut.* **2023**, *329*, 121709. [CrossRef]
18. Kerric, A.; Mazerolle, M.J.; Giroux, J.F.; Verreault, J. Halogenated flame retardant exposure pathways in urban-adapted gulls: Are atmospheric routes underestimated? *Sci. Total Environ.* **2023**, *860*, 160526. [CrossRef]
19. Capozzi, S.L.; Lehman, D.C.; Venier, M. Disentangling Source Profiles and Time Trends of Halogenated Flame Retardants in the Great Lakes. *Environ. Sci. Technol.* **2023**, *57*, 1309–1319. [CrossRef]
20. Eo, S.-M.; Cha, E.; Kim, D.-W. Effect of an inorganic additive on the cycling performances of lithium-ion polymer cells assembled with polymer-coated separators. *J. Power Sources* **2009**, *189*, 766–770. [CrossRef]
21. Katase, F.; Kajiyama, S.; Kato, T. Macromolecular templates for biomineralization-inspired crystallization of oriented layered zinc hydroxides. *Polym. J.* **2017**, *49*, 735–739. [CrossRef]
22. Sangian, D.; Ide, Y.; Bando, Y.; Rowan, A.E.; Yamauchi, Y. Materials Nanoarchitectonics Using 2D Layered Materials: Recent Developments in the Intercalation Process. *Small* **2018**, *14*, 1800551. [CrossRef] [PubMed]
23. Han, X.; Li, N.; Wu, B.; Li, D.; Pan, Q.; Wang, R. Microstructural characterization and corrosion resistance evaluation of chromate-phosphate/water-soluble resin composite conversion coating on Al surfaces. *Prog. Org. Coat.* **2022**, *173*, 107205. [CrossRef]
24. Gao, F.; Tong, L.F.; Fang, Z.P. Effect of a novel phosphorous-nitrogen containing intumescent flame retardant on the fire retardancy and the thermal behaviour of poly(butylene terephthalate). *Polym. Degrad. Stab.* **2006**, *91*, 1295–1299. [CrossRef]
25. Wang, D.-L.; Liu, Y.; Wang, D.-Y.; Zhao, C.-X.; Mou, Y.-R.; Wang, Y.-Z. A novel intumescent flame-retardant system containing metal chelates for polyvinyl alcohol. *Polym. Degrad. Stab.* **2007**, *92*, 1555–1564. [CrossRef]
26. Xue, M.; Zhang, X.; Wu, Z.; Wang, H.; Gu, Z.; Bao, C.; Tian, X. A Commercial Phosphorous-Nitrogen Containing Intumescent Flame Retardant for Thermoplastic Polyurethane. *J. Appl. Polym. Sci.* **2014**, *131*, 39772. [CrossRef]
27. Wang, C.; Wu, Y.; Li, Y.; Shao, Q.; Yan, X.; Han, C.; Wang, Z.; Liu, Z.; Guo, Z. Flame-retardant rigid polyurethane foam with a phosphorus-nitrogen single intumescent flame retardant. *Polym. Adv. Technol.* **2018**, *29*, 668–676. [CrossRef]
28. Anilkumar, Y.; Felipe, M.; de Souza, T.D.; Ram, K. Gupta, Recent Advancements in Flame-Retardant Polyurethane Foams: A Review. *Ind. Eng. Chem. Res.* **2022**, *61*, 15046–15065.
29. Jiang, Y.; Yang, H.; Lin, X.; Xiang, S.; Feng, X.; Wan, C. Surface Flame-Retardant Systems of Rigid Polyurethane Foams: An Overview. *Materials* **2023**, *16*, 2728. [CrossRef] [PubMed]
30. Li, F. Comprehensive Review of Recent Research Advances on Flame-Retardant Coatings for Building Materials: Chemical Ingredients, Micromorphology, and Processing Techniques. *Molecules* **2023**, *28*, 1842. [CrossRef]
31. Palacios, E.; Leret, P.; De La Mata, M.J.; Fernandez, J.F.; De Aza, A.H.; Rodriguez, M.A.; Rubio-Marcos, F. Self-Forming 3D Core-Shell Ceramic Nanostructures for Halogen-Free Flame Retardant Materials. *ACS Appl. Mater. Interfaces* **2016**, *8*, 9462–9471. [CrossRef]
32. Xue, B.; Niu, M.; Yang, Y.; Bai, J.; Song, Y.; Peng, Y.; Liu, X. Multi-functional carbon microspheres with double shell layers for flame retardant poly (ethylene terephthalate). *Appl. Surf. Sci.* **2018**, *435*, 656–665. [CrossRef]
33. Holdsworth, A.F.; Horrocks, A.R.; Kandola, B.K. Novel metal complexes as potential synergists with phosphorus based flame retardants in polyamide 6.6. *Polym. Degrad. Stab.* **2020**, *179*, 109220. [CrossRef]
34. Holdsworth, A.F.; Horrocks, A.R.; Kandola, B.K. Potential Synergism between Novel Metal Complexes and Polymeric Brominated Flame Retardants in Polyamide 6.6. *Polymers* **2020**, *12*, 1543. [CrossRef]
35. Horrocks, A.R. The Potential for Bio-Sustainable Organobromine-Containing Flame Retardant Formulations for Textile Applications—A Review. *Polymers* **2020**, *12*, 2160. [CrossRef] [PubMed]
36. Do, J.-L.; Friščić, T. Mechanochemistry: A Force of Synthesis. *ACS Cent. Sci.* **2017**, *3*, 13–19. [CrossRef]
37. Burmeister, C.F.; Kwade, A. Process engineering with planetary ball mills. *Chem. Soc. Rev.* **2013**, *42*, 7660–7667. [CrossRef] [PubMed]
38. Yan, H.; Dong, B.; Du, X.; Ma, S.; Wei, L.; Xu, B. Flame-Retardant Performance of Polystyrene Enhanced by Polyphenylene Oxide and Intumescent Flame Retardant. *Polym. Plast. Technol. Eng.* **2014**, *53*, 395–402. [CrossRef]
39. Zhang, T.; Kuga, S.; Wu, M.; Huang, Y. Chitin Nanofibril-Based Flame Retardant for Paper Application. *ACS Sustain. Chem. Eng.* **2020**, *8*, 12360–12365. [CrossRef]
40. Hwang, S.; Grätz, S.; Borchardt, L. A guide to direct mechanocatalysis. *Chem. Commun.* **2022**, *58*, 1661–1671. [CrossRef]
41. Zhang, Q.; Saito, F. A review on mechanochemical syntheses of functional materials. *Adv. Powder Technol.* **2012**, *23*, 523–531. [CrossRef]
42. Blumbergs, E.; Serga, V.; Shishkin, A.; Goljandin, D.; Shishko, A.; Zemcenkovs, V.; Markus, K.; Baronins, J.; Pankratov, V. Selective Disintegration-Milling to Obtain Metal-Rich Particle Fractions from E-Waste. *Metals* **2022**, *12*, 1468. [CrossRef]
43. Tsuzuki, T.; McCormick, P.G. Mechanochemical synthesis of nanoparticles. *J. Mater. Sci.* **2004**, *39*, 5143–5146. [CrossRef]
44. Nikolic, N.; Marinkovic, Z.; Sreckovic, T. The influence of grinding conditions on the mechanochemical synthesis of zinc stannate. *J. Mater. Sci.* **2004**, *39*, 5239–5242. [CrossRef]

45. Palazon, F.; Ajjouri, Y.E.; Sebastia-Luna, P.; Lauciello, S.; Manna, L.; Bolink, H.J. Mechanochemical synthesis of inorganic halide perovskites: Evolution of phase-purity, morphology, and photoluminescence. *J. Mater. Chem. C* **2019**, *7*, 11406–11410. [CrossRef]
46. Kim, M.-J.; Jeon, I.-Y.; Seo, J.-M.; Dai, L.; Baek, J.-B. Graphene Phosphonic Acid as an Efficient Flame Retardant. *ACS Nano* **2014**, *8*, 2820–2825. [CrossRef]
47. Jeon, I.-Y.; Shin, S.-H.; Choi, H.-J.; Yu, S.-Y.; Jung, S.-M.; Baek, J.-B. Heavily aluminated graphene nanoplatelets as an efficient flame-retardant. *Carbon* **2017**, *116*, 77–83. [CrossRef]
48. Chen, Y.; Wu, H.; Duan, R.; Zhang, K.; Meng, W.; Li, Y.; Qu, H. Graphene doped Sn flame retardant prepared by ball milling and synergistic with hexaphenoxy cyclotriphosphazene for epoxy resin. *J. Mater. Res. Technol.* **2022**, *17*, 774–788. [CrossRef]
49. Qiu, S.; Hou, Y.; Xing, W.; Ma, C.; Zhou, X.; Liu, L.; Kan, Y.; Yuen, R.K.K.; Hu, Y. Self-assembled supermolecular aggregate supported on boron nitride nanoplatelets for flame retardant and friction application. *Chem. Eng. J.* **2018**, *349*, 223–234. [CrossRef]
50. Chen, S.; Xu, R.; Liu, J.; Zou, X.; Qiu, L.; Kang, F.; Liu, B.; Cheng, H.-M. Simultaneous Production and Functionalization of Boron Nitride Nanosheets by Sugar-Assisted Mechanochemical Exfoliation. *Adv. Mater.* **2019**, *31*, 1804810. [CrossRef]
51. Han, G.; Zhao, X.; Feng, Y.; Ma, J.; Zhou, K.; Shi, Y.; Liu, C.; Xie, X. Highly flame-retardant epoxy-based thermal conductive composites with functionalized boron nitride nanosheets exfoliated by one-step ball milling. *Chem. Eng. J.* **2021**, *407*, 127099. [CrossRef]
52. Han, G.; Zhang, D.; Kong, C.; Zhou, B.; Shi, Y.; Feng, Y.; Liu, C.; Wang, D.-Y. Flexible, thermostable and flame-resistant epoxy-based thermally conductive layered films with aligned ionic liquid-wrapped boron nitride nanosheets via cyclic layer-by-layer blade-casting. *Chem. Eng. J.* **2022**, *437*, 135482. [CrossRef]
53. Guo, J.; Yang, L.; Zhang, L.; Li, C. Simultaneous exfoliation and functionalization of black phosphorus by sucrose-assisted ball milling with NMP intercalating and preparation of flame retardant polyvinyl alcohol film. *Polymer* **2022**, *255*, 125036. [CrossRef]
54. Duan, Z.; Wang, Y.; Bian, S.; Liu, D.; Zhang, Y.; Zhang, X.; He, R.; Wang, J.; Qu, G.; Chu, P.K.; et al. Size-dependent flame retardancy of black phosphorus nanosheets. *Nanoscale* **2022**, *14*, 2599–2604. [CrossRef] [PubMed]
55. He, L.; Wang, J.; Wang, B.; Wang, X.; Zhou, X.; Cai, W.; Mu, X.; Hou, Y.; Hu, Y.; Song, L. Large-scale production of simultaneously exfoliated and Functionalized MXenes as promising flame retardant for polyurethane. *Compos. B. Eng.* **2019**, *179*, 107486. [CrossRef]
56. Chen, X.; Zhao, Z.; Hao, M.; Wang, D. Research of hydrogen generation by the reaction of Al-based materials with water. *J. Power Sources* **2013**, *222*, 188–195. [CrossRef]
57. Xie, L.; Ding, Y.; Ren, J.; Xie, T.; Qin, Y.; Wang, X.; Chen, F. Improved Hydrogen Generation Performance via Hydrolysis of MgH_2 with Nb_2O_5 and CeO_2 Doping. *Mater. Trans.* **2021**, *62*, 880–886. [CrossRef]
58. Wang, R.; Zhu, Z.X.; Tan, S.F.; Guo, J.; Xu, Z.M. Mechanochemical degradation of brominated flame retardants in waste printed circuit boards by Ball Milling. *J. Hazard. Mater.* **2020**, *385*, 121509. [CrossRef]
59. Guo, X.; Geng, J.; Sun, B.; Xu, Q.; Li, Y.; Xie, S.; Xue, Y.; Yan, H. Great enhancement of efficiency of intumescent flame retardants by titanate coupling agent and polysiloxane. *Polym. Adv. Technol.* **2021**, *32*, 41–53. [CrossRef]
60. Lei, Y.; Bai, Y.; Shi, Y.; Liang, M.; Zou, H.; Zhou, S. Composite nanoarchitectonics of poly(vinylidene fluoride)/graphene for thermal and electrical conductivity enhancement via constructing segregated network structure. *J. Polym. Res.* **2022**, *29*, 213. [CrossRef]
61. Antonio Puertolas, J.; Jose Martinez-Morlanes, M.; Javier Pascual, F.; Morimoto, T. Influence of mechanical blending method and consolidation temperature on electrical properties of the prepared graphene nanoplatelet/UHMWPE composite. *J. Polym. Res.* **2023**, *30*, 21. [CrossRef]
62. Kumar, M.; Xiong, X.; Wan, Z.; Sun, Y.; Tsang, D.C.W.; Gupta, J.; Gao, B.; Cao, X.; Tang, J.; Ok, Y.S. Ball milling as a mechanochemical technology for fabrication of novel biochar nanomaterials. *Bioresour. Technol.* **2020**, *312*, 123613. [CrossRef] [PubMed]
63. Dudina, D.V.; Bokhonov, B.B. Materials Development Using High-Energy Ball Milling: A Review Dedicated to the Memory of M.A. Korchagin. *J. Compos. Sci.* **2022**, *6*, 188. [CrossRef]
64. Bocz, K.; Krain, T.; Marosi, G. Effect of Particle Size of Additives on the Flammability and Mechanical Properties of Intumescent Flame Retarded Polypropylene Compounds. *Int. J. Polym. Sci.* **2015**, *2015*, 493710. [CrossRef]
65. Bao, Y.; Li, X.; Tang, P.; Liu, C.; Zhang, W.; Ma, J. Attapulgite modified cotton fabric and its flame retardancy. *Cellulose* **2019**, *26*, 9311–9322. [CrossRef]
66. Üreyen, M.E.; Kaynak, E. Effect of Zinc Borate on Flammability of PET Woven Fabrics. *Adv. Polym. Technol.* **2019**, *2019*, 7150736. [CrossRef]
67. Jawaid, M.; Kian, L.K.; Alamery, S.; Saba, N.; Fouad, H.; Alothman, O.Y.; Sain, M. Development and characterization of fire retardant nanofiller from date palm biomass. *Biomass Convers. Bior.* **2022**. [CrossRef]
68. Azizi, H.; Ahmad, F.; Yusoff, P.S.M.M.; Zia-ul-Mustafa, M.I. Fire Performance, Microstructure and Thermal Degradation of an Epoxy Based Nano Intumescent Fire Retardant Coating for Structural Applications. In Proceedings of the 23rd Scientific Conference of Microscopy-Society-Malaysia (SCMSM), Univ Teknologi Petronas, Tronoh, Malaysia, 10–12 December 2014.
69. Andrikopoulos, K.S.; Bounos, G.; Lainioti, G.C.; Ioannides, T.; Kallitsis, J.K.; Voyiatzis, G.A. Flame Retardant Nano-Structured Fillers from Huntite/Hydromagnesite Minerals. *Nanomaterials* **2022**, *12*, 2433. [CrossRef]

70. Pate, R.; Klise, G.; Wu, B. Resource demand implications for US algae biofuels production scale-up. *Appl. Energy* **2011**, *88*, 3377–3388. [CrossRef]
71. Siebert, H.M.; Wilker, J.J. Deriving Commercial Level Adhesive Performance from a Bio-Based Mussel Mimetic Polymer. *ACS Sustain. Chem. Eng.* **2019**, *7*, 13315–13323. [CrossRef]
72. Gupta, I.; Gupta, O. Recent Advancements in the Recovery and Reuse of Organic Solvents Using Novel Nanomaterial-Based Membranes for Renewable Energy Applications. *Membranes* **2023**, *13*, 108. [CrossRef] [PubMed]
73. Wang, H.; Zhu, Z.; Yuan, J.; Wang, H.; Wang, Z.; Yang, F.; Zhan, J.; Wang, L. A new recycling strategy for preparing flame retardants from polyphenylene sulfide waste textiles. *Compos. Commun.* **2021**, *27*, 100852. [CrossRef]
74. Dragoman, M.; Dragoman, D. Graphene-based quantum electronics. *Prog. Quantum Electron.* **2009**, *33*, 165–214. [CrossRef]
75. Wang, H.; Wang, H.S.; Ma, C.; Chen, L.; Jiang, C.; Chen, C.; Xie, X.; Li, A.-P.; Wang, X. Graphene nanoribbons for quantum electronics. *Nat. Rev. Phys.* **2021**, *3*, 791–802. [CrossRef]
76. Julkapli, N.M.; Bagheri, S. Graphene supported heterogeneous catalysts: An overview. *Int. J. Hydrogen Energy* **2015**, *40*, 948–979. [CrossRef]
77. Alaf, M.; Tocoglu, U.; Kartal, M.; Akbulut, H. Graphene supported heterogeneous catalysts for LiO_2 batteries. *Appl. Surf. Sci.* **2016**, *380*, 185–192. [CrossRef]
78. Ratnikov, P.V.; Silin, A.P. Planar Graphene-Narrow-Gap Semiconductor-Graphene Heterostructure. *Bull. Lebedev Phys. Inst.* **2008**, *35*, 328–335. [CrossRef]
79. Ebrahimi, M.; Horri, A.; Sanaeepur, M.; Tavakoli, M.B. Tight-binding description of graphene-BCN-graphene layered semiconductors. *J. Comput. Electron.* **2020**, *19*, 62–69. [CrossRef]
80. Zabihi, O.; Ahmadi, M.; Li, Q.; Ferdowsi, M.R.G.; Mahmoodi, R.; Kalali, E.N.; Wang, D.-Y.; Naebe, M. A sustainable approach to scalable production of a graphene based flame retardant using waste fish deoxyribonucleic acid. *J. Clean. Prod.* **2020**, *247*, 119150. [CrossRef]
81. Yang, P.X.; Wu, H.G.; Yang, F.F.; Yang, J.; Wang, R.; Zhu, Z.G. A Novel Self-Assembled Graphene-Based Flame Retardant: Synthesis and Flame Retardant Performance in PLA. *Polymers* **2021**, *13*, 4216. [CrossRef]
82. Wang, C.; Wang, J.; Men, Z.; Wang, Y.; Han, Z. Thermal Degradation and Combustion Behaviors of Polyethylene/Alumina Trihydrate/Graphene Nanoplatelets. *Polymers* **2019**, *11*, 772. [CrossRef]
83. Tran, V.Q.; Doan, H.T.; Nguyen, N.T.; Do, C.V. Preparation of Graphene Nanoplatelets by Thermal Shock Combined with Ball Milling Methods for Fabricating Flame-Retardant Polymers. *J. Chem.* **2019**, *2019*, 5284160. [CrossRef]
84. Duan, R.; Wu, H.; Li, J.; Zhou, Z.; Meng, W.; Liu, L.; Qu, H.; Xu, J. Phosphor nitrile functionalized UiO-66-NH_2/graphene hybrid flame retardants for fire safety of epoxy. *Colloids Surf. A* **2022**, *635*, 128093. [CrossRef]
85. Li, J.; Lyu, Y.; Li, C.; Zhang, F.; Li, K.; Li, X.; Li, J.; Kim, K.-H. Development of strong, tough and flame-retardant phenolic resins by using Acacia mangium tannin-functionalized graphene nanoplatelets. *Int. J. Biol. Macromol.* **2023**, *227*, 1191–1202. [CrossRef] [PubMed]
86. Xu, J.; Jiang, Z.; Zhu, K.; Zhang, Y.; Zhu, M.; Wang, C.; Wang, H.; Ren, A. Highly flame-retardant and low toxic polybutylene succinate composites with functionalized BN@APP exfoliated by ball milling. *J. Appl. Polym. Sci.* **2022**, *139*, 52217. [CrossRef]
87. Gibertini, E.; Carosio, F.; Aykanat, K.; Accogli, A.; Panzeri, G.; Magagnin, L. Silica-encapsulated red phosphorus for flame retardant treatment on textile. *Surf. Interfaces* **2021**, *25*, 101252. [CrossRef]
88. Chen, X.; Lan, W.; Dou, W. Polystyrene nanospheres coated red phosphorus flame retardant for polyamide 66. *J. Appl. Polym. Sci.* **2022**, *139*, e52772. [CrossRef]
89. Yin, S.; Ren, X.; Lian, P.; Zhu, Y.; Mei, Y. Synergistic Effects of Black Phosphorus/Boron Nitride Nanosheets on Enhancing the Flame-Retardant Properties of Waterborne Polyurethane and Its Flame-Retardant Mechanism. *Polymers* **2020**, *12*, 1487. [CrossRef]
90. Qiu, S.; Yang, W.; Wang, X.; Hu, Y. Phthalocyanine zirconium diazo passivation of black phosphorus for efficient smoke suppression, flame retardant and mechanical enhancement. *Chem. Eng. J.* **2023**, *453*, 139759. [CrossRef]
91. Qu, Z.; Wu, K.; Xu, C.-A.; Li, Y.; Jiao, E.; Chen, B.; Meng, H.; Cui, X.; Wang, K.; Shi, J. Facile Construction of a Flexible Film with Ultrahigh Thermal Conductivity and Excellent Flame Retardancy for a Smart Fire Alarm. *Chem. Mater.* **2021**, *33*, 3228–3240. [CrossRef]
92. Qu, Z.; Wang, K.; Xu, C.-A.; Li, Y.; Jiao, E.; Chen, B.; Meng, H.; Cui, X.; Shi, J.; Wu, K. Simultaneous enhancement in thermal conductivity and flame retardancy of flexible film by introducing covalent bond connection. *Chem. Eng. J.* **2021**, *421*, 129729. [CrossRef]
93. Zou, B.; Qiu, S.; Qian, Z.; Wang, J.; Zhou, Y.; Xu, Z.; Yang, W.; Xing, W. Phosphorus/Nitrogen-Codoped Molybdenum Disulfide/Cobalt Borate Nanostructures for Flame-Retardant and Tribological Applications. *ACS Appl. Nano Mater.* **2021**, *4*, 10495–10504. [CrossRef]
94. Yang, Z.; Kang, X.; Lu, S.; Wang, J.; Fang, X.; Li, J.; Liu, B.; Ding, T.; Xu, Y. Synergistic effects of molybdenum disulfide on a novel intumescent flame retardant polyformaldehyde system. *J. Appl. Polym. Sci.* **2023**, *140*, e53385. [CrossRef]
95. Qiu, S.; Hu, Y.; Shi, Y.; Hou, Y.; Kan, Y.; Chu, F.; Sheng, H.; Yuen, R.K.K.; Xing, W. In situ growth of polyphosphazene particles on molybdenum disulfide nanosheets for flame retardant and friction application. *Compos. Part A Appl.* **2018**, *114*, 407–417. [CrossRef]

96. Wang, X.; Ji, H.; Wang, F.; Cui, X.; Liu, Y.; Du, X.; Lu, X. NiFe$_2$O$_4$-based magnetic covalent organic framework nanocomposites for the efficient adsorption of brominated flame retardants from water. *Microchim. Acta* **2021**, *188*, 161. [CrossRef] [PubMed]
97. Peng, H.; Mao, Y.; Wang, D.; Fu, S. B-N-P-linked covalent organic frameworks for efficient flame retarding and toxic smoke suppression of polyacrylonitrile composite fiber. *Chem. Eng. J.* **2022**, *430*, 133120. [CrossRef]
98. Mu, X.; Zhan, J.; Feng, X.; Yuan, B.; Qiu, S.; Song, L.; Hu, Y. Novel Melamine/o-Phthalaldehyde Covalent Organic Frameworks Nanosheets: Enhancement Flame Retardant and Mechanical Performances of Thermoplastic Polyurethanes. *ACS Appl. Mater. Interfaces* **2017**, *9*, 23017–23026. [CrossRef]
99. Mu, X.; Pan, Y.; Ma, C.; Zhan, J.; Song, L. Novel Co$_3$O$_4$/covalent organic frameworks nanohybrids for conferring enhanced flame retardancy, smoke and CO suppression and thermal stability to polypropylene. *Mater. Chem. Phys.* **2018**, *215*, 20–30. [CrossRef]
100. Ren, J.; Wang, Y.; Piao, J.; Cui, J.; Guan, H.; Jiao, C.; Chen, X. Facile construction of phosphorus-free and green organic-inorganic hybrid flame-retardant system: For improving fire safety of EP. *Prog. Org. Coat.* **2023**, *179*, 107489. [CrossRef]
101. Ren, J.; Wang, Y.; Piao, J.; Ou, M.; Lian, R.; Cui, J.; Guan, H.; Liu, L.; Jiao, C.; Chen, X. Facile construction of organic–inorganic hybrid flame-retardant system based on fully biomass: Improving the fire safety and mechanical property of epoxy resin. *Chem. Eng. J.* **2023**, *460*, 141775. [CrossRef]
102. Ren, J.; Piao, J.; Wang, Y.; Feng, T.; Liu, L.; Jiao, C.; Chen, X. Facile synthesis of bio-based phosphorus/nitrogen compound for high efficiency flame retardant finishing of cotton fabric. *Cellulose* **2023**, *30*, 1245–1264. [CrossRef]
103. Huang, S.-C.; Deng, C.; Chen, H.; Li, Y.-M.; Zhao, Z.-Y.; Wang, S.-X.; Wang, Y.-Z. Novel Ultrathin Layered Double Hydroxide Nanosheets with In Situ Formed Oxidized Phosphorus as Anions for Simultaneous Fire Resistance and Mechanical Enhancement of Thermoplastic Polyurethane. *ACS Appl. Polym. Mater.* **2019**, *1*, 1979–1990. [CrossRef]
104. Bi, X.; Meng, W.; Meng, Y.; Di, H.; Li, J.; Xie, J.; Xu, J.; Fang, L. Novel [BMIM]PF$_6$ modified flake-ANP flame retardant: Synthesis and application in epoxy resin. *Polym. Test.* **2021**, *101*, 107284. [CrossRef]
105. Ou, H.; Xu, J.; Liu, B.; Xue, H.; Weng, Y.; Jiang, J.; Xu, G. Study on synergistic expansion and flame retardancy of modified kaolin to low density polyethylene. *Polymer* **2021**, *221*, 123586. [CrossRef]
106. Wang, Y.; Piao, J.; Ren, J.; Feng, T.; Wang, Y.; Liu, W.; Dong, H.; Chen, W.; Jiao, C.; Chen, X. Simultaneously improving the hydrophobic property and flame retardancy of aluminum hypophosphite using rare earth based coupling agent for epoxy composites. *Polym. Adv. Technol.* **2023**, *34*, 1154–1169. [CrossRef]
107. Członka, S.; Kairytė, A.; Miedzińska, K.; Strąkowska, A. Polyurethane Hybrid Composites Reinforced with Lavender Residue Functionalized with Kaolinite and Hydroxyapatite. *Materials* **2021**, *14*, 415. [CrossRef]
108. Miedzińska, K.; Członka, S.; Strąkowska, A.; Strzelec, K. Vermiculite Filler Modified with Casein, Chitosan, and Potato Protein as a Flame Retardant for Polyurethane Foams. *Int. J. Polym. Sci.* **2021**, *22*, 10825. [CrossRef]
109. Członka, S.; Kairytė, A.; Miedzińska, K.; Strąkowska, A. Polyurethane Composites Reinforced with Walnut Shell Filler Treated with Perlite, Montmorillonite and Halloysite. *Int. J. Mol. Sci.* **2021**, *22*, 7304. [CrossRef]
110. Niu, L.; Xu, J.; Yang, W.; Zhao, J.; Su, J.; Guo, Y.; Liu, X. Research on nano-Sb$_2$O$_3$ flame retardant in char formation of PBT. *Ferroelectrics* **2018**, *523*, 14–21. [CrossRef]
111. Díez-Pascual, A.M.; Xu, C.; Luque, R. Development and characterization of novel poly(ether ether ketone)/ZnO bionanocomposites. *J. Mater. Chem. B* **2014**, *2*, 3065. [CrossRef]
112. Chen, Z.; Guo, Y.; Chu, Y.; Chen, T.; Zhang, Q.; Li, C.; Jiang, J.; Chen, T.; Yu, Y.; Liu, L. Solvent-free and electron transfer-induced phosphorus and nitrogen-containing heterostructures for multifunctional epoxy resin. *Compos. B Eng.* **2022**, *240*, 109999. [CrossRef]
113. Fiss, B.G.; Hatherly, L.; Stein, R.S.; Friščić, T.; Moores, A. Mechanochemical Phosphorylation of Polymers and Synthesis of Flame-Retardant Cellulose Nanocrystals. *ACS Sustain. Chem. Eng.* **2019**, *7*, 7951. [CrossRef]
114. Zhang, T.; Wu, M.; Kuga, S.; Ewulonu, C.M.; Huang, Y. Cellulose Nanofibril-Based Flame Retardant and Its Application to Paper. *ACS Sustain. Chem. Eng.* **2020**, *8*, 10222–10229. [CrossRef]
115. Xu, J.; Liu, X.; Yang, W.; Niu, L.; Zhao, J.; Ma, B.; Kang, C. Influence of montmorillonite on the properties of halogen–antimony flame retardant polypropylene composites. *Polym. Compos.* **2019**, *40*, 1930–1938. [CrossRef]
116. Liu, Y.; Li, J.; Wang, Q. Preparation of High Loading Magnesium Hydroxide Flame Retardant Polypropylene by Solid State Shear Milling. *J. Compos. Mater.* **2007**, *41*, 1995–2003. [CrossRef]
117. Prabhakar, M.N.; Rehman Shah, A.U.; Song, J.-I. Improved flame-retardant and tensile properties of thermoplastic starch/flax fabric green composites. *Carbohydr. Polym.* **2017**, *168*, 201–211. [CrossRef] [PubMed]
118. Pi, H.; Guo, S.; Ning, Y. Mechanochemical improvement of the flame-retardant and mechanical properties of zinc borate and zinc borate-aluminum trihydrate-filled poly(vinyl chloride). *J. Appl. Polym. Sci.* **2003**, *89*, 753–762. [CrossRef]
119. Gao, C.; Shi, Y.; Chen, Y.; Zhu, S.; Feng, Y.; Lv, Y.; Yang, F.; Liu, M.; Shui, W. Constructing segregated polystyrene composites for excellent fire resistance and electromagnetic wave shielding. *J. Colloid Interface Sci.* **2022**, *606*, 1193–1204. [CrossRef] [PubMed]
120. Luo, Y.; Xie, Y.; Chen, R.; Zheng, R.; Wu, H.; Sheng, X.; Xie, D.; Mei, A. A low-density polyethylene composite with phosphorus-nitrogen based flame retardant and multi-walled carbon nanotubes for enhanced electrical conductivity and acceptable flame retardancy. *Front. Chem. Sci. Eng.* **2021**, *15*, 1332–1345. [CrossRef]
121. Mio, H.; Kano, J.; Saito, F. Scale-up method of planetary ball mill. *Chem. Eng. Sci.* **2004**, *59*, 5906–5916. [CrossRef]

122. Santhanam, P.R.; Dreizin, E.L. Predicting conditions for scaled-up manufacturing of materials prepared by ball milling. *Powder Technol.* **2012**, *221*, 403–411. [CrossRef]
123. Holdsworth, A.F.; Eccles, H.; Halman, A.M.; Mao, R.; Bond, G. Low-Temperature Continuous Flow Synthesis of Metal Ammonium Phosphates. *Sci. Rep.* **2018**, *8*, 13547. [CrossRef] [PubMed]

Disclaimer/Publisher's Note: The statements, opinions and data contained in all publications are solely those of the individual author(s) and contributor(s) and not of MDPI and/or the editor(s). MDPI and/or the editor(s) disclaim responsibility for any injury to people or property resulting from any ideas, methods, instructions or products referred to in the content.

Article

Effect of Layered Aminovanadic Oxalate Phosphate on Flame Retardancy of Epoxy Resin

Po Hu, Weixi Li, Shuai Huang, Zongmian Zhang, Hong Liu, Wang Zhan, Mingyi Chen and Qinghong Kong *

School of Emergency Management, Jiangsu University, Zhenjiang 212013, China
* Correspondence: kongqh@ujs.edu.cn

Abstract: To alleviate the fire hazard of epoxy resin (EP), layered ammonium vanadium oxalate-phosphate (AVOPh) with the structural formula of $(NH_4)_2[VO(HPO_4)]_2(C_2O_4)\cdot 5H_2O$ is synthesized using the hydrothermal method and mixed into an EP matrix to prepare EP/AVOPh composites. The thermogravimetric analysis (TGA) results show that AVOPh exhibits a similar thermal decomposition temperature to EP, which is suitable for flame retardancy for EP. The incorporation of AVOPh nanosheets greatly improves the thermal stability and residual yield of EP/AVOPh composites at high temperatures. The residue of pure EP is 15.3% at 700 °C. In comparison, the residue of EP/AVOPh composites is increased to 23.0% with 8 wt% AVOPh loading. Simultaneously, EP/6 wt% AVOPh composites reach UL-94 V1 rating ($t_1 + t_2 = 16$ s) and LOI value of 32.8%. The improved flame retardancy of EP/ AVOPh composites is also proven by the cone calorimeter test (CCT). The results of CCT of EP/8 wt% AVOPh composites show that the peak heat release rate (PHHR), total smoke production (TSP), peak of CO production (PCOP), and peak of CO_2 production (PCO_2P) decrease by 32.7%, 20.4%, 37.1%, and 33.3% compared with those of EP, respectively. This can be attributed to the lamellar barrier, gas phase quenching effect of phosphorus-containing volatiles, the catalytic charring effect of transition metal vanadium, and the synergistic decomposition of oxalic acid structure and charring effect of phosphorus phase, which can insulate heat and inhibit smoke release. Based on the experimental data, AVOPh is expected to serve as a new high-efficiency flame retardant for EP.

Keywords: hydrothermal method; layered structure; phosphate; flame retardancy; vanadium; epoxy resin

Citation: Hu, P.; Li, W.; Huang, S.; Zhang, Z.; Liu, H.; Zhan, W.; Chen, M.; Kong, Q. Effect of Layered Aminovanadic Oxalate Phosphate on Flame Retardancy of Epoxy Resin. *Molecules* **2023**, *28*, 3322. https://doi.org/10.3390/molecules28083322

Academic Editors: Xin Wang, Weiyi Xing and Gang Tang

Received: 6 March 2023
Revised: 31 March 2023
Accepted: 7 April 2023
Published: 9 April 2023

Copyright: © 2023 by the authors. Licensee MDPI, Basel, Switzerland. This article is an open access article distributed under the terms and conditions of the Creative Commons Attribution (CC BY) license (https://creativecommons.org/licenses/by/4.0/).

1. Introduction

Epoxy resin (EP) is a class of polymer composed of large chains of hydrocarbons [1]. It has received much attention because of its excellent performance and convenient production. With the progress of technology, EP has been widely used in all aspects of production and life, such as coatings, architectural decoration, structural elements, and adhesives [2–4]. In the past decade, the global production of epoxy resin increased from 2.96 million tons in 2017 to 3.73 million tons in 2021, with a compound annual growth rate of 5.95%. As of 2021, the global total production capacity of epoxy resin was 5.37 million tons. However, EP also has disadvantages that limit its engineering applications. Due to the hydrocarbon structure of EP, it is flammable, resulting in instability at high temperatures, and easily cracks and burns [5,6]. At the same time, it releases a lot of heat and toxic gases, which could threaten life and property during a fire. To improve the safety of EP used in high-temperature environments and reduce its fire hazard, it is necessary to improve the flame retardancy of EP [7].

According to previous studies, two-dimensional (2D) layered phosphates have proven to be effective phosphate-containing flame retardants for EP [8–10]. Phosphate-based flame retardants are considered the most promising candidate to replace halogen-based flame retardants. Phosphate compounds have excellent flame retardancy in both gaseous and condensed phases. In the condensed phases, phosphorus is initially decomposed into phosphoric acid or dehydrated to form metaphosphate, which is polymerized to form polymetaphosphate. It has a catalytic effect on the dehydration of oxygen-containing polymers into char. In the gaseous phases, it acts as a free radical trapping agent that forms PO during combustion, which can combine with H· in the flame region to inhibit flame [11,12]. Materials with a 2D layered structure can provide a physical barrier effect, which can delay decomposition when they are added to the EP matrix. Phosphates containing transition metals, such as aluminum diethyl phosphate (ADP), copper phenylphosphate, and phosphor metal complex (CePn), are oxidized to form metal oxides at high temperatures [13–16]. Salen-diphenyl chloro-phosphate (Salen-DPCPs) metal compounds with phosphate structure and nickel components can reduce the flame hazards of EP. The results showed that Salen-DPCPs exhibited a good flame retardant effect for EP. The limiting oxygen index (LOI) value was up to 31.5%, and EP composites achieved a UL-94 V0 rating when 5 wt% Salen-DPCP-1 was incorporated [17]. It led to excellent catalytic properties, which accelerated cross-linking of EP to improve the charring properties [18,19]. Therefore, the combination of phosphorus and 2D metal compounds can reach a good synergistic effect, enhancing the compactness of the char layer.

The synergistic effect of nitrogen and phosphorus has a phosphorus-nitrogen structure-activity relationship, diluting the concentration of combustibles and oxygen around the combustion substrate and promoting flame extinction [20,21]. The results indicated that adding 7 wt% nitrogen/phosphorus modified lignin to epoxy resin achieved UL-94 V0 grade and limiting oxygen index of 31.4%, respectively. In addition, the residual char became denser [22]. $NiNH_4PO_4 \cdot H_2O$ nanoflakes have good flame retardancy in EP. When the addition amount was 5%, the peak heat release rate and peak smoke production of the composites were 69.1% and 36.5% lower than those of pure EP, respectively. $NiNH_4PO_4 \cdot H_2O$ promoted the formation of a stable carbon layer and released noncombustible gases. It prevented heat and oxygen transfer and diluted the concentration of combustible gases [21].

Inspired by the above work, 2D metal phosphonate with a layered structure and optimal thermal stability can effectively ameliorate the fire retardation and smoke inhibition performances of polymer composites. Additionally, the excellent synergistic flame retardant capability of gaseous phase flame retardancy, catalytic carbonation, and physical barrier can be achieved. In this work, ammonium vanadium oxalate-phosphate (AVOPh, $(NH_4)_2[VO(HPO_4)]_2(C_2O_4)5H_2O$) was successfully designed and synthesized, which was incorporated into an EP matrix as flame retardant filler for preparing EP/AVOPh composites. To better understand the effect of various components of AVOPh on the flame retardancy of EP composites, ammonium vanadium phosphates (AVPh, $(NH_4)VOPO_4 \cdot 1.5H_2O$) and vanadium phosphates (VPh, $VOPO_4 \cdot 2H_2O$) were selected as comparative samples. The results indicated that EP/AVOPh composites exhibited excellent thermal stability and flame retardancy, which were mainly attributed to the lamellar barrier, gas phase quenching of volatiles containing phosphorus, the catalytic charring of transition metal vanadium, and the synergistic decomposition of oxalic acid structure and phosphorus phase to participate in charring to form a stable char layer.

2. Results and Discussion
2.1. Structural Analysis of AVOPh

To characterize the structure of AVOPh, XRD, FTIR, TGA, and FESEM tests were carried out. As shown in Figure 1a, XRD results showed that AVOPh showed good crystallinity, and characteristic diffraction peaks appeared at 2θ = 13.2°, 26.4°, and 40.5°, corresponding to (100), (200), and (300) diffraction planes, respectively, indicating that AVOPh had a layered structure [23,24]. The characteristic peaks of VPh at 2θ = 11.9° and 24.0° correspond to (001) and (002) diffraction planes, respectively [24]. In the FTIR spectrum of Figure 1b, the absorption peaks near 3521 cm^{-1} and 3166 cm^{-1} are attributed to the vibration of N-H. The tensile vibration of V = O and V-O appears at 647 cm^{-1} and 523 cm^{-1}. The characteristic absorption bands of P-O appear at 1152 cm^{-1} and 952 cm^{-1} [25,26]. In addition, C-O bands at 1651 cm^{-1} and 900 cm^{-1} are assigned to the oxalate structure, and the peak at 1445 cm^{-1}

corresponds to the O-H bond [26]. The TGA was used to investigate the thermal stability of AVOPh, AVPh, VPh, and pure EP. From Figure 1c, the first stage of mass loss of AVOPh decreased by 8.2 wt% before 250 °C, mainly due to the desorption of absorbed water and the decomposition of a small amount of amine. The second significant weight loss occurs at 280 °C, mainly due to the thermal decomposition of the AVOPh structure. The mass loss is 21.5 wt% in the range of 280~500 °C. The initial decomposition temperature ($T_{5\%}$) of pure EP is 360 °C. From the results, the main decomposition temperature of AVOPh and EP is similar. After thermal decomposition, the final residual of AVOPh is 69.4 wt% at 700 °C. AVPh with a final residual of 62.5 wt% was less stable compared with AVOPh. The mass loss of VPh decreases by 17 wt% occurred before 150 °C, presumably due to the loss of adsorbed water, and the final residual was 82.5 wt% at 700 °C. The main decomposition temperatures of AVOPh and AVPh are lower than that of EP. According to the results of thermal analysis, the decomposition temperature of AVOPh is slightly lower than that of EP. AVOPh is suitable as a flame retardant for EP. Figure 1d–f are the SEM images of AVOPh, AVPh, and VPh, respectively. It can be seen that AVOPh, AVPh, and VPh are all layered structures, which is beneficial for the EP/AVOPh composites to play a lamellar barrier role in the flame retardant process.

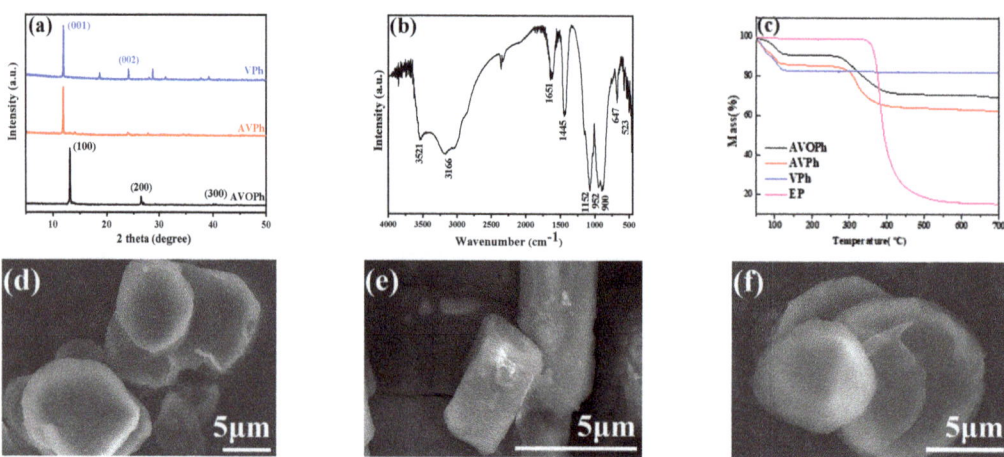

Figure 1. (a) XRD pattern of AVOPh, AVPh, and VPh; (b) FTIR spectrum of AVOPh; (c) TGA curves of AVOPh, AVPh, VPh, and EP; (d–f) SEM images of AVOPh, AVPh, and VPh.

2.2. Thermal Stability of EP Composites

To analyze the thermal stability of EP composites, a TGA measurement was carried out. The TGA curves of samples in an N_2 atmosphere are shown in Figure 2. The specific values are listed in Table 1. From the results, the initial decomposition temperature ($T_{5\%}$) of EP is 360 °C, and the initial decomposition temperature of EP/AVOPh composites is decreased. The $T_{5\%}$ of EP/8 wt% AVOPh composites are reduced to 339 °C, indicating that the vanadium group has a catalytic effect on EP. In addition, the temperature of 50% weight loss ($T_{50\%}$) and maximum decomposition temperature (T_{max}) of EP/AVOPh composites also decrease. After 280 °C, the decomposition of AVOPh makes the transition metal vanadium, oxalic acid structure, and phosphorus participate in the reactions of the condensation phase in the combustion, promoting the further char formation of the EP matrix. At 700 °C, the residue of EP is 15.3%. When the contents of AVOPh are 2, 4, 6, and 8 wt%, the residual yield of EP/AVOPh composites increases to 19.8%, 21.2%, 21.7%, and 23.0%, respectively. The results indicate that the addition of AVOPh improves the high-temperature thermal stability and residual yield of EP/AVOPh composites, mainly due to the catalytic capacity of the transition metal vanadium and the phosphorus compounds, which promote EP decomposition and cross-linking of decomposition products into a char layer [27]. Notably, the residue (22.6%) of EP/4 wt% VPh composites at 700 °C is higher than that of EP/4 wt% AVOPh composites, which is mainly due to the higher content of vanadium and phosphorus in VPh, exhibiting better catalytic ability [28]. The residual content of EP/4 wt% AVPh composites is only 17.8%, indicating that the existence of an oxalic acid structure can be beneficial to improve the residual char content and high thermal stability of EP composites.

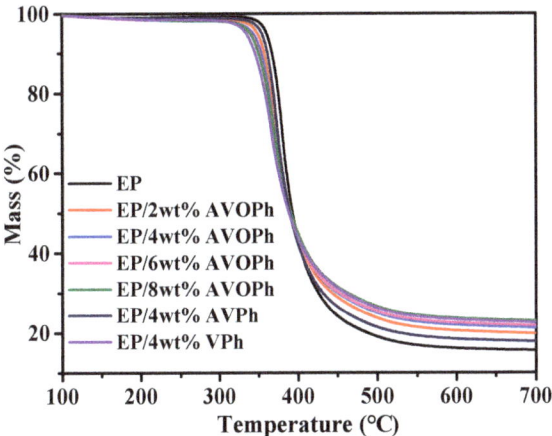

Figure 2. TGA curves of pure EP and its composites.

Table 1. TGA data of pure EP and its composites.

Samples	$T_{5\%}$ (°C)	$T_{50\%}$ (°C)	T_{max} (°C)	Residues (wt%, 700 °C)
EP	360	394	380	15.3
EP/2 wt% AVOPh	349	391	374	19.8
EP/4 wt% AVOPh	343	390	369	21.2
EP/6 wt% AVOPh	340	389	368	21.8
EP/8 wt% AVOPh	339	391	369	23.0
EP/4 wt% AVPh	355	389	370	17.8
EP/4 wt% VPh	334	388	366	22.6

2.3. Flame Retardancy of EP Composites

Promoting the self-extinguishing of combustible materials after ignition has potential value for reducing the fire hazard of polymers [29]. To explore the influence of AVOPh on the flame retardant performance of EP composites, vertical combustion and limiting oxygen index (LOI) tests were carried out. The results are listed in Table 2 and Figure 3. The UL-94 vertical combustion test shows that EP is no rating (NR). When 4, 6, and 8 wt% AVOPh are added, EP/AVOPh composites can reach the UL-94 V1 level. Specifically, the ($t_1 + t_2$) value of EP/6 wt% AVOPh composites is only 16 s, which is close to the UL-94 V0 level. As the content of AVOPh continues to increase, the ($t_1 + t_2$) value of EP/8 wt% AVOPh composites increases again, mainly due to the uneven dispersion of AVOPh in the EP matrix [30]. From the LOI results in Figure 4 and Table 2, the flame retardancy of EP composites was improved when the AVOPh was added to EP. When 2 wt% AVOPh is added to the EP matrix, the LOI value increases from 25.9% to 30.1%. After the addition of 4, 6, and 8 wt% AVOPh, the LOI values of EP/AVOPh composites increase to 31.7%, 32.8%, and 34.2%, respectively. The above results have positive significance for improving the fire safety of EP/AVOPh composites. This is the effect of AVOPh combined with gas phase and condensed phase flame retardancy during EP composite combustion. In comparison, although the EP/4 wt% AVPh composites can reach the V1 level, the ($t_1 + t_2$) value is as high as 41 s; EP/4 wt% VPh composites still do not have a rating of UL-94. Additionally, the LOI values of EP/4 wt% AVPh and EP/4 wt% VPh composites are 30.2% and 29.8%, respectively, which are inferior to that of EP/4 wt% AVOPh composite. It can be concluded that the presence of amino groups plays a positive role in promoting flame extinction. The improved flame retardancy of EP/AVOPh composites is mainly attributed to the release of phosphorus free radicals, NH_3, and water vapor in the gas phase during the decomposition process of AVOPh. They dilute the concentration of combustible and oxygen around the combustion matrix and promote flame extinction. In addition, the dense and strong char layer can isolate the external heat source and accelerate EP/AVOPh composite self-quenching [31,32].

Table 2. UL-94 and LOI results of EP and its composites.

Samples	LOI (vol%)	UL-94	
		$t_1 + t_2$ (s)	Rating
EP	25.9	>50	NR *
EP/2 wt% AVOPh	30.1	>50	NR
EP/4 wt% AVOPh	31.7	21	V1 **
EP/6 wt% AVOPh	32.8	16	V1
EP/8 wt% AVOPh	34.2	27	V1
EP/4 wt% AVPh	30.2	41	V1
EP/4 wt% VPh	29.8	>50	NR

* NR: no rating of UL-94 ($t_1 + t_2 > 50$ s). ** V1: UL-94 V1 (10 s < $t_1 + t_2$ < 50 s, no dripping or dripping does not ignite cotton).

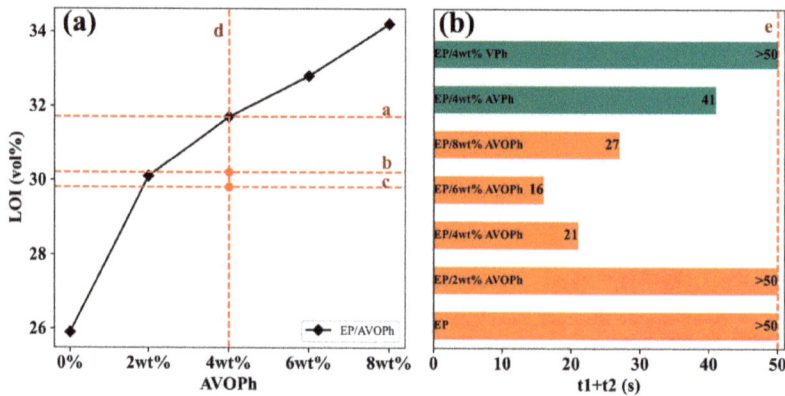

Figure 3. The LOI and vertical combustion curves of EP and EP composites: (**a**) LOI; (**b**) vertical combustion. a: the LOI of EP/4 wt% AVOPh. b: the LOI of EP/4 wt% AVPh. c: the LOI of EP/4 wt% VPh. d: additive ratio. e: V1: UL-94 V1 (10 s < $t_1 + t_2$ < 50 s, no dripping or dripping does not ignite cotton).

To further analyze the flame retardancy of EP/AVOPh composites, a cone calorimeter test (CCT) was carried out. Heat release rate (HRR), total smoke production (TSP), CO production (COP), and CO_2 production (CO_2P) are shown in Figure 4, and the specific values are listed in Table 3. The peak heat release rate (PHRR) of pure EP is 1004 kW·m^{-2}, indicating EP has great thermal hazard. The addition of AVOPh significantly reduces the PHRR values of EP composites. When 2, 4, 6, and 8 wt% AVOPh are added, the PHRR values of EP/AVOPh composites decrease to 931, 796, 795, and 675 kW·m^{-2}, respectively. The PHRR value of EP/8 wt% AVOPh composites decreases by 32.7% compared with that of pure EP. The PHRR values of EP/4 wt% AVPh and EP/4 wt% VPh composites are 877 kW·m^{-2} and 840 kW·m^{-2}, which are higher than that of EP/4 wt% AVOPh composites. Smoke production and toxicity are important indicators of fire hazard. The TSP value, peak CO production (PCOP), and peak CO_2 production (PCO$_2$P) of EP are 21.2 m^2, 0.035 g·s^{-1}, and 0.54 g·s^{-1}, respectively, indicating EP has a large production of smoke and toxic gas. When EP/AVOPh composites with 2, 4, 6, and 8 wt% AVOPh, TSP values decrease to 17.6, 17.4, 16.8, and 16.4 m^2, PCOP values decrease to 0.031, 0.026, 0.024, and 0.022 g·s^{-1}, and PCO$_2$P values decrease to 0.52, 0.44, 0.42, and 0.36 g·s^{-1}, respectively. Compared with EP, the TSP, PCOP, and PCO$_2$P values of EP/8 wt% AVOPh are decreased by 20.4%, 37.1%, and 33.3%, respectively. The results indicate that AVOPh can suppress smoke production and toxic gas release in EP composites, which provides more favorable conditions for rescue and escape in the fire. This can be attributed to the lamellar barrier and excellent charring of layered AVOPh, which act as heat insulation and gas isolation. Secondly, the phosphorus free radicals in the gas phase also play a key role in reducing the emission of smoke and toxic gases. In addition, the synergistic participation of oxalic acid structure and ammonium in the carbonization process forms a high-quality expanded char layer, thereby providing thermal insulation and smoke suppression [33,34].

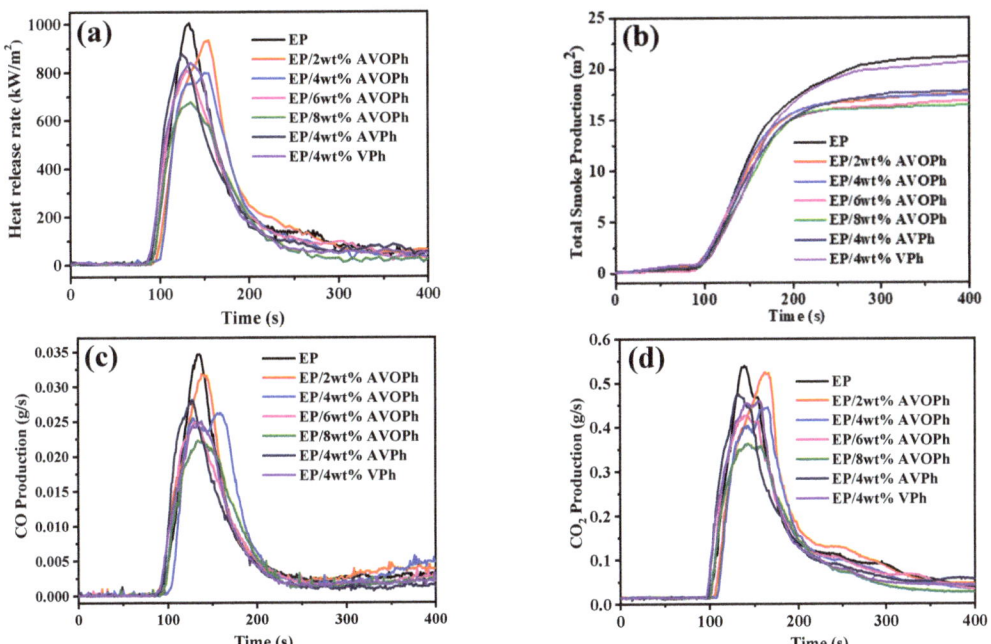

Figure 4. CCT curves of pure EP and its composites: (**a**) HRR curves; (**b**) TSP curves; (**c**) COP curves; (**d**) CO$_2$P curves.

Table 3. CCT data of pure EP and EP composites.

Samples	PHRR (kW/m^2)	TSP (m^2)	PCOP (g/s)	PCO$_2$P (g/s)	Mass Residue (%)
EP	1004	22.4	0.035	0.54	18.7
EP/2 wt% AVOPh	931	17.6	0.031	0.52	25.2
EP/4 wt% AVOPh	796	17.4	0.026	0.44	30.3
EP/6 wt% AVOPh	795	16.8	0.024	0.42	29.9
EP/8 wt% AVOPh	675	16.4	0.022	0.36	37.3
EP/4 wt% AVPh	877	17.8	0.028	0.47	25.0
EP/4 wt% VPh	840	19.7	0.025	0.45	24.3

To further confirm the positive influence of AVOPh on the char-forming performance of EP, the mass loss during the CCT is shown in Figure 5. The specific values are listed in Table 3. The mass residual of pure EP after the CCT is 18.7%. When the contents of AVOPh are 2, 4, 6, and 8 wt%, the residual amounts of EP/AVOPh composites are increased to 25.2%, 30.3%, 29.9%, and 37.3%. In addition, the residual amounts of EP/4 wt% AVPh composites and EP/4 wt% VPh composites are only 25.0% and 24.3%, indicating that the oxalic acid structure is involved in char formation during the decomposition of EP. The possible reason is that AVOPh promotes the formation of more expansive char on the surface of EP at the beginning of combustion, thereby isolating oxygen and heat from entering the EP interior. It provides a sufficient time of contact for small molecules decomposed from EP and vanadium, phosphides, and oxalic acid. In a word, AVOPh promotes the char formation of EP composites and improves the quality of the char layer after combustion, which further effectively isolates the internal and external mass exchange, improving the flame retardant performance of EP/AVOPh composites [35,36].

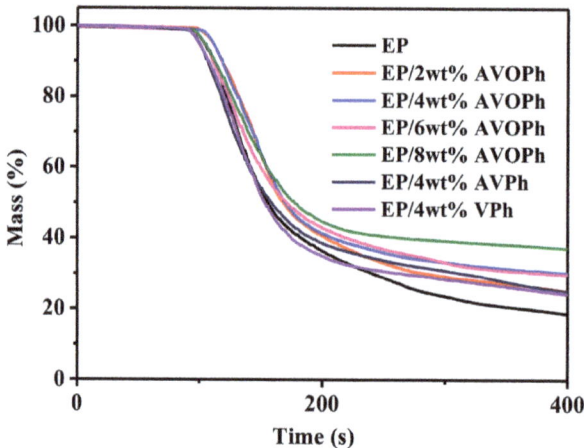

Figure 5. Mass loss curves of pure EP and its composites at CCT.

To observe the charring effect of AVOPh in EP composites, the SEM images of the inner and outer residue of pure EP and EP/AVOPh composites are shown in Figure 6. The outer and inner char of pure EP is shown in Figure 6a,b. The outer char has obvious collapse, with large and dense cracks. During the combustion process of the EP, it cannot act as a barrier during the combustion. There are large pores and cracks in the internal char. This is mainly caused by the intense combustion of EP and the rapid escape of a large amount of heat and smoke within a short time. The residue of EP/2 wt% AVOPh composites is shown in Figure 6c,d. The quality of the external char is obviously improved. Cracks of external char disappear, and the char is dense. This indicates that AVOPh plays a certain barrier role. At high temperatures, the transition metal vanadium and phosphorus compounds catalyze the dehydration and charring of the EP matrix, accelerating the formation of the char layer [37]. The large pores and cracks in the internal char are reduced, indicating that the intensity of combustion is reduced compared to pure EP. The residue of the EP/6 wt% AVOPh composites is shown in Figure 6e,f. The outer char is very dense and compact with a smooth surface. Although the outer char of the EP/6 wt% AVOPh composites becomes dense, the surface is relatively loose, which is crucial for improving the barrier property of the char. The inner char becomes thicker compared to that of EP/2 wt% AVOPh composites. The results indicate that AVOPh plays a significant catalytic role in forming compact char, improving the flame retardancy of EP/AVOPh composites [38].

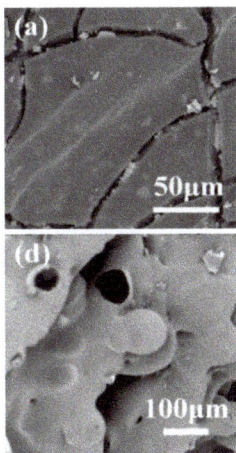
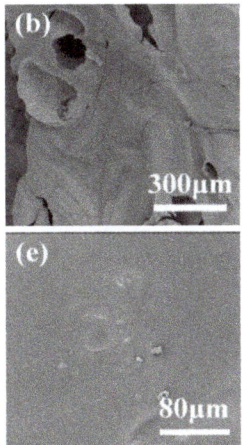
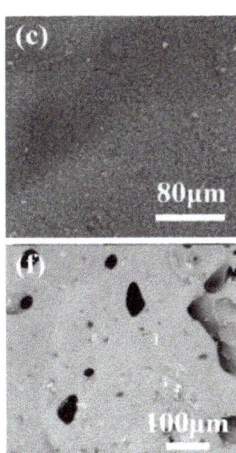

Figure 6. SEM images of char layer: (**a**) external and (**b**) internal char layer of pure EP; (**c**) external and (**d**) internal char layer of EP/2 wt% AVOPh composites; (**e**) external and (**f**) internal char layer of EP/6 wt% AVOPh composites.

2.4. Flame Retardant Mechanism of AVOPh in EP

According to the above analysis, the flame retardant mechanism of EP/AVOPh composites is discussed in Figure 7. AVOPh can promote flame extinguishment and inhibit heat release and smoke production, which improves the flame retardancy of EP/AVOPh composites. This can be attributed to the two-dimensional layered structure of AVOPh acting as a physical barrier effect during combustion and limiting energy transfer and gas escape [39–41]. Secondly, in the process of gas-phase flame retarding, AVOPh releases crystal water and NH_3, which dilute the concentration of combustible volatiles and oxygen around the combustion matrix. After the decomposition of the phosphorus compound, AVOPh releases HPO· and PO· free radicals to further capture and dilute high-energy free radicals, interfering with the combustion reaction and promoting flame self-extinguishing [42–44]. In addition, the transition metal vanadium oxidizes to form V_2O_5 and catalyzes the matrix to form char in the condensed phase. Meanwhile, oxalic acid structure decomposition participates in the char formation in this process. The quality of the char layer can be enhanced by the condensation of vanadium-based phosphate compounds to form H_3PO_4, $(VO)_2P_2O_7$, and $H_4P_2O_7$, leading to forming a dense and strong char layer to protect the internal unburned matrix and prevent further mass exchange [45,46].

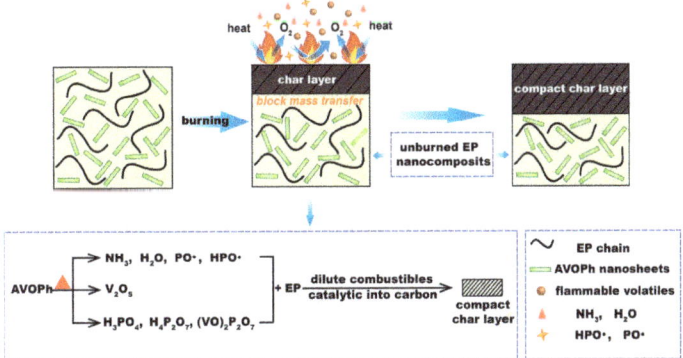

Figure 7. Schematic illustration of the mechanism for the enhanced flame resistance of EP/AVOPh composites.

3. Experimental Section

3.1. Materials

Phosphoric acid (H_3PO_4, 85% solution in H_2O), vanadium pentoxide (V_2O_5), ammonia solution (NH_4OH 30 wt% aqueous solution), and oxalic acid dihydrate ($C_2H_2O_4 \cdot 2H_2O$) were purchased from Sinopharm Chemical Reagent Co., Ltd. (Shanghai, China). Epoxy resin (NPEL 128) was purchased from NanYa Electronic Materials Co., Ltd. (Kunshan, China). 4, 4′-diaminodiphenyl methane (DDM) was purchased from Jiacheng Materials Co., Ltd (Dongguan, China).

3.2. Synthesis of AVOPh, AVPh, and VPh

Synthesis of AVOPh($(NH_4)_2[VO(HPO_4)]_2(C_2O_4) \cdot 5H_2O$): First, 6 mL deionized water and 0.6 mL H_3PO_4 were added to a 50 mL beaker. Then, 0.18 g V_2O_5 and 0.26 g $C_2H_2O_4 \cdot 2H_2O$ were added to the beaker. Then, 2.3 mL NH_4OH was dropped into the beaker and stirred continuously for 10 min. Finally, the above substances were transferred into a 50 mL Teflon reactor at 140 °C for three days. After the reaction, the green products were collected by vacuum extraction and filtrated, washed with deionized water three times, then dried in an oven at 80 °C. The product was labeled as AVOPh.

Synthesis of AVPh($(NH_4)VOPO_4 \cdot 1.5H_2O$): First, 0.90 g V_2O_5, 1.29 g $C_2H_2O_4 \cdot 2H_2O$, 2.1 mL H_3PO_4, 2.6 mL NH_4OH and 3 mL deionized water were added to a 100 mL beaker and stirred continuously for 10 min to make the various substances fully and evenly mixed. Then, the mixture was poured into a Teflon reactor with a capacity of 50 mL and placed in the oven at 140 °C for three days. After the reaction was completed, the green product was obtained. The product was dried and washed in the same way and labeled as AVPh.

Synthesis of VPh($VOPO_4 \cdot 2H_2O$): First, 115.4 mL deionized water was added into the 250 mL round-bottomed flask. Then 26.6 mL H_3PO_4 was slowly added into the round-bottomed flask and stirred at the speed of 500 rpm. Finally, 4.8 g V_2O_5 was added to the above solution. The round-bottomed flask was placed in an oil bath and refluxed at 110 °C for 16 h. The product was dried and washed in the same way and labeled as VPh.

3.3. Preparation of EP/AVOPh Composites

Firstly, AVOPh was uniformly dispersed in acetone through ultrasonicated dispersion. A total of 20 mL acetone is required per gram of AVOPh. Then, epoxy resin was added to the above dispersion liquid and the EP/AVOPh composites were prepared via ultrasonicated dispersion. The addition of AVOPh was controlled in 2, 4, 6, and 8 wt% in EP composites. Then the beaker was placed in the oil bath at 90 °C and stirred at 300 rpm for 4 h to remove the acetone. Curing agent DDM (EP:DDM = 4:1) was added and stirred continuously until DDM at the speed of 300 rpm. The mixture was put in a vacuum oven to remove bubbles at 80 °C for 5 min. Then the above-mentioned mixture was poured into preheated molds as quickly as possible. The samples were cured at 110 °C/2 h, 130 °C/2 h, and 150 °C/2 h, respectively. To further clarify the flame retardant effect of each component of AVOPh in epoxy resin, EP/4 wt% AVPh composites and EP/4 wt% VPh composites were prepared via the same method as comparative samples of EP/AVOPh. The formula of the EP/AVOPh composite is listed in Table 4.

Table 4. Ingredients of EP composites.

Samples	Components			
	EP (wt%)	AVOPh (wt%)	AVPh (wt%)	VPh (wt%)
EP	100	0	0	0
EP/2 wt% AVOPh	98	2	0	0
EP/4 wt% AVOPh	96	4	0	0

Table 4. *Cont.*

Samples	Components			
	EP (wt%)	AVOPh (wt%)	AVPh (wt%)	VPh (wt%)
EP/6 wt% AVOPh	94	6	0	0
EP/8 wt% AVOPh	92	8	0	0
EP/4 wt% AVPh	96	0	4	0
EP/4 wt% VPh	96	0	0	4

3.4. Characterization

X-ray diffraction (XRD) analysis was performed using Rigaku's MAX-RB diffractometer with a range of 5° to 80° at an operating voltage of 40 kV. Infrared spectroscopic characterization was conducted using 6700 spectrometers from Nicolet, USA, with the KBr tablet method in the range of 400 to 4000 cm^{-1}. The field emission scanning electron microscope (SEM) was characterized with an EVO MA15 scanning electron microscope from Zeiss Company, Aalen, Germany. The thermogravimetric analysis of materials was performed using a TG50 thermal analyzer (the same sample was tested two times, and the repeatability was good). The linear heating rate was 10 °C·min^{-1} under N_2. The limit oxygen index test (LOI) was performed using a JF-3 oxygen index meter from Jiangning Analytical Instrument Co., Ltd. (Nanjing, China) with a spline size of 130 × 6.5 × 3.2 mm^3 according to ASTM D 2863. The vertical combustion test (UL-94) was also carried out using a vertical combustion apparatus from Jiangning Analytical Instrument Co., Ltd. The spline size was 130 × 12.7 × 3.2 mm^3, and the test standard was carried out according to ASTM D 3801. The same sample was tested 15 times by LOI and vertical combustion. The cone calorimeter is an instrument from Stanton Redcroft, London, UK, with a continuous heat flux of 50 kW·m^{-2}, and the sample was wrapped in aluminum foil with a size of 100 × 100 × 3 mm^3. The same sample was tested three times with CCT.

4. Conclusions

In this work, layered AVOPh was synthesized using the hydrothermal method, which was used to prepare EP/AVOPh composites. The structure–activity relationship of phosphorus and nitrogen and the role of an oxalic acid structure in the flame retardancy of EP/AVOPh composites were investigated with TGA, LOI, vertical combustion, and CCT. Compared with pure EP, the residue of EP/8 wt% AVOPh composites increased from 15.3% to 23.0% at 700 °C in an N_2 atmosphere. The combustion results showed that the EP/4 wt% AVOPh composites passed a V1 rating, and the LOI value of EP/8 wt% AVOPh composites increased from 25.9% to 34.2% compared with pure EP. The PHRR, TSP, PCOP, and PCO$_2$P values of EP/8 wt% AVOPh composites decreased by 32.7%, 26.8%, 37.1%, and 33.3%, respectively. After the CCT, the residue of EP/4 wt% AVOPh composites was high at 30.3%, compared with 25.0% and 24.3% of EP/4 wt% AVPh and EP/4 wt% VPh composites, indicating that the oxalic acid structure was involved in the char formation during the combustion. The improved flame retardancy was mainly due to the synergistic effect of different components in AVOPh. In the condensed phase, the two-dimensional layered structure of AVOPh acted as a physical barrier effect during combustion and limited energy transfer and gas escape. Phosphoric acid and oxalic acid promoted the dehydration and carbonization of EP. The transition metal vanadium oxidized the catalyzed decomposition product of EP to form char. In the gas phase, water and NH_3 diluted combustible gas are released by AVOPh. The decomposition of AVOPh released HPO· and PO· free radicals to further capture and dilute high-energy free radicals, interfering with the combustion reaction and promoting flame self-extinguishing.

Author Contributions: Conceptualization, Q.K. and H.L.; methodology, P.H. and S.H.; investigation, W.L. and Z.Z.; major experimental work and writing—original draft preparation P.H.; writing—review and editing, Q.K.; resources, Q.K.; supervision, Z.Z.; revision, Q.K., W.Z. and M.C. All authors have read and agreed to the published version of the manuscript.

Funding: This work was supported by the Natural Science of Foundation of China (Grant No. 5220421), the Innovation and Entrepreneurship Training Plan for College Students in Jiangsu Province (202210299148Y), the Education Reform Research and Talent Training Project of Jiangsu University School of Emergency Management, and Jiangsu Collaborative Innovation Center of Technology and Material of Water Treatment, China.

Institutional Review Board Statement: This study was approved by the Regional Ethics Committee of our hospital and all patients signed informed consents.

Informed Consent Statement: Written informed consent has been obtained from the patient(s) to publish this paper.

Data Availability Statement: Not applicable.

Conflicts of Interest: The authors declare no conflict of interest.

Sample Availability: Samples of the compounds are not available from the authors.

References

1. Capricho, J.C.; Fox, B.; Hameed, N. Multifunctionality in Epoxy Resins. *Polym. Rev.* **2020**, *60*, 1–41. [CrossRef]
2. Zhang, H.; Li, K.; Wang, M.; Zhang, J. The preparation of a composite flame retardant of layered double hydroxides and α-zirconium phosphate and its modification for epoxy resin. *Mater. Today Commun.* **2021**, *28*, 102711. [CrossRef]
3. Yuan, Y.; Pan, Y.; Zhang, Z.; Zhang, W.; Li, X.; Yang, R. Nickle nanocrystals decorated on graphitic nanotubes with broad channels for fire hazard reduction of epoxy resin. *J. Hazard. Mater.* **2021**, *402*, 123880. [CrossRef] [PubMed]
4. Kong, Q.; Zhu, H.; Fan, J.; Zheng, G.; Zhang, C.; Wang, Y.; Zhang, J. Boosting flame retardancy of epoxy resin composites through incorporating ultrathin nickel phenylphosphate nanosheets. *J. Appl. Polym. Sci.* **2020**, *138*, 50265. [CrossRef]
5. Ding, J.; Zhang, Y.; Zhang, X.; Kong, Q.; Zhang, J.; Liu, H.; Zhang, F. Improving the flame-retardant efficiency of layered double hydroxide with disodium phenylphosphate for epoxy resin. *J. Therm. Anal. Calorim.* **2020**, *140*, 149–156. [CrossRef]
6. Kong, Q.; Sun, Y.; Zhang, C.; Guan, H.; Zhang, J.; Wang, D.; Zhang, F. Ultrathin iron phenyl phosphonate nanosheets with appropriate thermal stability for improving fire safety in epoxy. *Compos. Sci. Technol.* **2019**, *182*, 107748. [CrossRef]
7. Huo, S.; Song, P.; Yu, B.; Ran, S.; Chevali, V.; Liu, L.; Fang, Z.; Wang, H. Phosphorus-containing flame retardant epoxy thermosets: Recent advances and future perspectives. *Prog. Polym. Sci.* **2021**, *114*, 101366. [CrossRef]
8. Wang, X.; Chen, T.; Hong, J.; Luo, W.; Zeng, B.; Yuan, C.; Xu, Y.; Chen, G.; Dai, L. In-situ growth of metal-organophosphorus nanosheet/nanorod on graphene for enhancing flame retardancy and mechanical properties of epoxy resin. *Compos. Part B Eng.* **2020**, *200*, 108271. [CrossRef]
9. Qin, Z.; Yang, R.; Zhang, W.; Li, D.; Jiao, Q. Synergistic barrier effect of aluminum phosphate on flame retardant polypropylene based on ammonium polyphosphate/dipentaerythritol system. *Mater. Des.* **2019**, *181*, 107913. [CrossRef]
10. Hou, Y.; Hu, W.; Hu, Y. Preparation of layered organic-inorganic aluminum phosphonate for enhancing fire safety of polystyrene. *Mater. Chem. Phys.* **2017**, *196*, 109–117. [CrossRef]
11. Wang, C.; Wu, Y.; Li, Y.; Shao, Q.; Yan, X.; Han, C.; Wang, Z.; Liu, Z.; Guo, Z. Flame-retardant rigid polyurethane foam with a phosphorus-nitrogen single intumescent flame retardant. *Polym. Adv. Technol.* **2018**, *29*, 668–676. [CrossRef]
12. Wang, X.; Zhou, S.; Guo, W.-W.; Wang, P.-L.; Xing, W.; Song, L.; Hu, Y. Renewable Cardanol-Based Phosphate as a Flame Retardant Toughening Agent for Epoxy Resins. *ACS Sustain. Chem. Eng.* **2017**, *5*, 3409–3416. [CrossRef]
13. Zhan, Z.; Xu, M.; Li, B. Synergistic effects of sepiolite on the flame retardant properties and thermal degradation behaviors of polyamide 66/aluminum diethylphosphinate composites. *Polym. Degrad. Stab.* **2015**, *117*, 66–74. [CrossRef]
14. Gu, L.; Qiu, J.; Sakai, E. Thermal stability and fire behavior of aluminum diethylphosphinate-epoxy resin nanocomposites. *J. Mater. Sci. Mater. Electron.* **2017**, *28*, 18–27. [CrossRef]
15. Kong, Q.; Zhu, H.; Huang, S.; Wu, T.; Zhu, F.; Zhang, Y.; Wang, Y.; Zhang, J. Influence of multiply modified FeCu-montmorillonite on fire safety and mechanical performances of epoxy resin nanocomposites. *Thermochim. Acta* **2022**, *707*, 179112. [CrossRef]
16. Yang, Y.; Sun, P.; Sun, J.; Wen, P.; Zhang, S.; Kan, Y.; Liu, X.; Tang, G. Enhanced flame retardancy of rigid polyurethane foam via iron tailings and expandable graphite. *J. Mater. Sci.* **2022**, *57*, 18853–18873. [CrossRef]
17. Cui, J.; Zhang, Y.; Wang, L.; Liu, H.; Wang, N.; Yang, B.; Guo, J.; Tian, L. Phosphorus-containing Salen-Ni metal complexes enhancing the flame retardancy and smoke suppression of epoxy resin composites. *J. Appl. Polym. Sci.* **2020**, *137*, 48734. [CrossRef]
18. Wang, P.; Chen, L.; Xiao, H. Flame retardant effect and mechanism of a novel DOPO based tetrazole derivative on epoxy resin. *J. Anal. Appl. Pyrolysis.* **2019**, *139*, 104–113. [CrossRef]

19. Kong, Q.; Wu, T.; Zhang, J.; Wang, D.-Y. Simultaneously improving flame retardancy and dynamic mechanical properties of epoxy resin nanocomposites through layered copper phenylphosphate. *Compos. Sci. Technol.* **2018**, *154*, 136–144. [CrossRef]
20. Liu, J.; Qi, P.; Meng, D.; Li, L.; Sun, J.; Li, H.; Gu, X.; Jiang, S.; Zhang, S. Eco-friendly flame retardant and smoke suppression coating containing boron compounds and phytic acids for nylon/cotton blend fabrics. *Ind. Crops Prod.* **2022**, *186*, 115239. [CrossRef]
21. Yang, Y.; Yang, S.; Jiang, S.; Liu, M.; Liu, X.; Tang, G.; Cai, R.; Tan, W. Facile synthesis of zinc-containing mesoporous silicate from iron tailings and enhanced Fire retardancy of rigid polyurethane foam. *Macromol. Mater. Eng.* **2022**. [CrossRef]
22. Zhou, S.; Tao, R.; Dai, P.; Luo, Z.; He, M. Two-step fabrication of lignin-based flame retardant for enhancing the thermal and fire retardancy properties of epoxy resin composites. *Polym. Compos.* **2020**, *41*, 2025–2035. [CrossRef]
23. Zhu, H.; Chen, Y.; Huang, S.; Wang, Y.; Yang, R.; Chai, H.; Zhu, F.; Kong, Q.; Zhang, Y.; Zhang, J. Suppressing fire hazard of poly(vinyl alcohol) based on $(NH_4)_2[VO(HPO_4)]_2(C_2O_4)\cdot 5H_2O$ with layered structure. *J. Appl. Polym. Sci.* **2021**, *138*, 51345. [CrossRef]
24. Wang, J.; Tan, S.; Xiong, F.; Yu, R.; Wu, P.; Cui, L.; An, Q. $VOPO_4 \cdot 2H_2O$ as a new cathode material for rechargeable Ca-ion batteries. *Chem. Commun.* **2020**, *56*, 3805–3808. [CrossRef]
25. Do, J.; Bontchev, R.P.; Jacobson, A.J. A Hydrothermal Investigation of the $1/2V_2O_5-H_2C_2O_4/H_3PO_4/NH_4OH$ System: Synthesis and Structures of $(NH_4)VOPO_4 \cdot 1.5H_2O$, $(NH_4)_{0.5}VOPO_4 \cdot 1.5H_2O$, $(NH_4)_2[VO(H_2O)_3]_2[VO(H_2O)][VO(PO_4)_2]_2 \cdot 3H_2O$, and $(NH_4)_2[VO(HPO_4)]_2(C_2O_4) \cdot 5H_2O$. *Inorg. Chem.* **2000**, *39*, 3230–3237. [CrossRef] [PubMed]
26. Bircsk, Z.; Harrison, W.T.A. $NH_4VOPO_4 \cdot H_2O$, a New One-Dimensional Ammonium Vanadium(IV) Phosphate Hydrate. *Inorg. Chem.* **1998**, *37*, 5387–5389. [CrossRef]
27. Zheng, J.; Liu, X.; Dian, Y.; Chen, L.; Zhang, X.; Feng, X.; Chen, W.; Zhao, Y. Stable cross-linked gel terpolymer electrolyte containing methyl phosphonate for sodium ion batteries. *J. Membr. Sci.* **2019**, *583*, 163–170. [CrossRef]
28. Liu, Q.; Wang, D.; Li, Z.; Li, Z.; Peng, X.; Liu, C.; Zhang, Y.; Zheng, P. Recent Developments in the Flame-Retardant System of Epoxy Resin. *Materials* **2020**, *13*, 2145. [CrossRef]
29. Ou, M.; Lian, R.; Cui, J.; Guan, H.; Liu, L.; Jiao, C.; Chen, X. Co-curing preparation of flame retardant and smoke-suppressive epoxy resin with a novel phosphorus-containing ionic liquid. *Chemosphere* **2023**, *311*, 137061. [CrossRef]
30. Chai, H.; Li, W.; Wan, S.; Liu, Z.; Zhang, Y.; Zhang, Y.; Zhang, J.; Kong, Q. Amino Phenyl Copper Phosphate-Bridged Reactive Phosphaphenanthrene to Intensify Fire Safety of Epoxy Resins. *Molecules* **2023**, *28*, 623. [CrossRef]
31. Sag, J.; Goedderz, D.; Kukla, P.; Greiner, L.; Schoenberger, F.; Doering, M. Phosphorus-Containing Flame Retardants from Biobased Chemicals and Their Application in Polyesters and Epoxy Resins. *Molecules* **2019**, *24*, 3764. [CrossRef] [PubMed]
32. Kong, Q.; Li, L.; Zhang, M.; Chai, H.; Li, W.; Zhu, F.; Zhang, J. Improving the Thermal Stability and Flame Retardancy of Epoxy Resins by Lamellar Cobalt Potassium Pyrophosphate. *Polymers* **2022**, *14*, 4927. [CrossRef] [PubMed]
33. Qian, Z.; Zou, B.; Xiao, Y.; Qiu, S.; Xu, Z.; Yang, Y.; Jiang, G.; Zhang, Z.; Song, L.; Hu, Y. Targeted modification of black phosphorus by MIL-53(Al) inspired by "Cannikin's Law" to achieve high thermal stability of flame retardant polycarbonate at ultra-low additions. *Compos. Part B Eng.* **2022**, *238*, 109943. [CrossRef]
34. Li, L.; Hua, F.; Xi, H.; Yang, J.; Xiao, T.; Zuo, R.; Xu, Y.; Yang, Z.; Lei, Z. Synthesis of Phosphorous Phenanthrene/L-Tryptophan Flame Retardant for Enhanced Flame Retardancy of Epoxy Resins. *Macromol. Res.* **2022**, *30*, 937–946. [CrossRef]
35. He, T.; Guo, J.; Qi, C.; Feng, R.; Yang, B.; Zhou, Y.; Tian, L.; Cui, J. Core-shell structure flame retardant Salen-PZN-Cu@Ni-Mof microspheres enhancing fire safety of epoxy resin through the synergistic effect. *J. Polym. Res.* **2022**, *29*, 27. [CrossRef]
36. Feng, X.; Wang, B.; Wang, X.; Wen, P.; Cai, W.; Hu, Y.; Liew, K.M. Molybdenum disulfide nanosheets as barrier enhancing nanofillers in thermal decomposition of polypropylene composites. *Chem. Eng. J.* **2016**, *295*, 278–287. [CrossRef]
37. Sang, L.; Cheng, Y.; Yang, R.; Li, J.; Kong, Q.; Zhang, J. Polyphosphazene-wrapped Fe-MOF for improving flame retardancy and smoke suppression of epoxy resins. *J. Therm. Anal. Calorim.* **2021**, *144*, 51–59. [CrossRef]
38. Yu, C.; Wu, T.; Yang, F.; Wang, H.; Rao, W.; Zhao, H.-B. Interfacial engineering to construct P-loaded hollow nanohybrids for flame-retardant and high-performance epoxy resins. *J. Colloid Interface Sci.* **2022**, *628*, 851–863. [CrossRef] [PubMed]
39. Huang, C.; Fang, G.; Tao, Y.; Meng, X.; Lin, Y.; Bhagia, S.; Wu, X.; Yong, Q.; Ragauskas, A.J. Nacre-inspired hemicelluloses paper with fire retardant and gas barrier properties by self-assembly with bentonite nanosheets. *Carbohydr. Polym.* **2019**, *225*, 115219. [CrossRef]
40. Zhao, H.; Ding, J.; Yu, H. Phosphorylated Boron Nitride Nanosheets as Highly Effective Barrier Property Enhancers. *Ind. Eng. Chem. Res.* **2018**, *57*, 14096–14105. [CrossRef]
41. Shi, Y.; Yu, B.; Duan, L.; Gui, Z.; Wang, B.; Hu, Y.; Yuen, R.K.K. Graphitic carbon nitride/phosphorus-rich aluminum phosphinates hybrids as smoke suppressants and flame retardants for polystyrene. *J. Hazard. Mater.* **2017**, *332*, 87–96. [CrossRef]
42. Suparanon, T.; Phetwarotai, W. Fire-extinguishing characteristics and flame retardant mechanism of polylactide foams: Influence of tricresyl phosphate combined with natural flame retardant. *Int. J. Biol. Macromol.* **2020**, *158*, 1090–1101. [CrossRef] [PubMed]
43. Nguyen, C.; Lee, M.; Kim, J. Relationship between structures of phosphorus compounds and flame retardancies of the mixtures with acrylonitrile–butadiene–styrene and ethylene–vinyl acetate copolymer. *Polym. Adv. Technol.* **2011**, *22*, 512–519. [CrossRef]
44. Zhang, J.; Kong, Q.; Yang, L.; Wang, D. Few layered $Co(OH)_2$ ultrathin nanosheet-based polyurethane nanocomposites with reduced fire hazard: From eco-friendly flame retardance to sustainable recycling. *Green. Chem.* **2016**, *18*, 3066–3074. [CrossRef]

45. Gong, K.; Cai, L.; Shi, C.; Gao, F.; Yin, L.; Qian, X.; Zhou, K. Organic-inorganic hybrid engineering MXene derivatives for fire resistant epoxy resins with superior smoke suppression. *Compos. Part A Appl. Sci. Manuf.* **2022**, *161*, 107109. [CrossRef]
46. Wang, H.; Li, X.; Su, F.; Xie, J.; Xin, Y.; Zhang, W.; Liu, C.; Yao, D.; Zheng, Y. Core-Shell ZIF67@ZIF8 Modified with Phytic Acid as an Effective Flame Retardant for Improving the Fire Safety of Epoxy Resins. *ACS Omega* **2022**, *7*, 21664–21674. [CrossRef] [PubMed]

Disclaimer/Publisher's Note: The statements, opinions and data contained in all publications are solely those of the individual author(s) and contributor(s) and not of MDPI and/or the editor(s). MDPI and/or the editor(s) disclaim responsibility for any injury to people or property resulting from any ideas, methods, instructions or products referred to in the content.

Article

Amino Phenyl Copper Phosphate-Bridged Reactive Phosphaphenanthrene to Intensify Fire Safety of Epoxy Resins

Huiyu Chai [1], Weixi Li [1], Shengbing Wan [2], Zheng Liu [2], Yafen Zhang [2], Yunlong Zhang [1], Junhao Zhang [3] and Qinghong Kong [1,*]

1 School of Emergency Management, Jiangsu University, Zhenjiang 212013, China
2 Zhejiang Jiamin New Materials Co., Ltd., Jiaxing 314027, China
3 School of Environmental and Chemical Engineering, Jiangsu University of Science and Technology, Zhenjiang 212003, China
* Correspondence: kongqh@ujs.edu.cn

Abstract: To improve the compatibility between flame retardant and epoxy resin (EP) matrix, amino phenyl copper phosphate-9, 10-dihydro-9-oxygen-10-phospha-phenanthrene-10-oxide (CuPPA-DOPO) is synthesized through surface grafting, which is blended with EP matrix to prepare EP/CuPPA-DOPO composites. The amorphous structure of CuPPA-DOPO is characterized by X-ray diffraction and Fourier-transform infrared spectroscopy. Scanning electron microscope (SEM) images indicate that the agglomeration of hybrids is improved, resisting the intense intermolecular attractions on account of the acting force between CuPPA and DOPO. The results of thermal analysis show that CuPPA-DOPO can promote the premature decomposition of EP and increase the residual amount of EP composites. It is worth mentioning that EP/6 wt% CuPPA-DOPO composites reach UL-94 V-1 level and limiting oxygen index (LOI) of 32.6%. Meanwhile, their peak heat release rate (PHRR), peak smoke production release (PSPR) and CO_2 production (CO_2P) are decreased by 52.5%, 26.1% and 41.4%, respectively, compared with those of EP. The inhibition effect of CuPPA-DOPO on the combustion of EP may be due to the release of phosphorus and ammonia free radicals, as well as the catalytic charring ability of metal oxides and phosphorus phases.

Keywords: amino phenyl metal phosphate; organic–inorganic hybrid; reactive flame retardant; epoxy resin

Citation: Chai, H.; Li, W.; Wan, S.; Liu, Z.; Zhang, Y.; Zhang, Y.; Zhang, J.; Kong, Q. Amino Phenyl Copper Phosphate-Bridged Reactive Phosphaphenanthrene to Intensify Fire Safety of Epoxy Resins. *Molecules* **2023**, *28*, 623. https://doi.org/10.3390/molecules28020623

Academic Editors: Xin Wang, Weiyi Xing and Gang Tang

Received: 9 December 2022
Revised: 3 January 2023
Accepted: 4 January 2023
Published: 7 January 2023

Copyright: © 2023 by the authors. Licensee MDPI, Basel, Switzerland. This article is an open access article distributed under the terms and conditions of the Creative Commons Attribution (CC BY) license (https:// creativecommons.org/licenses/by/ 4.0/).

1. Introduction

Epoxy resin (EP) has been widely used in many fields because of its excellent performance [1,2]. Nevertheless, its highly flammable and asphyxiating fumes cannot meet the needs of sophisticated fields [3,4]. Therefore, improving the fire resistance of epoxy resin is an important goal in the field of intrinsic safety [5,6].

In recent years, two-dimensional layered metallic phosphates have attracted extensive attention owing to their regular layered structures and excellent flame-retardant properties [7,8]. Phenyl metal phosphonates, as a class of layered metallic phosphates, have been a research focus in the field of flame-retardant materials because of their stable chemical properties, suitable thermal stability and good compatibility with matrix [9,10]. For example, UL-94 of V-0 grade was observed with 1 wt% layered lanthanum phenyl phosphate (LaHPP) and 7 wt% decabromodiphenyl oxide (DBDPO) in a polycarbonate formulation, which was attributed to the lamellar structure and condensation phase reinforcement of lanthanum phenyl phosphate [11]. It was reported that the flue gas and toxic gas release of EP composites with the addition of copper phenyl phosphate (CuPP) was reduced [12]. However, phenyl metal phosphonates are difficult to uniformly disperse in polymer matrix due to their unique laminated structure, which has a great influence on the flame retardancy of polymer composites [13,14]. Therefore, it is necessary to improve the dispersion of phenyl metal phosphonate in polymer matrices.

Reactive flame retardants can effectively solve the problem of poor dispersion in polymer composites [15]. DOPO, as a reactive flame retardant, has good oxidation resistance and easily modifiable property [16]. The inherent P-H bond in the structure of DOPO has high reactivity, and so it can be reacted with unsaturated bonds and epoxy groups to form various derivatives, which can be cross-linked with EP matrix [17]. Furthermore, nitrogenous DOPO compounds synthesized by the addition reaction of DOPO and imine compounds are the most common solidified DOPO flame retardants [18]. The advantages of DOPO-imine flame retardants lie in their simple synthesis conditions, adjustable reaction process, high product yield and molecular structure containing both phosphorus and nitrogen flame-retardant elements [19,20]. The application of DOPO combined with inorganic nanomaterials through the amino group in flame-retardant EP has been reported. The PHRR and total heat release (THR) of EP/7 wt% zirconium hybrid polyhedron oligomeric polysiloxane derivative composites decreased by 37.1% and 16.7%, respectively, compared with those of EP [21]. A piperazinyl DOPO derivative was developed with the aim of functionalizing graphene oxide to form a hybrid, which reduced the mechanical cracking of epoxy resins and improved their biphasic compatibility [22]. Therefore, reactive DOPO-imine derivatives as flame retardants for EP are worthy of further investigation.

Given this context, it can be said that reactive DOPO-imine derivatives can effectively alleviate the compatibility disadvantage of phenyl metal phosphonate, and can further improve the flame-retardance efficiency. Hence, in this work, an amino phenyl copper phosphate (CuPPA) was successfully synthesized. Then, CuPPA was further grafted with DOPO to form complex compounds. Inorganic nanoparticles were combined with an organic high-efficiency phosphorus flame retardant through this molecular structural design. Meanwhile, reactive flame-retardant technology was introduced to promote the formation of uniform and stable EP composites due to the presence of imines. The results indicated that the uniform EP composites exhibited remarkable flame retardancy when CuPPA-DOPO was introduced into the epoxy resin combustion system.

2. Results and Discussion

2.1. Structural Analysis of CuPPA-DOPO

Figure 1a shows the XRD patterns of CuPPA and CuPPA-DOPO. The diffuse scattering peak of CuPPA occurs at about 26°, indicating that its degree of crystallinity is not high and revealing its amorphous structure [23]. The diffuse scattering peak decreased slightly and the crystallinity did not change significantly after grafting DOPO onto CuPPA. FTIR spectra (Figure 1b) reveal the structures of CuPPA and CuPPA-DOPO. The absorption peaks at 1148 cm^{-1} and 1054 cm^{-1} are the stretching vibrations of P=O and P-O bonds, and the peaks at 615 cm^{-1} and 533 cm^{-1} are the characteristic peaks of C-P bonds [12]. The absorption peak at 1092 cm^{-1} corresponds to the stretching vibrations of Cu-O-P bonds [24]. The absorption peaks at 1501 cm^{-1} and 834 cm^{-1} correspond to the stretching vibrations of C=C and C-N-C bonds. The absorption peaks at 3330 cm^{-1} and 3445 cm^{-1} belong to the stretching vibrations of -NH$_2$ bonds, which are consistent with the molecular formula of CuPPA [25,26]. The vibrational band occurring at 758 cm^{-1} is attributable to the absorption of P-O-Ph, and the vibrational band of P-H in DOPO disappears. The double peaks of -NH$_2$ in CuPPA shift to a single peak (3335 cm^{-1}) of -NH in CuPPA-DOPO [27]. The above analyses prove that DOPO was successfully introduced into the molecular chain of CuPPA. The microstructures of CuPPA and CuPPA-DOPO can be directly observed in the SEM images shown in Figure 1c,d. A rod-like nanostructure with tiny nanoparticles was clearly visible after grafting DOPO onto CuPPA, which was different from the stacked lamellar structure of CuPPA. The nanorods had a diameter of about 500 nm and a length of a few microns.

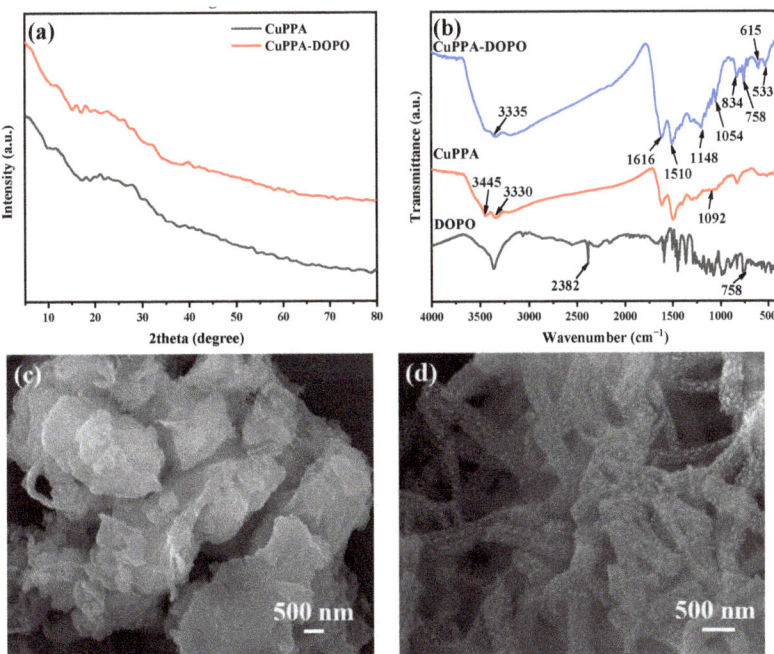

Figure 1. (a) XRD patterns of CuPPA and CuPPA-DOPO; (b) FTIR spectra of CuPPA and CuPPA-DOPO; (c) SEM image of CuPPA; (d) SEM image of CuPPA-DOPO.

The thermal decomposition behavior of CuPPA, DOPO and CuPPA-DOPO was obtained by TGA, as plotted in Figure 2. The decomposition of CuPPA-DOPO has three stages of weightlessness. The first weight loss (5.8 wt%) occurred at 30–105 °C, and was mainly the desorption of adsorbed water and a small amount of organic solvents [28]. The next weight loss was 14.2 wt% after 105–306 °C, accompanied by the decomposition of phosphaphenanthrene into small molecules [29]. The most intense weight loss was 52.0 wt% after 306 °C, attributable to the decomposition of metal compounds and the oxidation of organic molecules such as phosphonates, the phase of which coincided with the decomposition temperature of EP [30].

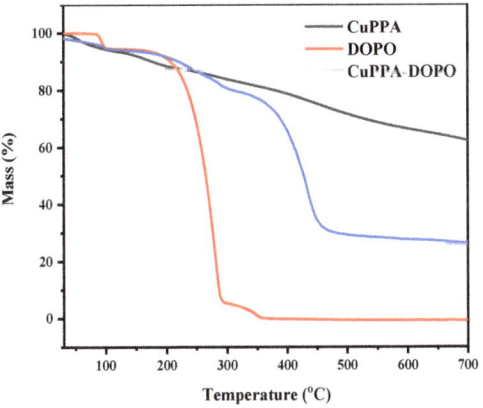

Figure 2. TGA curves of CuPPA, DOPO and CuPPA-DOPO.

2.2. Thermal Properties of EP Composites

The thermal stability of EP nanocomposites is an important evaluation index of fire safety. TGA and DTG curves and relevant data of EP nanocomposites are shown in Figure 3a,b and Table 1. The initial decomposition temperature ($T_{5\%}$) of EP was 360 °C, while EP/CuPPA-DOPO composites displayed lower $T_{5\%}$. Among EP composites, $T_{5\%}$ and T_{max} (the temperature at maximum pyrolysis rate) of EP/6 wt% CuPPA-DOPO composites were decreased to 333 °C and 368 °C, respectively, demonstrating that CuPPA-DOPO promoted the earlier decomposition of EP composites [31]. However, the promoting effect of CuPPA was not obvious, with $T_{5\%}$ and T_{max} of 350 °C and 376 °C. The decrease of the maximum decomposition rate of EP/6 wt% CuPPA-DOPO composites also indicates that the presence of CuPPA-DOPO delayed the degradation of the EP at a higher temperature [32]. Moreover, the incorporation of CuPPA-DOPO improved the char-forming ability of the EP composites. The residues of EP composites were increased to 17.4%, 23.0%, 24.9% and 26.1%, respectively, when the amounts of CuPPA-DOPO were 2, 4, 6 and 8 wt%. The residue of EP/4 wt% CuPPA composites at 700 °C was 19.8%, which is higher than that (14.9%) of pure EP, but lower than that of EP/4 wt% CuPPA-DOPO. The outstanding char-forming ability of EP is attributed to the polyphosphoric substances and metallic oxide produced in the early decomposition of CuPPA-DOPO [33,34].

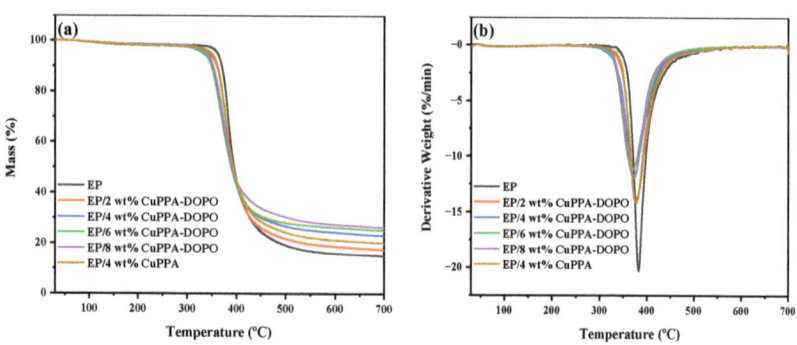

Figure 3. (a) TGA and (b) DTG curves of pure EP and EP composites.

Table 1. TGA data of pure EP and EP composites.

Samples	$T_{5\%}$ (°C)	T_{max} (°C)	Residues (wt%, 700 °C)
EP	360	382	14.9
EP/2 wt% CuPPA-DOPO	343	376	17.4
EP/4 wt% CuPPA-DOPO	333	373	23.0
EP/6 wt% CuPPA-DOPO	333	368	24.9
EP/8 wt% CuPPA-DOPO	342	372	26.1
EP/4 wt% CuPPA	350	376	19.8

2.3. Flame Retardancy of EP Composites

The results of UL-94 vertical burning and LOI test of EP/CuPPA-DOPO composites are shown in Table 2. EP composites all reached a UL-94 V-1 rating, and the LOI values increased to 28.8%, 30.8%, 32.6% and 31.4% when adding 2, 4, 6 and 8 wt% CuPPA-DOPO, respectively, while EP was evaluated as no grade and the LOI value increased only to 25.9%. EP/6 wt% CuPPA-DOPO composites had the best flame-retardant effect. The flame retardancy of EP/8 wt% CuPPA-DOPO composites was slightly poor, which may have been due to the poor dispersibility caused by the excessive dosage, degrading the fluidity of EP/CuPPA-DOPO composites and greatly promoting the combustion of EP composites [35,36]. However, UL-94 reached grade V-1 with 45.2 s, and the LOI value was only 28.2% when 4 wt% CuPPA was added, indicating that the flame retardancy of CuPPA was not as good as that of CuPPA-DOPO. In addition, some of the following features can be seen from the LOI char image in Figure 4. The residue of EP/4 wt% CuPPA-DOPO composites was dense and not easily broken, and more than that of EP/4 wt% CuPPA composites, while EP had less residue. In the system of EP/CuPPA-DOPO composites, the phosphoric acid and metal oxides generated by decomposition could also promote the formation of char layer, insulate the heat transfer, limit oxygen exchange, and thus inhibit the further combustion of EP [37,38].

Table 2. LOI data and UL-94 results of EP composites.

Samples	LOI (vol%)	UL-94			
		t_1 (s)	t_2 (s)	$t_1 + t_2$ (s)	Rating
EP	25.9 ± 0.2	-	-	>50	NR
EP/2 wt% CuPPA-DOPO	28.8 ± 0.2	37.0	6.3	43.3	V-1
EP/4 wt% CuPPA-DOPO	30.8 ± 0.3	20.5	7.0	27.5	V-1
EP/6 wt% CuPPA-DOPO	32.6 ± 0.3	13.6	5.0	18.6	V-1
EP/8 wt% CuPPA-DOPO	31.4 ± 0.3	14.5	9.0	23.5	V-1
EP/4 wt% CuPPA	28.2 ± 0.2	36.0	9.3	45.3	V-1

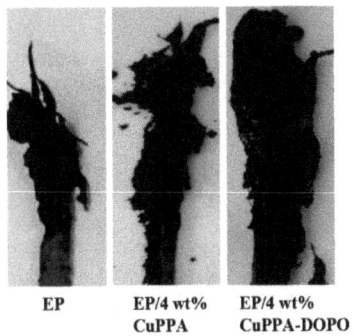

Figure 4. Photographs of pure EP and EP composites after LOI test.

In order to more intuitively and accurately reflect the combustion behavior of EP composites, CCT was conducted to obtain the heat release rate (HRR), smoke production rate (SPR), total smoke release (TSR) and CO_2 production (CO_2P), as displayed in Figure 5a–d and Table 3. The PHRRs of EP composites declined to 739, 617, 390 and 597 kW/m^2 with the addition of 2, 4, 6 and 8 wt% CuPPA-DOPO, respectively, representing decreases of 10.1%, 24.9%, 52.5% and 27.4% compared with that (822 kW/m^2) of pure EP. The PHRR of EP/4 wt% CuPPA composites was 784 kW/m^2, 4.6% lower than that of pure EP, which indicates that CuPPA-DOPO had a better inhibitory effect on the combustion of EP than CuPPA. The decrease of PHRR can be interpreted as being a result of H_2O, CO_2 and NH_2 released by the decomposition of CuPPA-DOPO diluting the concentration of combustible volatile matter in the gas phase and absorbing part of the combustion heat [39,40].

Notably, the PSPR of pure EP was 0.23 m^2/s, and that of EP/4 wt% CuPPA composites was 0.21 m^2/s. The PSPR of EP composites reduced to 0.21, 0.20, 0.17 and 0.20 m^2/s when adding 2, 4, 6 and 8 wt% CuPPA-DOPO, respectively, representing decreases of 8.7%, 13.0%, 26.1% and 13.0% compared with pure EP. Meanwhile, the TSR of EP composites decreased by 28.3%, 20.3%, 13.4% and 14.5% with the addition of 2, 4, 6 and 8 wt% CuPPA-DOPO, respectively, compared with that (2438 m^2/m^2) of pure EP. The TSR of EP/4 wt% CuPPA composites was 1585 m^2/m^2, 35% lower than pure EP and 18.4% lower than EP/4 wt% CuPPA-DOPO, which is because small phospho-oxygen molecules released after CuPPA-DOPO combustion became parts of the flue gas and entered the flame, thus increasing the TSR value and degenerating the smoke suppression performance [41–43].

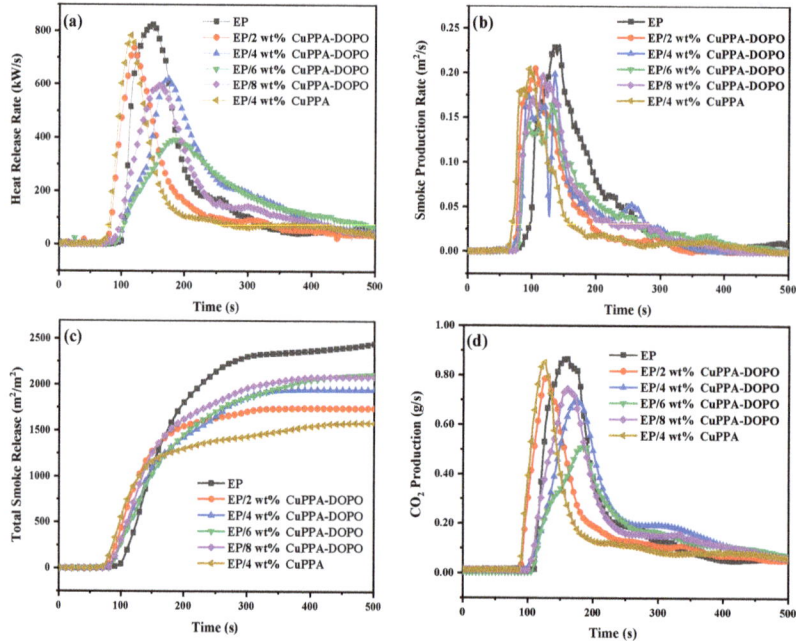

Figure 5. (a) HRR, (b) SPR, (c) TSR and (d) CO_2P curves of pure EP and EP composites.

Table 3. CCT data of pure EP and EP composites.

Samples	PHRR (kW/m^2)	PSPR (m^2/s)	TSR (m^2/m^2)	CO_2P (g/s)
EP	822	0.23	2438	0.87
EP/2 wt% CuPPA-DOPO	739	0.21	1746	0.79
EP/4 wt% CuPPA-DOPO	617	0.20	1942	0.70
EP/6 wt% CuPPA-DOPO	390	0.17	2112	0.51
EP/8 wt% CuPPA-DOPO	597	0.20	2085	0.75
EP/4 wt% CuPPA	784	0.21	1585	0.85

High levels of carbon dioxide in fires are a major cause of asphyxia and poisoning. The CO_2 production (CO_2P) curves of EP composites show that upon adding 2, 4, 6 and 8 wt% CuPPA-DOPO, the CO_2P of EP composites decreases to 0.79, 0.70, 0.51 and 0.75 g/s, respectively, representing decreases of 9.2%, 19.5%, 41.4% and 13.8% compared with that (0.87 g/s) of pure EP. However, the CO_2P of EP/4 wt% CuPPA composites was 0.85 g/s, slightly lower than that of pure EP and higher than that of EP/4 wt% CuPPA-DOPO. The main reason for the reduction of CO_2P is that the combustion of CuPPA-DOPO can produce CuO and phosphorous flame retardant, catalyzing the formation of a dense carbon layer to protect the surface of the composite materials. The PO• produced by CuPPA-DOPO combustion exerts flame retardancy in the gas phase and inhibits the complete combustion of the gas-phase products [44,45].

2.4. Flame-Retardance Mechanism

The microstructures of the outer (Figure 6a–c) and inner (Figure 6d–f) char layers of EP composites were characterized by SEM to obtain more detailed information on the flame-retardance mechanism. The outer surface of the char layer of pure EP was obviously lumpy, with large and dense cracks, which could not play a barrier role in combustion. In addition, there were many large pores and a large number of cracks on the inner surface of the char layer of pure EP, which were caused by intense burning, with a large amount of heat and smoke escaping. More dense and hard expanded char appeared on the outer surface of EP/4 wt% CuPPA-DOPO composites' char layer. Its inner surface char layer was thicker and rougher, with bonded particles compared with EP/4 wt% CuPPA composites. On the one hand, phosphorous groups contributed to the formation of char. On the other hand, the cured cross-linked network structure enhanced the strength of the residual coke, which could play the role of insulation barrier, reducing the transfer of heat, mass and oxygen between internal and external of EP composites [46,47].

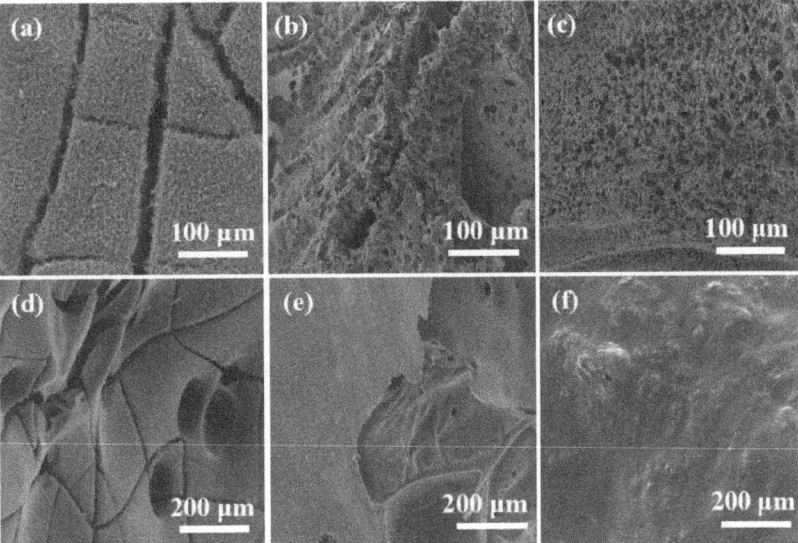

Figure 6. SEM images of external char layer: (**a**) EP; (**b**) EP/4 wt% CuPPA; (**c**) EP/4 wt% CuPPA-DOPO. SEM images of internal char layer: (**d**) EP; (**e**) EP/4 wt% CuPPA; (**f**) EP/4 wt% CuPPA-DOPO.

The flame-retardance mechanism of EP/CuPPA-DOPO composites was inferred by analyzing the above SEM results of the EP composites, as shown in Figure 7. In the condensed phase, the exposed copper-metal centers can catalyze organic reactions such as dehydrogenation and esterification [48,49]. In addition, phosphoric acid derivatives also promote the formation of skeleton-stable polycyclic aromatic hydrocarbons, thereby reducing heat and mass transfer and further inhibiting combustion [50–52]. In the gas phase, the HPO• and PO• radicals generated during the decomposition of CuPPA-DOPO interrupt the chain reaction [53]. Meanwhile, the non-flammable gases dilute the combustible gases on the surface of EP composites, which is conducive to retarding flame spread [54,55].

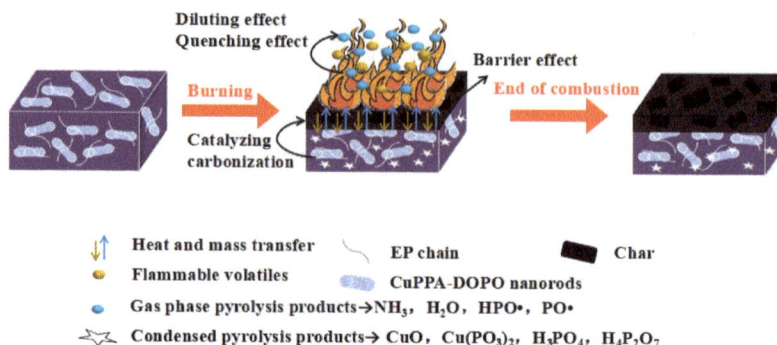

Figure 7. Flame-retardance mechanism diagram of EP/CuPPA-DOPO composites.

3. Materials and Methods

3.1. Materials

Para-phenylene diamine ($C_6H_8N_2$), potassium iodide (KI) and 2-chloroethyl phosphate ($C_2H_6ClO_3P$) were supplied by Macklin Co., Ltd. (Shanghai, China). Additionally, 9,10-dihydro-9-oxygen-10-phospha-phenanthrene-10-oxide (DOPO) (phosphorus content, ≥14% by weight), copper chloride dihydrate ($CuCl_2·2H_2O$), paraformaldehyde (HCHO), tetrahydrofuran (C_4H_8O), sodium hydroxide (NaOH) and diaminodiphenylmethane (DDM) (viscosity at 25 °C, 2.5–4.0 Pa·s, amine value, 480 mg KOH g^{-1}) were acquired from Sinopharm Chemical Reagent Co., Ltd. (Shanghai, China). EP (NFEL128) (viscosity, 12–15 Pa·s, epoxy equivalent, 184–190 g/mol) was bought from Nanya Electronic Materials (Kunshan) Co., Ltd. (Kunshan, China).

3.2. Preparation of Amino Phenyl Copper Phosphate

The schematic diagram of CuPPA synthesis is shown in Figure 8a. First, 400 mL deionized water, 7.3 g 2-chloroethyl phosphoric acid, 5.4 g p-phenylenediamine and 0.8 g KI were added into a 500 mL beaker and stirred until the mixture was completely dissolved. Next, 40 mL cold NaOH (1 mol/L) solution was dropped into a beaker and stirred for 30 min. After reacting at 18 °C for 48 h, 6.8 g $CuCl_2·2H_2O$ was added into the mixture and the mixed system was continuously stirred for 4 h at 70 °C under magnetic agitation. Finally, the flask was aged for 24 h. The precipitates were filtered and washed with deionized water, then followed by vacuum drying at 80 °C for 24 h.

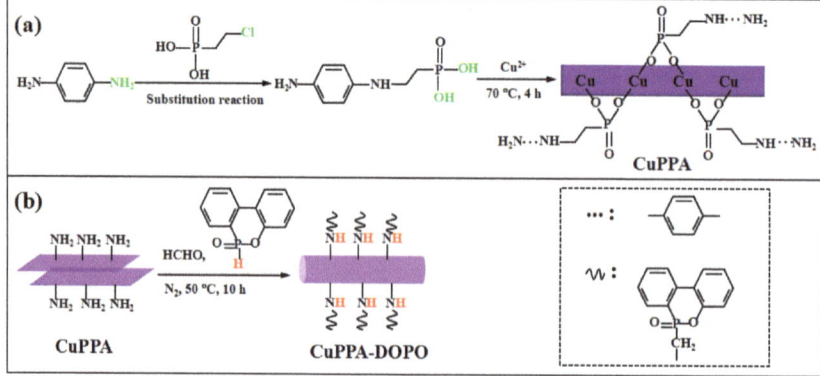

Figure 8. Synthetic route of (**a**) amino phenyl copper phosphate; (**b**) DOPO-amino phenyl copper phosphate.

3.3. Preparation of DOPO-Amino Phenyl Copper Phosphate

The reaction process of CuPPA-DOPO is illustrated in Figure 8b. First, 100 mL tetrahydrofuran and 5 g CuPPA were added into a 250 mL three-necked flask and stirred for 0.5 h. Next, 2.5 g DOPO and 0.5 g paraformaldehyde were added into the mixture and stirred vigorously at 50 °C for 10 h, keeping the condensation reflux under nitrogen atmosphere. The products were washed three times and dried under vacuum at 80 °C for 24 h.

3.4. Preparation of EP/CuPPA-DOPO Composites

Firstly, CuPPA-DOPO was dispersed in acetone and sonicated for 30 min. The preheated EP matrix was added in the mixture for 30 min of ultrasonic. Next, the mixture was stirred at 90 °C for 4 h until the acetone was completely removed. Then, DDM (mass ratio of EP matrix to DDM was 4:1) was added to the dispersion and stirred until DDM was completely dissolved. Then, the mixture was quickly poured into the preheated silicone rubber mold and cured at 100 °C for 2 h. Finally, the splines were transferred to the drying oven and cured at 110 °C/2 h, 130 °C/2 h, 150 °C/2 h. The specific contents and curing reaction of EP composites are shown in Table 4 and Figure 9.

Table 4. Ingredients of EP composites.

Samples	Components			
	EP (wt%)	DDM (wt%)	CuPPA-DOPO (wt%)	CuPPA (wt%)
EP	80.0	20.0	0	0
EP/2 wt% CuPPA-DOPO	78.4	19.1	2	0
EP/4 wt% CuPPA-DOPO	76.8	17.1	4	0
EP/6 wt% CuPPA-DOPO	75.2	15.7	6	0
EP/8 wt% CuPPA-DOPO	73.6	14.2	8	0
EP/4 wt% CuPPA	76.8	19.2	0	4

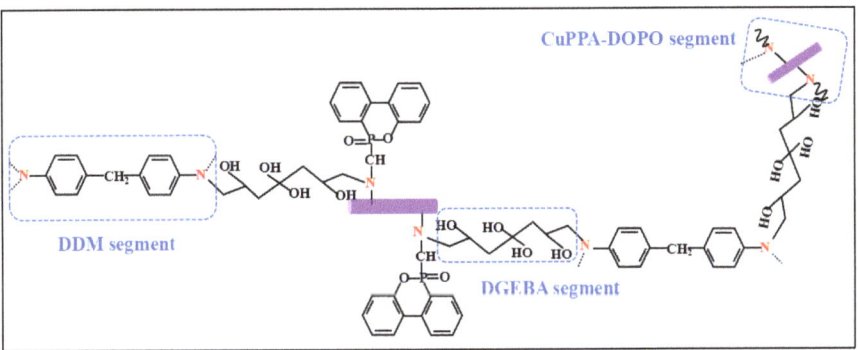

Figure 9. Schematic diagram of the curing reaction of EP composites.

3.5. Measurements

X ray diffraction (XRD) patterns of CuPPA-DOPO and CuPPA were measured with a MAX-Rb instrument produced by Rigaku Co., Ltd. (Matsumoto, Japan), which had a scanning range of 5° to 80°, a scanning speed of 5°/min and a step length of 0.02°. Fourier-transform infrared (FTIR) spectra were obtained with a 6700 spectrometer produced by Nicolet Instruments Co., Ltd. (Madison, WI, USA). The scanning electron microscopy (SEM) instrument was a JSM-7001F microscope produced by JEOL (Tokyo, Japan), and gold spraying was carried out before testing. Vertical burning (UL-94) tests were performed with a CZF-1 vertical combustion instrument produced by Jiang Ning Analytical Instrument Co., Ltd. (Nanjing, China). The spline size was 130 × 13 × 3.2 mm^3, and the test standard was in accordance with ASTM D 3801. Limiting oxygen index (LOI) was tested by a JF-3 oxygen index instrument produced by Jiang Ning Analytical Instrument Co., Ltd. (Nanjing, China). The spline size was $130 \times 6.5 \times 3.2$ mm^3 and the test standard was in accordance with ASTM D 2863. Thermogravimetric analysis (TGA) was conducted using a thermogravimetric analyzer produced by Mettler Toledo, with a sample quantity of about 10 mg at a ramp rate of 10 °C/min in nitrogen environment. Burning data of samples were obtained using the cone calorimeter test (CCT) with sheet dimensions of $100 \times 100 \times 3.0$ mm^3 according to ISO 5660-1. The spline was wrapped in aluminum foil and placed at a horizontal heat flux of 35 kW/m^2.

4. Conclusions

In summary, DOPO-amino phenyl copper phosphate was successfully synthesized, which was incorporated into EP matrix for preparing uniformly dispersed EP/CuPPA-DOPO composites. FTIR and SEM characterizations showed that CuPPA-DOPO was an amorphous rod-like nanoparticle. TGA results showed that the residue of EP/8 wt% CuPPA-DOPO composites was increased to 26.1%, about 11.2% higher than that of pure EP. Moreover, the obtained EP nanocomposites exhibited noteworthy flame-retardant properties with low additives. EP composites reached a UL-94 V-1 rating with a dosage of CuPPA-DOPO as low as 2 wt%. The PHRR, PSPR, TSR and CO_2P of EP/CuPPA-DOPO were all decreased compared with those of EP. The inhibition of CuPPA-DOPO on EP combustion was attributed to a char barrier in the condensed phase, quenching by releasing some phosphorus molecular debris in the gas phase as well as the synergy of phosphorous and nitrogen to dilute combustible gases. Therefore, the multi-synergistic flame-retardance mechanism of CuPPA-DOPO is of great significance in the preparation of organic and inorganic hybrid flame retardants.

Author Contributions: Conceptualization, Q.K., S.W. and J.Z.; methodology, H.C. and Z.L.; investigation, W.L. and Y.Z. (Yafen Zhang); major experimental work and writing—original draft preparation, H.C.; writing—review and editing, Q.K., J.Z. and Y.Z. (Yunlong Zhang); resources, Q.K. and J.Z.; supervision, Y.Z. (Yulong Zhang). All authors have read and agreed to the published version of the manuscript.

Funding: This work was supported by Natural Science of Foundation of China (Grant No.5220421), Innovation and Entrepreneurship Training Plan for College Students in Jiangsu Province (202210299148Y) and the Education Reform Research and Talent Training Project of Jiangsu University School of Emergency Management (KY-B-06).

Institutional Review Board Statement: Not applicable.

Informed Consent Statement: Not applicable.

Data Availability Statement: All data has been provided in this article.

Conflicts of Interest: The authors declare no conflict of interest.

Sample Availability: Samples of the compounds are not available from the authors.

References

1. De Farias, M.A.; Coelho, L.A.F.; Pezzin, S.H. Hybrid Nanocomposites Based on Epoxy/silsesquioxanes Matrices Reinforced with Multi-walled Carbon Nanotubes. *Mater. Res.* **2015**, *18*, 1304–1312. [CrossRef]
2. Lin, C.H.; Chen, J.C.; Huang, C.M.; Jehng, J.M.; Chang, H.C.; Juang, T.Y.; Su, W.C. Side-chain phenol-functionalized poly (ether sulfone) and its contribution to high-performance and flexible epoxy thermosets. *Polymer* **2013**, *54*, 6936–6941. [CrossRef]
3. Lv, Q.; Huang, J.-Q.; Chen, M.-J.; Zhao, J.; Tan, Y.; Chen, L.; Wang, Y.-Z. An Effective Flame Retardant and Smoke Suppression Oligomer for Epoxy Resin. *Ind. Eng. Chem. Res.* **2013**, *52*, 9397–9404. [CrossRef]
4. Zhang, J.; Kong, Q.; Wang, D.-Y. Simultaneously improving the fire safety and mechanical properties of epoxy resin with Fe-CNTs via large-scale preparation. *J. Mater. Chem. A* **2018**, *6*, 6376–6386. [CrossRef]
5. Zhou, S.; Tao, R.; Dai, P.; Luo, Z.Y.; He, M. Two-step fabrication of lignin-based flame retardant for enhancing the thermal and fire retardancy properties of epoxy resin composites. *Polym. Compos.* **2020**, *41*, 2025–2035. [CrossRef]

6. Kong, Q.; Zhu, H.; Fan, J.; Zheng, G.; Zhang, C.; Wang, Y.; Zhang, J. Boosting flame retardancy of epoxy resin composites through incorporating ultrathin nickel phenylphosphate nanosheets. *J. Appl. Polym. Sci.* **2021**, *138*, 50265. [CrossRef]
7. Mostovoy, A.S.; Nurtazina, A.S.; Kadykova, Y.A.; Bekeshev, A.Z. Highly Efficient Plasticizers-Antipirenes for Epoxy Polymers. *Inorg. Mater. Appl. Res.* **2019**, *10*, 1135–1139. [CrossRef]
8. Bekeshev, A.; Mostovoy, A.; Kadykova, Y.; Akhmetova, M.; Tastanova, L.; Lopukhova, M. Development and Analysis of the Physicochemical and Mechanical Properties of Diorite-Reinforced Epoxy Composites. *Polymers* **2021**, *13*, 2421. [CrossRef]
9. Bao, X.; Wu, F.; Wang, J. Thermal Degradation Behavior of Epoxy Resin Containing Modified Carbon Nanotubes. *Polymers* **2021**, *13*, 3332. [CrossRef]
10. Chi, Z.; Guo, Z.; Xu, Z.; Zhang, M.; Li, M.; Shang, L.; Ao, Y. A DOPO-based phosphorus-nitrogen flame retardant bio-based epoxy resin from diphenolic acid: Synthesis, flame-retardant behavior and mechanism. *Polym. Degrad. Stab.* **2020**, *176*, 109151. [CrossRef]
11. Ran, S.; Ye, R.; Cai, Y.; Shen, H.; He, Y.; Fang, Z.; Guo, Z. Synergistic Flame Retardant Mechanism of Lanthanum Phenylphosphonate and Decabromodiphenyl Oxide in Polycarbonate. *Polym. Compos.* **2019**, *40*, 986–999. [CrossRef]
12. Kong, Q.; Wu, T.; Zhang, J.; Wang, D. Simultaneously improving flame retardancy and dynamic mechanical properties of epoxy resin nanocomposites through layered copper phenylphosphate. *Compos. Sci. Technol.* **2018**, *154*, 136–144. [CrossRef]
13. Jiang, T.; Liu, L.; Liu, L.; Hong, J.; Dong, M.; Deng, X. Synergistic flame retardant properties of a layered double hydroxide in combination with zirconium phosphonate in polypropylene. *RSC Adv.* **2016**, *94*, 91720–91727. [CrossRef]
14. Wang, X.; Zhang, P.; Huang, Z.; Xing, W.; Song, L.; Hu, Y. Effect of aluminum diethylphosphinate on the thermal stability and flame retardancy of flexible polyurethane foams. *Fire Saf. J.* **2019**, *106*, 72–79. [CrossRef]
15. Zhu, H.; Chen, Y.; Huang, S.; Wang, Y.; Yang, R.; Chai, H.; Zhu, F.; Kong, Q.; Zhang, Y.; Zhang, J. Suppressing fire hazard of poly(vinyl alcohol) based on $(NH_4)_2[VO(HPO_4)]_2(C_2O_4)\cdot 5H_2O$ with layered structure. *J. Appl. Polym. Sci.* **2021**, *138*, 51345. [CrossRef]
16. Hou, Y.; Xu, Z.; Chu, F.; Gui, Z.; Song, L.; Hu, Y.; Hu, W. A review on metal-organic hybrids as flame retardants for enhancing fire safety of polymer composites. *Compos. Part B Eng.* **2021**, *221*, 109014. [CrossRef]
17. Ding, J.; Zhang, Y.; Zhang, X.; Kong, Q.; Zhang, H.; Zhang, F. Improving the flame-retardant efficiency of layered double hydroxide with disodium phenylphosphate for epoxy resin. *J. Therm. Anal. Calorim.* **2020**, *140*, 149–156. [CrossRef]
18. Ran, S.; Guo, Z.; Chen, C.; Zhao, L.; Fang, Z. Carbon nanotube bridged cerium phenylphosphonate hybrids, fabrication and their effects on the thermal stability and flame retardancy of the HDPE/BFR composite. *J. Mater. Chem. A* **2014**, *2*, 2999–3007. [CrossRef]
19. Liu, W.; Nie, L.; Luo, L.; Yue, J.; Gan, L.; Lu, J.; Huang, J.; Liu, C. Enhanced dispersibility and uniform distribution of iron phosphonate to intensify its synergistic effect on polypropylene-based intumescent flame-retardant system. *J. Appl. Polym. Sci.* **2020**, *137*, 49552. [CrossRef]
20. Chen, C.; Guo, Z.; Ran, S.; Fang, Z. Synthesis of cerium phenylphosphonate and its synergistic flame retardant effect with decabromodiphenyl oxide in glass-fiber reinforced poly(ethylene terephthalate). *Polym. Compos.* **2014**, *35*, 539–547. [CrossRef]
21. Zhang, F.; Bao, Y.; Ma, S.; Liu, L.; Shi, X. Hierarchical flower-like nickel phenylphosphonate microspheres and their calcined derivatives for supercapacitor electrodes. *J. Mater. Chem. A* **2017**, *5*, 7474–7481. [CrossRef]
22. Cui, J.; Zhang, Y.; Wang, L.; Liu, H.; Wang, N.; Yang, B.; Guo, J.; Tian, L. Phosphorus-containing Salen-Ni metal complexes enhancing the flame retardancy and smoke suppression of epoxy resin composites. *J. Appl. Polym. Sci.* **2020**, *137*, 48734. [CrossRef]
23. Guo, X.; Duan, M.; Zhang, Y.; Xi, B.; Li, M.; Yin, R.; Zheng, X.; Liu, Y.; Cao, F.; An, X.; et al. A general self-assembly induced strategy for synthesizing two-dimensional ultrathin cobalt-based compounds toward optimizing hydrogen evolution catalysis. *Adv. Funct. Mater.* **2022**, *32*, 2209397. [CrossRef]
24. He, Z.; Honeycutt, C.W.; Zhang, T.; Bertsch, P.M. Preparation and FT-IR characterization of metal phytate compounds. *J. Environ. Qual.* **2006**, *35*, 1319–1328. [CrossRef] [PubMed]
25. Guo, W.; Yu, B.; Yuan, Y.; Song, L.; Hu, Y. In situ preparation of reduced graphene oxide/DOPO-based phosphonamidate hybrids towards high-performance epoxy nanocomposites. *Compos. Part B Eng.* **2017**, *123*, 154–164. [CrossRef]
26. Tang, H.; Zhu, Z.; Chen, R.; Wang, J.; Zhou, H. Synthesis of DOPO-based pyrazine derivative and its effect on flame retardancy and thermal stability of epoxy resin. *Polym. Adv. Technol.* **2019**, *30*, 2331–2339. [CrossRef]
27. Yang, S.; Wang, J.; Huo, S.; Wang, M.; Wang, J. Preparation and flame retardancy of a compounded epoxy resin system composed of phosphorus/nitrogen-containing active compounds. *Polym. Degrad. Stab.* **2015**, *121*, 398–406. [CrossRef]
28. Chen, J.; Guo, X.; Gao, M.; Wang, J.; Sun, S.; Xue, K.; Zhang, S.; Liu, Y.; Zhang, J. Free-supporting dual-confined porous Si@c-ZIF@carbon nanofibers for high-performance lithium-ion batteries. *Chem. Commun.* **2021**, *57*, 10580–10583. [CrossRef]
29. Gnanasekar, P.; Feng, M.; Yan, N. Facile Synthesis of a Phosphorus-Containing Sustainable Biomolecular Platform from Vanillin for the Production of Mechanically Strong and Highly Flame-Retardant Resins. *ACS Sustain. Chem. Eng.* **2020**, *8*, 17417–17426. [CrossRef]
30. Kong, Q.; Sun, Y.; Zhang, C.; Guan, H.; Zhang, J.; Wang, D.; Zhang, F. Ultrathin iron phenyl phosphonate nanosheets with appropriate thermal stability for improving fire safety in epoxy. *Compos. Sci. Technol.* **2019**, *182*, 107748. [CrossRef]
31. Luo, Q.; Sun, Y.; Yu, B.; Li, C.; Song, J.; Tan, D.; Zhao, J. Synthesis of a novel reactive type flame retardant composed of phenophosphazine ring and maleimide for epoxy resin. *Polym. Degrad. Stab.* **2019**, *165*, 137–144. [CrossRef]

32. Bifulco, A.; Varganici, C.D.; Rosu, L.; Mustata, F.; Rosu, D.; Gaan, S. Recent advances in flame retardant epoxy systems containing non-reactive DOPO based phosphorus additives. *Polym. Degrad. Stab.* **2022**, *200*, 109962. [CrossRef]
33. Wang, P.; Cai, Z. Highly efficient flame-retardant epoxy resin with a novel DOPO-based triazole compound: Thermal stability, flame retardancy and mechanism. *Polym. Degrad. Stab.* **2017**, *137*, 138–150. [CrossRef]
34. Wang, P.; Fu, X.; Kan, Y.; Wang, X.; Hu, Y. Two high-efficient DOPO-based phosphonamidate flame retardants for transparent epoxy resin. *High Perform. Polym.* **2019**, *31*, 249–260. [CrossRef]
35. Jin, S.; Qian, L.; Qiu, Y.; Chen, Y.; Xin, F. High-efficiency flame retardant behavior of bi-DOPO compound with hydroxyl group on epoxy resin. *Polym. Degrad. Stab.* **2019**, *166*, 344–352. [CrossRef]
36. Wang, J.; Ma, C.; Wang, P.; Qiu, S.; Cai, W.; Hu, Y. Ultra-low phosphorus loading to achieve the superior flame retardancy of epoxy resin. *Polym. Degrad. Stab.* **2018**, *149*, 119–128. [CrossRef]
37. Chen, Y.; Huang, S.; Zhao, H.; Yang, R.; He, Y.; Zhao, T.; Zhang, Y.; Kong, Q.; Sun, S.; Zhang, J. Influence of beta-cyclodextrin functionalized tin phenylphosphonate on the thermal stability and flame retardancy of epoxy composites. *J. Renew. Mater.* **2022**, *10*, 3121–3130. [CrossRef]
38. Kong, Q.; Li, L.; Zhang, M.; Chai, H.; Li, W.; Zhu, F.; Zhang, J. Improving the thermal stability and flame retardancy of epoxy resin by lamellar cobalt potassium pyrophosphate. *Polymers* **2022**, *14*, 4927. [CrossRef]
39. Salmeia, K.A.; Gaan, S. An overview of some recent advances in DOPO-derivatives: Chemistry and flame retardant applications. *Polym. Degrad. Stab.* **2015**, *113*, 119–134. [CrossRef]
40. Klinkowski, C.; Wagner, S.; Ciesielski, M.; Doring, M. Bridged phosphorylated diamines: Synthesis, thermal stability and flame retarding properties in epoxy resins. *Polym. Degrad. Stab.* **2014**, *106*, 122–128. [CrossRef]
41. Zhang, Y.; Yu, B.; Wang, B.; Liew, K.M.; Song, L.; Wang, C.; Hu, Y. Highly effective P–P synergy of a novel DOPO-based flame retardant for epoxy resin. *Ind. Eng. Chem. Res.* **2017**, *56*, 1245–1255. [CrossRef]
42. Wei, Z.; Wu, J.; Liu, Z.; Gu, Y.; Luan, G.; Sun, H.; Yu, Q.; Zhang, S.; Wang, Z. Effect of ethyl-bridged diphenylphosphine oxide on flame retardancy and thermal properties of epoxy resin. *Polym. Adv. Technol.* **2020**, *31*, 1426–1436. [CrossRef]
43. Yang, S.; Wang, J.; Huo, S.; Wang, M.; Cheng, L. Synthesis of a phosphorus/nitrogen-containing additive with multifunctional groups and its flame retardant effect in epoxy resin. *Ind. Eng. Chem. Res.* **2015**, *54*, 7777–7786. [CrossRef]
44. Xu, M.; Xu, G.; Leng, Y.; Li, B. Synthesis of a novel flame retardant based on cyclotriphosphazene and DOPO groups and its application in epoxy resins. *Polym. Degrad. Stab.* **2016**, *123*, 105–114. [CrossRef]
45. Fang, F.; Song, P.; Ran, S.; Guo, Z.; Wang, H.; Fang, Z. A facile way to prepare phosphorus-nitrogen-functionalized graphene oxide for enhancing the flame retardancy of epoxy resin. *Compos. Commun.* **2018**, *10*, 97–102. [CrossRef]
46. Kong, Q.; Zhu, H.; Huang, S.; Wu, T.; Zhu, F.; Zhang, Y.; Wang, Y.; Zhang, J. Influence of multiply modified FeCu-montmorillonite on fire safety and mechanical performances of epoxy resin nanocomposites. *Thermochim. Acta* **2022**, *707*, 179112. [CrossRef]
47. Wang, X.; Hu, Y.; Song, L.; Xing, W.; Lu, H.; Lv, P.; Jie, G. Flame retardancy and thermal degradation mechanism of epoxy resin composites based on a DOPO substituted organophosphorus oligomer. *Polymers* **2010**, *51*, 2435–2445. [CrossRef]
48. Guo, X.; Liu, S.; Wan, X.; Zhang, J.; Liu, Y.; Zheng, X.; Kong, Q.; Jin, Z. Controllable solid-phase fabrication of $Fe_2O_3/Fe_5C_2/Fe$-N-C electrocatalyst towards optimizing the oxygen reduction reaction in zinc-air batteries. *Nano Lett.* **2022**, *22*, 4879–4887. [CrossRef]
49. Wang, H.; Li, S.; Yuan, Y.; Liu, X.; Sun, T.; Wu, Z. Study of the epoxy/amine equivalent ratio on thermal properties, cryogenic mechanical properties, and liquid oxygen compatibility of the bisphenol A epoxy resin containing phosphorus. *High Perform. Polym.* **2019**, *32*, 429–430. [CrossRef]
50. Sai, T.; Ran, S.; Guo, Z.; Yan, H.; Zhang, Y.; Song, P.; Zhang, T.; Wang, H.; Fang, Z. Deposition growth of Zr-based MOFs on cerium phenylphosphonate lamella towards enhanced thermal stability and fire safety of polycarbonate. *Compos. Part B Eng.* **2020**, *197*, 108064. [CrossRef]
51. Gao, M.; Xue, Y.; Zhang, Y.; Zhu, C.; Yu, H.; Guo, X.; Sun, S.; Xiong, S.; Kong, Q.; Zhang, J. Growing Co-Ni-Se nanosheets on 3D carbon frameworks as advanced dualfunctional electrodes for supercapacitors and sodium ion batteries. *Inorg. Chem. Front.* **2022**, *9*, 3933–3942. [CrossRef]
52. Tang, G.; Liu, X.; Zhou, L.; Zhang, P.; Deng, D.; Jiang, H. Steel slag waste combined with melamine pyrophosphate as a flame retardant for rigid polyurethane foams. *Adv. Powder Technol.* **2020**, *31*, 279–286. [CrossRef]
53. Chen, Y.; Lu, Q.; Zhong, G.; Zhang, H.; Chen, M.; Liu, C. DOPO-based curing flame retardant of epoxy composite material for char formation and intumescent flame retardance. *J. Appl. Polym. Sci.* **2020**, *138*, 49918. [CrossRef]
54. Tang, G.; Liu, X.; Yang, Y.; Chen, D.; Zhang, H.; Zhou, L.; Zhang, P.; Jiang, H.; Deng, D. Phosphorus-containing silane modified steel slag waste to reduce fire hazards of rigid polyurethane foams. *Adv. Powder Technol.* **2020**, *31*, 1420–1430. [CrossRef]
55. Peng, W.; Xu, Y.; Nie, S.; Yang, W. A bio-based phosphaphenanthrene-containing derivative modified epoxy thermosets with good flame retardancy, high mechanical properties and transparency. *RSC Adv.* **2021**, *11*, 30943–30954. [CrossRef]

Disclaimer/Publisher's Note: The statements, opinions and data contained in all publications are solely those of the individual author(s) and contributor(s) and not of MDPI and/or the editor(s). MDPI and/or the editor(s) disclaim responsibility for any injury to people or property resulting from any ideas, methods, instructions or products referred to in the content.

Article

Flame-Retarded Rigid Polyurethane Foam Composites with the Incorporation of Steel Slag/Dimelamine Pyrophosphate System: A New Strategy for Utilizing Metallurgical Solid Waste

Mingxin Zhu [1,*], Sujie Yang [2], Zhiying Liu [1], Shunlong Pan [1] and Xiuyu Liu [2]

[1] College of Environmental Science and Engineering, Nanjing Tech University, Nanjing 211816, China
[2] School of Architecture and Civil Engineering, Anhui University of Technology, Ma'anshan 243002, China
* Correspondence: zmx@njtech.edu.cn

Abstract: Rigid polyurethane (RPUF) was widely used in external wall insulation materials due to its good thermal insulation performance. In this study, a series of RPUF and RPUF-R composites were prepared using steel slag (SS) and dimelamine pyrophosphate (DMPY) as flame retardants. The RPUF composites were characterized by thermogravimetric (TG), limiting oxygen index (LOI), cone calorimetry (CCT), and thermogravimetric infrared coupling (TG-FTIR). The results showed that the LOI of the RPUF-R composites with DMPY/SS loading all reached the combustible material level (22.0 vol%~27.0 vol%) and passed UL-94 V0. RPUF-3 with DMPY/SS system loading exhibited the lowest pHRR and THR values of 134.9 kW/m^2 and 16.16 MJ/m^2, which were 54.5% and 42.7% lower than those of unmodified RPUF, respectively. Additionally, PO· and PO$_2$· free radicals produced by pyrolysis of DMPY could capture high energy free radicals, such as H·, O·, and OH·, produced by degradation of RPUF matrix, effectively blocking the free radical chain reaction of composite materials. The metal oxides in SS reacted with the polymetaphosphoric acid produced by the pyrolysis of DMPY in combustion. It covered the surface of the carbon layer, significantly insulating heat and mass transport in the combustion area, endowing RPUF composites with excellent fire performance. This work not only provides a novel strategy for the fabrication of high-performance RPUF composites, but also elucidates a method of utilizing metallurgical solid waste.

Keywords: rigid polyurethane; steel slag; dimelamine pyrophosphate; solid waste utilization; flame retardant

1. Introduction

Rigid polyurethane (RPUF) is widely used in thermal insulation, refrigeration, architectural decoration, amongst others, due to its excellent thermal insulation, mechanical properties, and environmental resistance [1–5]. However, due to the porous structure and organic skeleton of RPUF, it is also easily ignited and releases toxic gases, such as HCN, CO, etc. The limiting oxygen index (LOI) of RPUF is only approximately 19.0 vol%, with intense dripping in fire [6]. The above disadvantages significantly limit the further application of RPUF. Thus, how to reduce the toxic gases generated during the combustion process of RPUF and enhance its fire performance has attracted much scholarly attention [7].

Steelmaking as a basic industry and steel products have penetrated all areas of life. According to statistics, the global crude steel production in 2020 was as high as 1.878 billion tons [8]. Steel slag (SS), as an unavoidable solid waste in the steelmaking process, is partially stored in piles, causing excessive resource waste and environmental pollution. Thus, exploring the potential value of SS and improving its comprehensive utilization rate is the current focus of researchers in the solid waste industry. Additionally, exploring the high value-added utilization pathway of steel slag is an important step to accelerating the improvement of the green low-carbon circular economy system. In recent years, many studies on the utilization of SS have been conducted, for example, in cement fields, mine backfilling, and soil remediation, amongst others [9–12].

Steel slag (SS) is rich in metal oxides such as CaO, SiO_2, and Al_2O_3. These compounds exhibit excellent catalytic carbonization and smoke suppression effects in combustion and thus have potential applications in fire-retarding fields [13,14]. Xu et al. [15] prepared melamine-cyanuric acid fumed silica (MCA-SiO_2) by reacting melamine (ME) and cyanuric acid (CA) in aqueous suspension and deposited it on fumed silica. Subsequently, GF-PP and IFR containing MCA-SiO_2 were introduced to prepare GF-PP/IFR-(MCA-SiO_2) composites via a melt blending technology. The results showed that the heat release rate (HRR) of GF-PP/IFR-MCA composites decreased, and the flame-retardant property increased with the addition of SiO_2. When the content of SiO_2 was 20 wt%, the flame-retardant property of the composites reached a UL-94 V-0 rating, and the limiting oxygen index (LOI) increased to 32.4 vol%. Tang et al. [16] modified steel slag (SS) by DOPO-derived silanes through a sol–gel reaction, synergizing the modified steel slag (mSS) with expanded graphite (EG) in rigid polyurethane foam. The results showed that when the mSS/EG additions reached 20 wt%, the RPUF composite exhibited the best overall performance. The peak heat release rate (pHRR) and total heat release (THR) of the composite were decreased by 55% and 47%, respectively, and ultimate oxygen index increased to 24.0 vol% and UL-94 V0 rating in flame-retardant tests.

Dimelamine pyrophosphate (DMPY) is an environmentally friendly non-halogen phosphorus–nitrogen flame retardant commonly used in polyolefins, coatings, and fibers [17]. DMPY molecules contain triazine rings, which have excellent charring ability [18–20]. Thus, DMPY has attracted much scholarly attention and use alone or as a compound with other flame retardants. Liu et al. [21] synthesized DMPY using sodium pyrophosphate and melamine as raw materials, which was further introduced into thermoplastic epoxy resin (EP) composites to enhance its fire performance. It was observed that the flame retardancy and mechanical properties of EP/DMPY composites were significantly improved compared to pure EP, and the ultimate oxygen index of the composites was increased to 28.7 vol% when 9 wt% DMPY was incorporated. Liu et al. [22] combined DMPY and aluminum diethylphosphinate (ADP) as a synergistic system (FRs), which was further used in UPR thermoset composites. The results showed that the decomposition of FRs produced pyrophosphate, polyphosphate, and metaphosphate, which catalyzed the degradation of the UPR matrix to form coherent, partially graphitized, dense carbon layers.

Currently, the direct addition of flame retardants is the simplest strategy to improve the flame-retardant properties of polyurethane [23,24]. However, there are no reports on the applications of DMPY in RPUF. Thus, in this study, a series of RPUF and RPUF-R composites were prepared with DMPY and SS as flame retardants, and the flame-retardant properties, combustion properties, and morphological changes of the composites were investigated using limiting oxygen index (LOI), vertical combustion (UL-94), cone calorimetry (CCT), and scanning electron microscopy (SEM) test methods. This work provides a novel strategy for the high value-added utilization of SS and fabricating high-performance RPUF composites.

2. Results and Discussion

2.1. Bubble Structure of RPUF and RPUF-R Composites

SEM was a useful tool for investigation particle size and micromorphology of materials [25,26]. The morphology of RPUF and RPUF-R composites were analyzed by SEM. As observed from Figure 1a, the pure RPUF had a bubble structure with uniform pores structure, which was consistent with a previous report [27]. Additionally, as shown in red circles in Figure 1a, some big bubble structures were observed, which may come from the uneven distribution of water (blowing agent) in the foaming process. As presented in Figure 1a1, pure RPUF was endowed with a thin wall for bubble structure. As can been found in Figure 1b, the big bubble structures were also observed when 20 wt% DMPY was added. Figure 1b1 confirmed that the wall of blister in RPUF-1 thickened and the density increased significantly. Meanwhile, it also could be found that DMPY particles remained on the surface of the composite and some of the blister holes were broken (as the yellows arrow point), which may be due to the excessive amount of DMPY increasing viscosity and promoting the agglomeration of flame-retardant particles. When SS replaced part of DMPY, it could be observed that the agglomeration of the RPUF-3 composite was reduced, but the hole wall was still partially damaged (points by red circle), indicating that the addition of inorganic SS was conducive to improving the compatibility of DMPY and RPUF matrix. When SS was added to the RPUF matrix completely instead of DMPY, it could be observed from Figure 1d1 that the fracture surface of the vesicle pore wall of the RPUF-1 composite was smooth and the wall layer was thickened (points by red circle), which indicated that the inorganic SS particles possessed excellent compatibility with the RPUF matrix. It can be found from Figure 1b–d that when SS, DMPY, and their combination were introduced into RPUF, the pore size of the composites were smaller than that of unmodified RPUF, which was mainly due to the nucleation effect of powdery flame retardant.

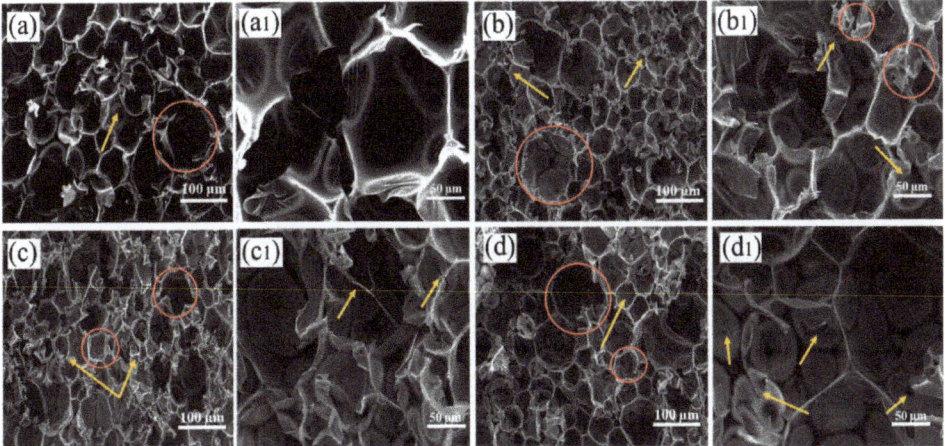

Figure 1. SEM images of RPUF and RPUF-R composites: (**a,a1**) RPUF; (**b,b1**) RPUF-1; (**c,c1**) RPUF-3; (**d,d1**) RPUF-5.

2.2. Physical Properties of RPUF and RPUF-R Composites

Apparent density was a very important physical property index for composites [28,29]. As shown in Table 1, pure RPUF exhibited density of 45.28 kg/m^3, with thermal conductivity of 0.0373 W/m·k, and compressive strength of 0.325 KPa. When 20 wt% DMPY was added, the apparent density of RPUF-1 was increased to 71.84 kg/m^3, which was significantly higher than that of unmodified RPUF. This may be due to the fact that the density of DMPY was much higher than that of the RPUF matrix. Additionally, the apparent density of RPUF-1 was increased, which may come from the enhancement of nucleation ability and the decrease in cell size after DMPY loading [30]. With the increase in the SS and decrease in DMPY, the apparent density of the RPUF-R composites were gradually decreased. This may come from the fact that SS contained free f-CaO and f-MgO components. When SS met water, it underwent hydration reaction and generated Ca(OH)$_2$ and Mg(OH)$_2$, resulting in volume expansion. The total mass of the composite was a constant, with the increase of SS addition, the expansion rate of the composites gradually increased, thus resulting in decrease in the apparent density for the composites [31]. It was observed that the thermal conductivity of RPUF-R composites was higher than that of unmodified RPUF. This was mainly ascribed to the damage of part of the cell structure caused by the addition of powdery flame-retardant particles. Compression strength was another important index to characterize performance of rigid polyurethane foam [32]. In general, the compressive strength of polyurethane rigid foam increased with the increase in density. RPUF-R composites exhibited lower compressive strength than that of pure RPUF, which was related to the excessive addition of powdery flame retardants and the destruction of hydrogen bonds in RPUF composites.

Table 1. Apparent density, thermal conductivity, and compressive strength of RPUF and RPUF-R composites.

Sample	Thermal Conductivity/(W/m·k)	Compressive Strength/(MPa)	Apparent Density/(kg/m^3)
RPUF	0.0373	0.325	45.28
RPUF-1	0.0392	0.315	71.84
RPUF-2	0.0380	0.289	63.80
RPUF-3	0.0389	0.273	61.44
RPUF-4	0.0381	0.257	61.20
RPUF-5	0.0369	0.240	59.25

2.3. Flame Retardant Properties of RPUF and RPUF-R Composites

LOI and UL-94 were important methods for evaluating the flame retardancy of polymer materials [33]. As shown in Table 2, the LOI value of pure RPUF was only 19.0 vol%, which was a flammable material and exhibited molten dripping in combustion. When 20 wt% DMPY was added, the LOI value of RPUF-1 increased to 24.8 vol% with UL-94 V-0 rating, and the dripping phenomenon disappeared. This was due to the existence of DMPY, which produced polymetaphosphate in the decomposition process, and promoted the carbonization of the RPUF matrix. At the same time, polymetaphosphate was a viscous glass material. After dehydration and carbonization, the glass material could attach to the surface of the carbon layer to form a liquid film and achieve the purpose of flame retardancy [34]. With the replacement of DMPY by SS, the LOI values of the composites were in the range of 22.0 vol%~25 vol%, which reached the combustible material rating (22.0 vol%~27.0 vol%) compared with flammable material (<22.0 vol%) of pure RPUF. Additionally, the RPUF composites with DMPY and SS loading all could pass UL-94 V0 flame retardant rating [35]. This may be due to the fact that SS contained Al_2O_3, MgO, and CaO, which combined with DMPY to form a protective layer and prevent the further combustion of the underlying composites. When 20 wt% of SS was added to completely replace DMPY, the LOI value of RPUF-5 composite was 20.4 vol%, which was 1.4 vol% higher compared to unmodified RPUF and failed to pass UL-94 V-0 flame retardant rating. It was also found that the melt dripping phenomenon disappeared. The above results indicated that the addition of SS alone has some effect on RPUF flame retardancy, but the combination of SS with DMPY can be effective for flame retardancy enhancement of RPUF-R composites.

Table 2. LOI, UL-94 data for RPUF and RPUF-R composites.

Sample	LOI/vol%	UL-94. 3.2 mm Bar		
		t_1/t_2[a] (s)	Dripping	Rating
RPUF	19.0	BC[b]	Y	NR[c]
RPUF-1	24.8	4.37/0	N	V-0
RPUF-2	23.7	6.42/0	N	V-0
RPUF-3	23.4	6.84/0	N	V-0
RPUF-4	22.4	8.72/0	N	V-0
RPUF-5	20.4	BC	N	NR

Note: a t_1, t_2—the first, second ignition after the burning time; b BC—burn to fixture; c NR—no rating.

2.4. Thermal Stability of RPUF and RPUF-R Composites

Thermogravimetry was a useful tool for investigating the thermal stability of polymer composites [36]. Herein, the thermal stability of RPUF and RPUF-R composites was characterized by thermogravimetry (TGA). Figure 2 showed the thermogravimetric (TG) and thermal weight loss (DTG) curves of RPUF and RPUF-R composites, and the related data were listed in Table 3. As shown in the table, the initial degradation temperature ($T_{-5wt\%}$) of unmodified RPUF was 271 °C. The $T_{-5wt\%}$ of RPUF-1, RPUF-3, and RPUF-5 composites were higher than that of pure RPUF, indicating that the addition of DMPY and SS inhibited the initial decomposition of RPUF composites. The whole pyrolysis process of the composites was divided into two stages. The first stage in the interval of 300–350 °C corresponded to the degradation of the hard segments of the polyurethane molecular chain, and the second stage in the interval of 450–500 °C could be ascribed to the degradation of the soft segments in the polyurethane molecular chain, accompanied by the generation of CO [37,38]. It can also be noted from Table 4 that the maximum degradation temperature (T_{max2}) of RPUF-1, RPUF-3, and RPUF-5 in the second stage were higher than that of pure RPUF, which may be due to the fact that the polymetaphosphate generated by pyrolysis of DMPY promoted the dehydration and carbonization of RPUF composites, and free PO· radicals captured high-energy radicals such as H·, O·, and OH·, which could effectively block the free radical chain reaction in combustion [39]. At the same time, the metal oxides in SS chelated with the polyphosphoric acid generated by DMPY pyrolysis to form metal salts. It could be retained on the surface of the carbon layer, which was a benefit for isolating oxygen and heat, preventing the further combustion of the composites and improving the thermal stability of the composites [40]. The residual carbon of RPUF-1, RPUF-3, and RPUF-5 composites at 700 °C were higher than those of pure RPUF. All these results indicated that the addition of DMPY and SS could improve the high temperature thermal stability of RPUF-R composites.

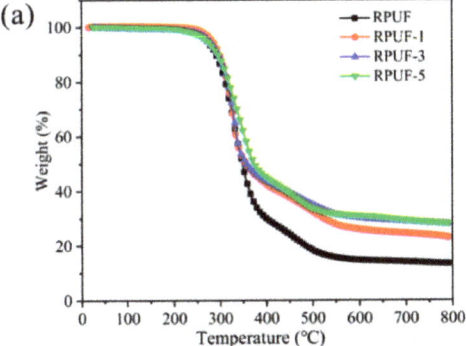
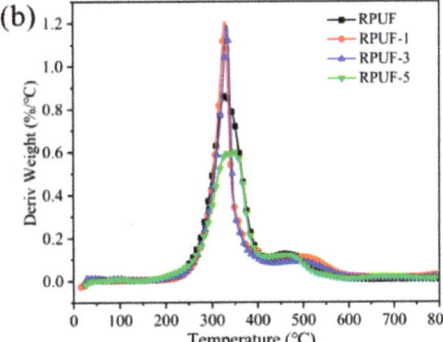

Figure 2. (a) thermogravimetric (TG) and (b) thermal weight loss (DTG) curves of RPUF and RPUF-R composites.

Table 3. Thermogravimetric data of RPUF and RPUF-R composites (nitrogen atmosphere).

Sample	$T_{-5wt\%}$/°C	T_{max1}/°C	T_{max2}/°C	700 °C Carbon Residue/wt%
RPUF	271	330	460	14.0
RPUF-1	283	328	490	24.4
RPUF-3	277	333	470	28.8
RPUF-5	272	350	463	29.3

Table 4. Cone calorimetric data of RPUF, RPUF-R composites.

Sample	RPUF	RPUF-1	RPUF-3	RPUF-5
TTI (s)	5	3	4	3
Tp (s)	72	68	72	64
Td (s)	390	415	516	441
pHRR (kW/m^2)	296.5	145.1	134.9	188.6
THR (MJ/m^2)	28.18	18.95	16.16	21.52
FPI (m^2·s/kW)	0.0169	0.0207	0.0297	0.0159
FGI (kW/m^2·s)	4.12	2.13	1.87	2.95
CY (wt%)	17.27	25.22	24.77	23.29

Note: TTI, time to ignition; Tp, time to pHRR; Td, continuous combustion time; pHRR, peak of release rate; THR, total heat release; FPI, fire performance index; FGI, fire growth rate index; CY, char yield.

2.5. Combustion Properties of RPUF and RPUF-R Composites

Cone calorimeter was widely used in evaluating the combustion performance of materials [41]. Figures 3 and 4 showed the conical calorimetric data plots of RPUF and RPUF-R composites, and the related data were listed in Table 4. The pure RPUF reached the maximum heat release rate (pHRR) at 72 s, with the highest value of 296.5 kW/m^2. The pHRR values of RPUF-1, RPUF-3, and RPUF-5 composites were 145.1 kW/m^2, 134.9 kW/m^2, and 188.6 kW/m^2, respectively, which were 51.1%, 54.5%, and 36.4% lower than those of pure RPUF. The HRR values of RPUF-R composites decreased rapidly after reaching pHRR, which were lower than that of RPUF. RPUF-3 was endowed with the lowest pHRR value, suggesting the synergistic effect between SS and DMPY in RPUF composites. It can be seen from Figure 3b that the THR value of pure RPUF was 28.18 MJ/m^2, and THR values of RPUF-R were lower than that of pure RPUF. Among them, the decrease in the RPUF-3 composite was the most obvious, with THR value of 16.16 MJ/m^2, which was 42.7% lower than that of pure RPUF, implying that DMPY/SS could synergistically reduce the heat release of RPUF composite in combustion. This result was highly consistent with the flame retardant test.

Furthermore, typical gaseous products and mass vs. time curves were given in Figure 4. CO was a deadly toxic gas produced in a fire accident [42]. As observed in Figure 4, the CO production of RPUF-R composites showed a decreasing trend, which was mainly due to the possible adsorption of CO by the metal ions contained in SS. Additionally, part of the CO could be oxidized into CO_2 by oxidation reaction under high temperature conditions [43]. Fire Performance Index (FPI) and Fire Spread Index (FGI) were often used to evaluate the fire safety of flame-retardant polymer composites, where FPI was the ratio of PHRR to peak burning time of the composite materials (T_P), and FGI was the ratio of TTI to PHRR. The higher the FPI value, the smaller the FGI value, and the smaller the fire hazard [44,45]. As shown in Table 5, the FPI values of RPUF, RPUF-1, RPUF-3, and RPUF-5 composites were 0.0169 m^2·s/kW, 0.0207 m^2·s/kW, 0.0297 m^2·s/kW, and 0.0159 m^2·s/kW, respectively, and the FGI values were 4.12 kW/m^2·s, 2.13 kW/m^2·s, and 1.87 kW/m^2·s, and 2.95 kW/m^2·s, respectively. Compared with pure RPUF, the fire hazards of RPUF-R composites were all significantly reduced, in which RPUF-3 exhibited the most significant effect, indicating that the addition of DMPY/SS synergistically reduced the fire safety of RPUF composites.

Table 5. Composition of RPUF and RPUF-R composites.

Sample	LY-4110 /g	PM-200 /g	LC /g	AK-8805 /g	A33 /g	TEOA /g	Water /g	SS /g	DMPY /g
RPUF	100	135	0.5	2	1	3	2	0	0
RPUF-1	100	135	0.5	2	1	3	2	60.9	0
RPUF-2	100	135	0.5	2	1	3	2	40.6	20.3
RPUF-3	100	135	0.5	2	1	3	2	30.45	30.45
RPUF-4	100	135	0.5	2	1	3	2	20.3	40.6
RPUF-5	100	135	0.5	2	1	3	2	0	60.9

Note: LY-4110, polyether polyol; PM-200, polyaryl polymethylene isocyanate; LC, dibutyltin dilaurate, AK-8805, silicone surfactant; A33, triethylenediamine; TEOA, triethanolamine, Water, blowing agent; SS, steel slag; DMPY, dimelamine pyrophosphate.

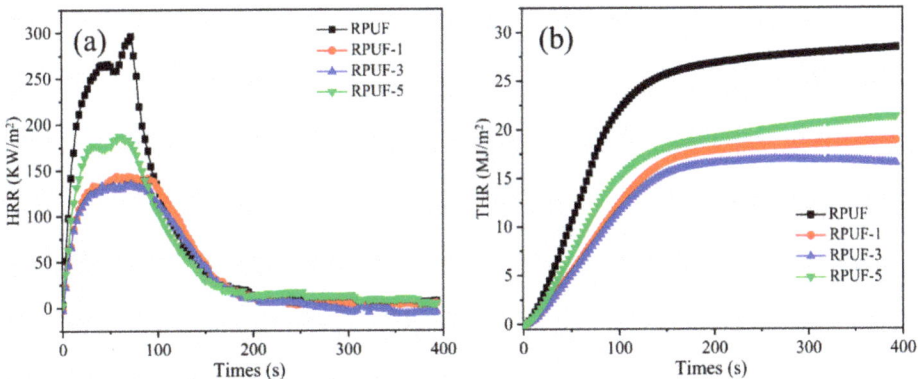

Figure 3. (a) HRR and (b) THRcurves of RPUF and RPUF-R composites.

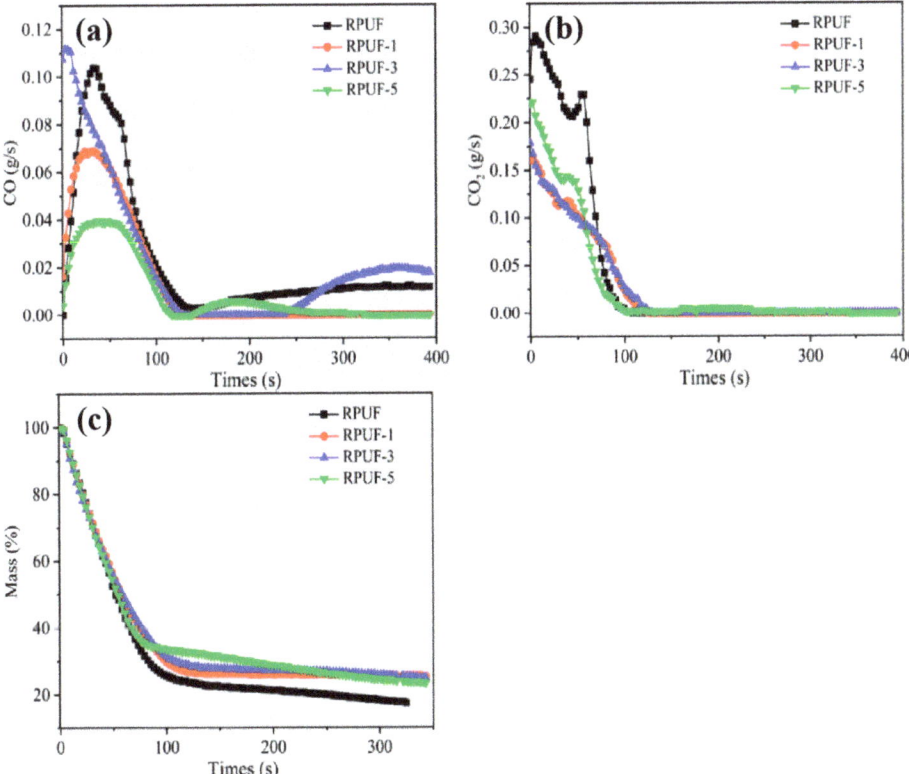

Figure 4. (a) CO, (b) CO_2, and (c) Mass curves of RPUF and RPUF-R composites.

2.6. Analysis of Gas Phase Products of RPUF and RPUF-R Composites

Thermogravimetric infrared coupling (TG-FTIR) was a common method to analyze the gas phase products and work mechanism of flame-retarding composites [46]. Figure 5 provided TG-FTIR 3D Spectra of RPUF and RPUF-R Composites. It can be observed from the figure that the gas phase products of RPUF and RPUF-R composites were mainly distributed in the bands of 3500–4000 cm^{-1}, 2500–3400 cm^{-1}, 2100–2400 cm^{-1}, 1500–2000 cm^{-1}, 700–1300 cm^{-1}. Figure 6 showed the FTIR spectra of the gas products of RPUF and RPUF-R composites at different times. The characteristic peaks at around 3818 cm^{-1} and 2930 cm^{-1} corresponded to the stretching vibration of the N–H bond and the stretching vibration of the C–H bond in hydrocarbons, respectively [47]. The characteristic peaks of CO_2 and -NCO were 2384 cm^{-1} and 2306 cm^{-1}, respectively. The characteristic peaks at around 1604 cm^{-1} and 1517 cm^{-1} were attributed to aromatic compounds and esters [48]. The characteristic peak of HCN was found at 721 cm^{-1}, which was consistent with a previous report. From Figure 6a, it could be found that the release intensity of CO_2 and -NCO from the RPUF-3 composite was significantly reduced compared to the other three composites. Additionally, the release intensity of HCN was reduced compared to pure RPUF, indicating that the chelation reaction between DMPY and SS inhibited the release of toxic and hazardous gases associated with the decomposition of RPUF matrix. Figure 6b–e showed the variation curves of typical gaseous products intensity for RPUF and RPUF-R composites with time. As observed from Figure 6b, the CO_2 release of pure RPUF started at around 12 min, while the CO_2 generation time of the other three composites was delayed, implying that the addition of DMPY and SS inhibited the initial degradation of the composites and improved their thermal stability, which was consistent with the thermogravimetric test.

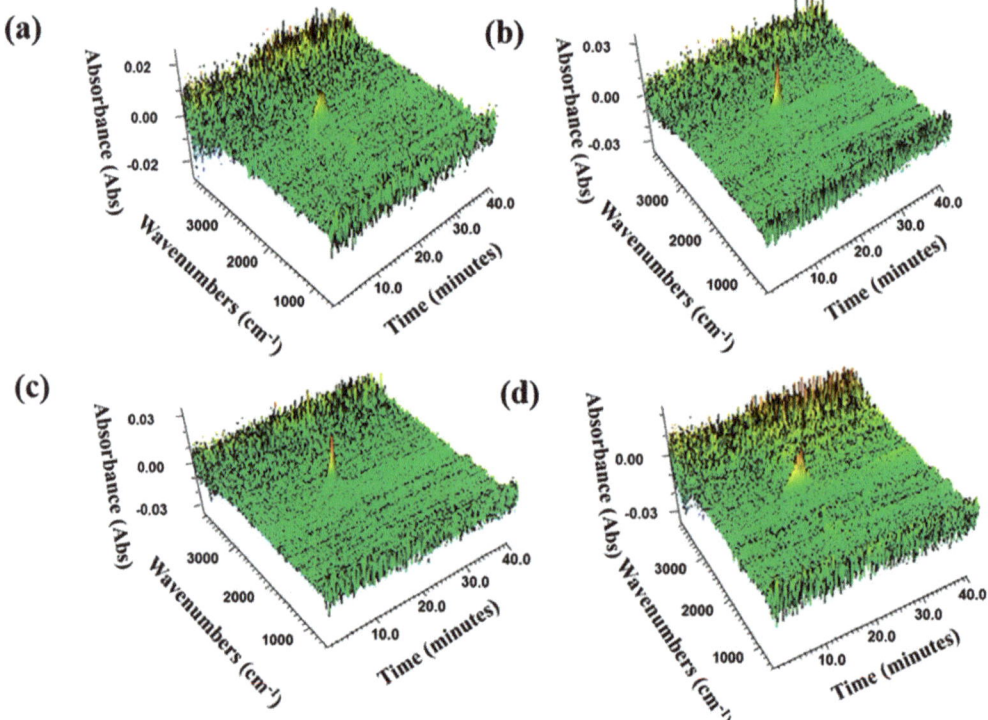

Figure 5. TG-FTIR 3D images of RPUF and RPUF-R composites: (**a**) RPUF; (**b**) RPUF-1; (**c**) RPUF-3; (**d**) RPUF-5.

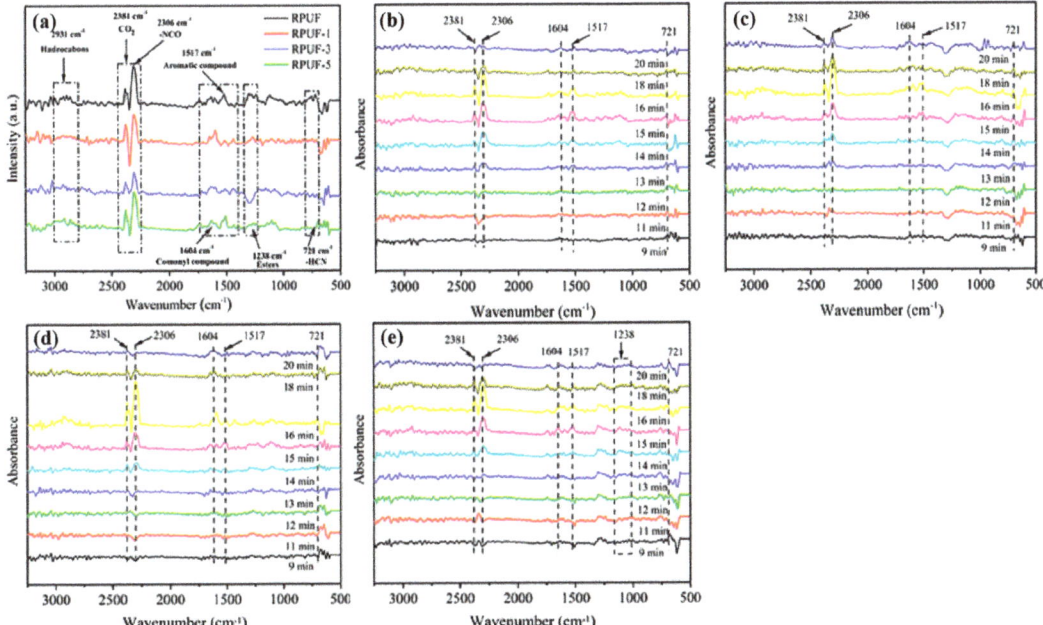

Figure 6. FTIR spectra of thermal decomposition products of RPUF and RPUF-R composites: (**a**) FTIR spectra at T_{max}; (**b**) RPUF; (**c**) RPUF-1; (**d**) RPUF-3; (**e**) RPUF-5.

2.7. Carbon Slag Analysis of RPUF and RPUF-R Composites

Carbon residue in RPUF and RPUF-R composites was obtained after calcination in muffle furnace, which were further investigated by SEM [49]. As shown in Figure 7a, the carbon layer of pure RPUF exhibited a thin and smooth structure after calcination (as shown by red circle), and the char layer was lax, which could not inhibit the mass and heat transport in combustion. As shown in Figure 7b, when 20 wt% DMPY was added, the carbon layer compactness of RPUF-1 composite was significantly improved compared with that of unmodified RPUF, but some pore structures (as observed from the red circle) was observed, which may be due to the poor compatibility between DMPY and RPUF matrix and the uneven dispersion caused by excessive addition of DMPY. When 20 wt% of SS was added, the compactness of carbon layer for RPUF-5 was also improved. It was also found that the surface of the carbon layer was loaded with a large amount of metal oxides and SS particles (pointed by yellow arrow), which acted as a barrier to prevent the transport of heat and mass. When 20 wt% DMPY/SS (1:1) was added, many obvious worm structures and dense carbon layers were observed in RPUF-3 (as shown by red circle). This was mainly due to the combination of the catalytic carbonization effect of polyphosphate resulting from DMPY decomposition, and the crosslinking effect of metal ions that came from SS during combustion. Based on the above investigation, we came to the conclusion that the synergistic effect of DMPY/SS system effectively inhibited the heat and mass transfer in combustion region, endowing RPUF-3 with excellent fire-retarding performance.

Raman spectroscopy is a common method to measure the graphitization degree of carbon materials [50]. Figure 8 shows the Raman spectra of RPUF and RPUF-R composites. As shown in the figure, two broadband bands were present in all the samples. The D band at around 1360 cm^{-1} corresponded to the amorphous phase consisting of disordered carbon atoms, and the G band at 1590 cm^{-1} was attributed to crystalline phase consisting of graphited carbon atoms [51]. The area ratio of I_D/I_G was usually used to represent the graphitization degree of the sample. The smaller the ratio of I_D/I_G, the higher graphitization degree of carbon residue with better fire resistance [52,53]. As observed from Figure 8, the I_D/I_G values of the carbon residues for RPUF-R composites were all lower than that of unmodified RPUF, suggesting that the usage, either alone or combination, of SS and DMPY could endow RPUF-R composites with enhanced fire-retarding performance.

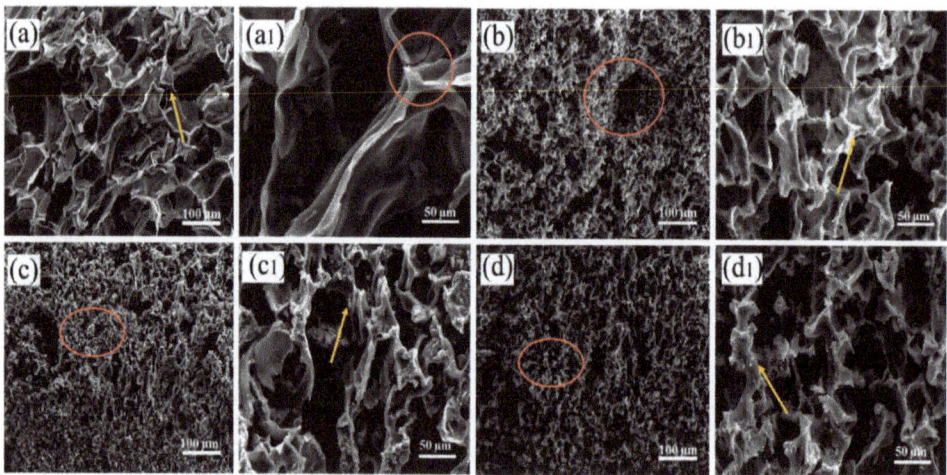

Figure 7. SEM images of RPUF and RPUF-R composite carbon slag: (**a,a1**) RPUF; (**b,b1**) RPUF-1; (**c,c1**) RPUF-3; (**d,d1**) RPUF-5.

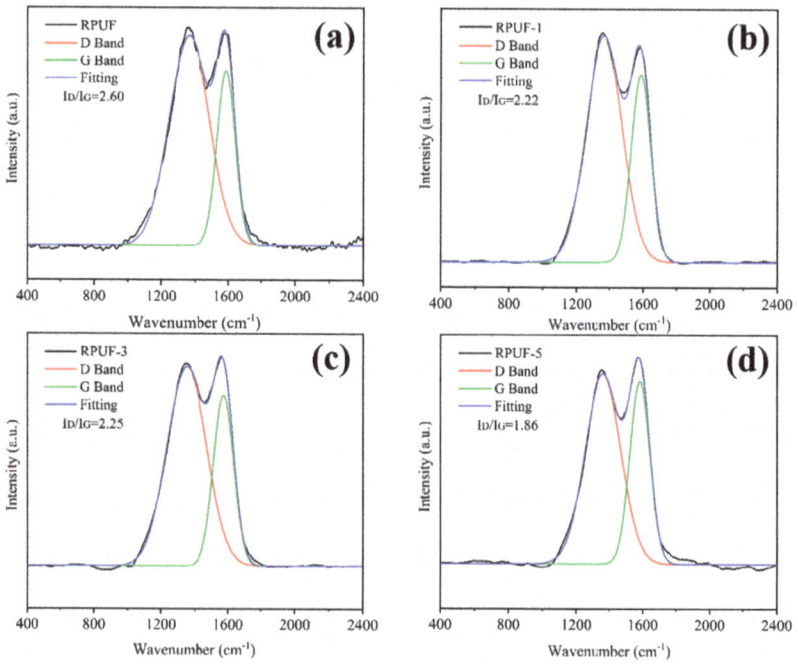

Figure 8. Raman spectra of RPUF and RPUF-R composites: (**a**) RPUF; (**b**) RPUF-1; (**c**) RPUF-3; (**d**) RPUF-5.

3. Experiment

3.1. Experimental Materials

Polyether polyols (LY-4110), triethylenediamine solution (A33), industrial grade, were all provided by Jiangsu Lvyuan New Materials Technology Co., Ltd. (Nantong, China). Trimethylene polyphenyl polyisocyanate (PM-200, industrial grade) was purchased from Wanhua Chemical Group Co., Ltd. (Yantai, China). Silicone oil foam stabilizer (AK-8805, industrial grade) was purchased from Jining Hengtai Chemical Co., Ltd. (Jining, China). Triethanolamine (TEOA, Chemically Pure), was purchased from Chemical Reagent Co., Ltd. (Shanghai, China). Dibutyltin dilaurate (LC, chemical purity) was purchased from Air Chemical Co., Ltd. (Delaware, Ameica). Dimelamine pyrophosphate (DMPY) was purchased from Dongguan Shengde New Materials Co., Ltd. (Dongguan, China). Steel slag (SS) was kindly provided by Masteel Group. None of the above chemicals have been further purified.

3.2. Sample Foaming

A series of RPUF and RPUF-R composites were prepared by one-step aqueous foaming process. The composition ratio of the composites was shown in Table 5. Firstly, LY-4110, LC, A33, AK-8005, TEA, SS, DMPY, and distilled water were added into a 500 mL beaker and stirred at 2000 rpm for 1 min. Then, the weighed PM-200 was further added into the beaker with intense stirring for another 10 s until the mixture turned white, which was then poured into the mold quickly. The samples were further reacted at 80 °C for 6 h to complete the polymerization process. Then, the obtained sample was cut into suitable sizes for the following testing.

3.3. Sample Characterization

Limit oxygen index (LOI): JF-3 oxygen index analyzer (Nanjing Jiangning District Instrument Analysis Factory) was applied to test the limit oxygen index of the composites with sample size of 127 mm × 10 mm × 10 mm. Vertical combustion (UL-94): Horizontal vertical combustion tester (CZF-2) was introduced to test combustion rating of RPUF and RPUF composites with sample size of 127 mm × 13 mm × 10 mm. Mechanical properties: The DCS-5000 universal material testing machine was applied to test the compressive strength of the RPUF and RPUF-R composites with sample size of 50 mm × 50 mm × 40 mm, and each sample was tested for five times to obtain average value. Thermal conductivity test: TC3000E thermal conductivity meter was used to test the thermal conductivity of the composites with sample size of 50 mm × 50 mm × 25 mm, and the average value was obtained by test for three measurements. Cone calorimeter (CCT): 6180 (Siemens analyzer) cone calorimetry was used to test the heat release rate and smoke release of the composites, where the radiation flux was 35 kW/m^2 and the sample size was 100 mm × 100 mm × 25 mm. Scanning electron microscopy (SEM): The morphology of the composites and carbon slags were investigated by JSM 6490LV scanning electron microscope. In order to enhance the electrical conductivity of the samples, the samples were sprayed with conductive layer. Laser confocal Raman spectroscopy (Roman): The carbon slag was obtained by calcining the composites in a muffle furnace at 600 °C for 10 min, and then tested by laser confocal Raman spectroscopy (LabRAM HR Evolution, HORIBA Scientific) to investigate the graphitization degree of the carbon slag for the composites.

4. Conclusions

In this work, a series of RPUF and RPUF-R composites were fabricated by one-step all-water foaming method using DMPY and SS as flame retardants. Additionally, the effects of DMPY and SS on the microscopic morphology, thermal stability, flame retardancy, combustion properties, gas phase products, and carbon residue of the composites were systematically investigated. The LOI test confirmed that RPUF/DMPY and RPUF/DMPY/SS composites all passed UL 94 V0 rating, and the LOI values reached the combustible material grade (22.0 vol%~27.0 vol%). The RPUF/SS composite with SS alone cannot pass UL 94

V0 rating with relatively low LOI value of 20.4 vol%. RPUF-R composites were observed dripping phenomenon disappeared, implying the enhanced fire hazard of the composites. TG test showed that the residual carbon value and $T_{-5wt\%}$ of RPUF-1, RPUF-3, and RPUF-5 composites at 700 °C were higher than those of unmodified RPUF, confirming that the addition of DMPY and SS could significantly improve the high temperature thermal stability of the composites. Cone colorimetry test showed that RPUF-3 composite possessed the lowest pHRR and THR values of 134.9 kW/m^2 and 16.16 MJ/m^2, indicating that the combination of DMPY and SS could inhibit the heat release of the composites in combustion. The carbon slag investigation showed that the DMPY/SS synergistic system improved the denseness and graphitization degree of the carbon layer for RPU composites, thus inhibiting the heat and mass transport of the composites in combustion. The above results show that DMPY/SS can be a feasible strategy for fabricating fire-retardant RPUF composites, which also paves a new way for solid waste utilization.

Author Contributions: Conceptualization, M.Z. and X.L.; methodology, S.Y. and Z.L.; validation, S.P.; formal analysis, M.Z. and S.Y.; investigation, M.Z. and X.L.; data curation, S.P. and Z.L.; writing—original draft preparation, M.Z. and S.Y.; writing—review and editing, X.L. and Z.L.; visualization, S.P.; supervision, M.Z. and X.L. All authors have read and agreed to the published version of the manuscript.

Funding: This work was supported by National Natural Science Foundation of China (No. 52000102), Provincial Nature Science Foundation of Jiangsu Province (No. BK20190689) and Nature Science Foundation of Anhui Province (No. 2108085ME178).

Institutional Review Board Statement: Not applicable.

Informed Consent Statement: Not applicable.

Data Availability Statement: The data will be available on request.

Acknowledgments: National Natural Science Foundation of China (No. 52000102), Provincial Nature Science Foundation of Jiangsu Province (No. BK20190689) and Nature Science Foundation of Anhui Province (No. 2108085ME178).

Conflicts of Interest: The authors declare that they have no conflict of interest with this work.

References

1. Huang, Y.B.; Jiang, S.H.; Liang, R.C.; Sun, P.; Hai, Y.; Zhang, L. Thermal-triggered insulating fireproof layers: A novel fire-extinguishing MXene composites coating. *Chem. Eng. J.* **2020**, *391*, 123621. [CrossRef]
2. Dong, M.Y.; Li, Q.; Liu, H.; Liu, C.T.; Wujcik, E.K.; Shao, Q.; Ding, T.; Mai, X.M.; Shen, C.Y.; Guo, Z.H. Thermoplastic polyurethane-carbon black nanocomposite coating: Fabrication and solid particle erosion resistance. *Polymer* **2018**, *158*, 381–390. [CrossRef]
3. Tang, G.; Zhou, L.; Zhang, P.; Hang, Z.Q.; Chen, D.P.; Liu, X.Y.; Zhou, Z.J. Effect of aluminum diethylphosphinate on flame retardant and thermal properties of rigid polyurethane foam composites. *J. Therm. Anal. Calorim.* **2020**, *140*, 625–636. [CrossRef]
4. Tan, C.J.; Lee, J.J.L.; Ang, B.C.; Andriyana, A.; Chagnon, G.; Sukiman, M.S. Design of polyurethane fibers: Relation between the spinning technique and the resulting fiber topology. *J. Appl. Polym. Sci.* **2019**, *136*, 47706. [CrossRef]
5. Tang, G.; Liu, M.R.; Deng, D.; Zhao, R.Q.; Liu, X.L.; Yang, Y.D.; Yang, S.J.; Liu, X.Y. Phosphorus-containing soybean oil-derived polyols for flame-retardant and smoke-suppressant rigid polyurethane foams. *Polym. Degrad. Stab.* **2021**, *191*, 109701. [CrossRef]
6. Naldzhiev, D.; Mumovic, D.; Strlic, M. Polyurethane insulation and household products-a systematic review of their impact on indoor environmental quality. *Build. Environ.* **2020**, *169*, 106559. [CrossRef]
7. Yuan, Y.; Wang, W.; Shi, Y.Q.; Song, L.; Ma, C.; Hu, Y. The influence of highly dispersed Cu$_2$O-anchored MoS$_2$ hybrids on reducing smoke toxicity and fire hazards for rigid polyurethane foam. *J. Hazard. Mater.* **2020**, *382*, 121028. [CrossRef]
8. Gajdzik, B. Polish crude steel production in pandemic year of 2020 compared to previous five years. *Metalurgija* **2022**, *61*, 137–140.
9. Yüksel, İ. A review of steel slag usage in construction industry for sustainable development. *Environ. Dev. Sustain.* **2016**, *19*, 369–384. [CrossRef]
10. Yi, H.; Xu, G.; Cheng, H.; Wang, J.; Wan, Y.; Chen, H. An overview of utilization of steel slag. *Procedia Environ. Sci.* **2012**, *16*, 791–801. [CrossRef]
11. Gencel, O.; Karadag, O.; Oren, O.H.; Bilir, T. Steel slag and its applications in cement and concrete technology: A review. *Constr. Build. Mater.* **2021**, *283*, 122783. [CrossRef]
12. Guo, J.; Bao, Y.; Wang, M. Steel slag in China: Treatment, recycling, and management. *Waste Manag.* **2018**, *78*, 318–330. [CrossRef] [PubMed]

13. Wang, Z.Y.; Li, J.F.; He, X.L.; Yang, G.; Qi, J.Y.; Zhao, C. Organic pollutants removal performance and enhanced mechanism investigation of surface-modified steel slag particle electrode. *Environ. Prog. Sustain. Energy* **2019**, *38*, S7–S14. [CrossRef]
14. Thomas, C.; Rosales, J.; Polanco, J.A.; Agrela, F. *New Trends in Eco-Efficient and Recycled Concrete*; Woodhead Publishing: Sawston, UK, 2019; pp. 169–190.
15. Xu, J.Y.; Li, K.D.; Deng, H.M.; Lv, S.; Fang, P.K.; Liu, H.; Shao, Q.; Guo, Z.H. Preparation of MCA-SiO_2 and its flame retardant effects on glass fiber reinforced polypropylene. *Fibers Polym.* **2019**, *20*, 120–128. [CrossRef]
16. Tang, G.; Liu, X.L.; Yang, Y.D.; Chen, D.P.; Zhang, H.; Zhou, L.; Zhang, P.; Deng, D. Phosphorus-containing silane modified steel slag waste to reduce fire hazards of rigid polyurethane foams. *Adv. Powder Technol.* **2020**, *31*, 1420–1430. [CrossRef]
17. Levchik, S.V.; Balabanovich, A.I.; Levchik, G.F.; Costa, L. Effect of melamine and its salts on combustion and thermal decomposition of polyamide 6. *Fire Mater.* **1997**, *21*, 75–83. [CrossRef]
18. Sut, A.; Metzsch-Zilligen, E.; Großhauser, M.; Pfaendner, R.; Schartel, B. Synergy between melamine cyanurate, melamine polyphosphate and aluminum diethylphosphinate in flame retarded thermoplastic polyurethane. *Polym. Test.* **2019**, *74*, 196–204. [CrossRef]
19. Xu, M.; Ma, Y.; Hou, M.; Bourbigot, S. Synthesis of a cross-linked triazine phosphine polymer and its effect on fire retardancy, thermal degradation and moisture resistance of epoxy resins. *Polym. Degrad. Stab.* **2015**, *119*, 14–22. [CrossRef]
20. Lai, X.; Tang, S.; Li, H.; Zeng, X. Flame-retardant mechanism of a novel polymeric intumescent flame retardant containing caged bicyclic phosphate for polypropylene. *Polym. Degrad. Stab.* **2015**, *113*, 22–31. [CrossRef]
21. Liu, L.; Xu, Y.; Xu, M.; Li, Z.; Hu, Y.; Li, B. Economical and facile synthesis of a highly efficient flame retardant for simultaneous improvement of fire retardancy, smoke suppression and moisture resistance of epoxy resins. *Compos. Part B-Eng.* **2019**, *167*, 422–433. [CrossRef]
22. Liu, L.B.; Xu, Y.; Xu, M.J.; He, Y.T.; Li, S.; Li, B. An efficient synergistic system for simultaneously enhancing the fire retardancy, moisture resistance and electrical insulation performance of unsaturated polyester resins. *Mater. Des.* **2020**, *187*, 108302. [CrossRef]
23. Akdogan, E.; Erdem, M.; Ureyen, M.E.; Kaya, M. Rigid polyurethane foams with halogen-free flame retardants: Thermal insulation, mechanical, and flame retardant properties. *J. Appl. Polym. Sci.* **2020**, *137*, 47611. [CrossRef]
24. Chai, H.; Duan, Q.; Jiang, L.; Sun, J. Effect of inorganic additive flame retardant on fire hazard of polyurethane exterior insulation material. *J. Therm. Anal. Calorim.* **2018**, *135*, 2857–2868. [CrossRef]
25. Almessiere, M.A.; Slimani, Y.; Güngüneş, H.; Korkmaz, A.D.; Zubar, T.; Trukhanov, S.; Trukhanov, A.; Manikandan, A.; Alahmari, F.; Baykal, A. Influence of Dy^{3+} ions on the microstructures and magnetic, electrical, and microwave properties of $[Ni_{0.4}Cu_{0.2}Zn_{0.4}](Fe_2\text{-xDy}x)O_4$ $(0.00 \leq x \leq 0.04)$ spinel ferrites. *ACS Omega* **2021**, *6*, 10266–10280. [CrossRef] [PubMed]
26. Darwish, M.A.; Zubar, T.I.; Kanafyev, O.D.; Zhou, D.; Trukhanova, E.L.; Trukhanov, S.V.; Trukhanov, A.V.; Henaish, A.M. Combined effect of microstructure, surface energy, and adhesion force on the friction of PVA/ferrite spinel nanocomposites. *Nanomaterials* **2022**, *12*, 1998. [CrossRef] [PubMed]
27. Chen, K.X.; Yang, D.; Shi, Y.Q.; Feng, Y.Z.; Fu, L.B.; Liu, C.; Chen, M.; Yang, F.Q. Synergistic function of N-P-Cu containing supermolecular assembly networks in intumescent flame retardant thermoplastic polyurethane. *Polym. Adv. Technol.* **2021**, *32*, 4450–4463. [CrossRef]
28. Yang, S.; Zhang, B.; Liu, M.; Yang, Y.; Liu, X.; Chen, D.; Wang, B.; Tang, G.; Liu, X. Fire performance of piperazine phytate modified rigid polyurethane foam composites. *Polym. Adv. Technol.* **2021**, *32*, 4531–4546. [CrossRef]
29. Yang, R.; Wang, B.; Xu, L.; Zhao, C.X.; Zhang, X.; Li, J.C. Synthesis and characterization of rigid polyurethane foam with dimer fatty acid-based polyols. *Polym. Bull.* **2019**, *76*, 3753–3768. [CrossRef]
30. Pang, X.; Xin, Y.; Shi, X.Z.; Xu, J.Z. Effect of different size–modified expandable graphite and ammonium polyphosphate on the flame retardancy, thermal stability, physical, and mechanical properties of rigid polyurethane foam. *Polym. Eng. Sci.* **2019**, *59*, 1381–1394. [CrossRef]
31. Tang, G.; Liu, X.L.; Zhou, L.; Zhang, P.; Deng, D.; Jiang, H.H. Steel slag waste combined with melamine pyrophosphate as a flame retardant for rigid polyurethane foams. *Adv. Powder Technol.* **2020**, *31*, 279–286. [CrossRef]
32. Kaya, H.; Özdemir, E.; Kaynak, C.; Hacaloglu, J. Effects of nanoparticles on thermal degradation of polylactide/aluminium diethylphosphinate composites. *J. Anal. Appl. Pyrolysis* **2016**, *118*, 115–122. [CrossRef]
33. Chambhare, S.U.; Lokhande, G.P.; Jagtap, R.N. Design and UV-curable behaviour of boron based reactive diluent for epoxy acrylate oligomer used for flame retardant wood coating. *Des. Monomers Polym.* **2016**, *20*, 125–135. [CrossRef] [PubMed]
34. Liu, L.B.; Xu, Y.; He, Y.T.; Xu, M.J.; Shi, Z.X.; Hu, H.C.; Yang, Z.C.; Li, B. An effective mono-component intumescent flame retardant for the enhancement of water resistance and fire safety of thermoplastic polyurethane composites. *Polym. Degrad. Stab.* **2019**, *167*, 146–156. [CrossRef]
35. Wang, X.; Song, L.; Pornwannchai, W.; Hu, Y.; Kandola, B. The effect of graphene presence in flame retarded epoxy resin matrix on the mechanical and flammability properties of glass fiber-reinforced composites. *Compos. Part A Appl. Sci. Manuf.* **2013**, *53*, 88–96. [CrossRef]
36. Zhang, J.H.; Kong, Q.H.; Yang, L.W.; Wang, D.Y. Few layered $Co(OH)_2$ ultrathin nanosheet-based polyurethane nanocomposites with reduced fire hazard: From eco-friendly flame retardance to sustainable recycling. *Green Chem.* **2016**, *18*, 3066–3074. [CrossRef]
37. Chattopadhyay, D.K.; Webster, D.C. Thermal stability and flame retardancy of polyurethanes. *Prog. Polym. Sci.* **2009**, *34*, 1068–1133. [CrossRef]

38. Chen, M.J.; Chen, C.R.; Tan, Y.; Huang, J.Q.; Wang, X.L.; Chen, L.; Wang, Y.Z. Inherently flame-retardant flexible polyurethane foam with low content of phosphorus-containing cross-linking agent. *Ind. Eng. Chem. Res.* **2014**, *53*, 1160–1171. [CrossRef]
39. Liu, C.; Zhang, P.; Shi, Y.; Rao, X.; Cai, S.; Fu, L.; Feng, Y.; Wang, L.; Zheng, X.; Yang, W. Enhanced fire safety of rigid polyurethane foam via synergistic effect of phosphorus/nitrogen compounds and expandable graphite. *Molecules* **2020**, *25*, 4741. [CrossRef]
40. Yang, S.; Mo, L.; Deng, M. Effects of ethylenediamine tetra-acetic acid (EDTA) on the accelerated carbonation and properties of artificial steel slag aggregates. *Cem. Concr. Compos.* **2021**, *118*, 103948. [CrossRef]
41. Chen, X.L.; Ma, C.Y.; Jiao, C.M. Synergistic effects between iron-graphene and melamine salt of pentaerythritol phosphate on flame retardant thermoplastic polyurethane. *Polym. Adv. Technol.* **2016**, *27*, 1508–1516. [CrossRef]
42. Chen, H.B.; Shen, P.; Chen, M.-J.; Zhao, H.B.; Schiraldi, D.A. Highly efficient flame retardant polyurethane foam with alginate/clay aerogel coating. *ACS Appl. Mater. Interfaces* **2016**, *8*, 32557–32564. [CrossRef] [PubMed]
43. Zhang, B.; Feng, Z.H.; Han, X.X.; Wang, B.B.; Yang, S.J.; Chen, D.P.; Peng, J.W.; Yang, Y.D.; Liu, X.Y.; Tang, G. Effect of ammonium polyphosphate/cobalt phytate system on flame retardancy and smoke & toxicity suppression of rigid polyurethane foam composites. *J. Polym. Res.* **2021**, *28*, 407. [CrossRef]
44. Wu, Z.H.; Wang, Q.; Fan, Q.X.; Cai, Y.J.; Zhao, Y.Q. Synergistic effect of Nano-ZnO and intumescent flame retardant on flame retardancy of polypropylene/ethylene-propylene-diene monomer composites using elongational flow field. *Polym. Compos.* **2019**, *40*, 2819–2833. [CrossRef]
45. Qian, L.J.; Li, L.J.; Chen, Y.J.; Xu, B.; Qiu, Y. Quickly self-extinguishing flame retardant behavior of rigid polyurethane foams linked with phosphaphenanthrene groups. *Compos. Part B Eng.* **2019**, *175*, 107186. [CrossRef]
46. Tang, G.; Zhao, R.; Deng, D.; Yang, Y.; Chen, D.; Zhang, B.; Liu, X.; Liu, X. Self-extinguishing and transparent epoxy resin modified by a phosphine oxide-containing bio-based derivative. *Front. Chem. Sci. Eng.* **2021**, *15*, 1269–1280. [CrossRef]
47. Shi, X.X.; Jiang, S.H.; Zhu, J.Y.; Li, G.H.; Peng, X.F. Establishment of a highly efficient flame-retardant system for rigid polyurethane foams based on bi-phase flame-retardant actions. *RSC Adv.* **2018**, *8*, 9985–9995. [CrossRef]
48. Zhou, K.Q.; Gui, Z.; Hu, Y.; Jiang, S.H.; Tang, G. The influence of cobalt oxide–graphene hybrids on thermal degradation, fire hazards and mechanical properties of thermoplastic polyurethane composites. *Compos. Part A Appl. Sci. Manuf.* **2016**, *88*, 10–18. [CrossRef]
49. Chen, Y.J.; Li, L.S.; Qi, X.Q.; Qian, L.J. The pyrolysis behaviors of phosphorus-containing organosilicon compound modified APP with different polyether segments and their flame retardant mechanism in polyurethane foam. *Compos. Part B Eng.* **2019**, *173*, 106784. [CrossRef]
50. Gong, K.L.; Zhou, K.Q.; Yu, B. Superior thermal and fire safety performances of epoxy-based composites with phosphorus-doped cerium oxide nanosheets. *Appl. Surf. Sci.* **2019**, *504*, 144314. [CrossRef]
51. Liu, C.; Yao, A.; Chen, K.; Shi, Y.; Feng, Y.; Zhang, P.; Yang, F.; Liu, M.; Chen, Z. MXene based core-shell flame retardant towards reducing fire hazards of thermoplastic polyurethane. *Compos. Part B Eng.* **2021**, *226*, 109363. [CrossRef]
52. Zhou, K.Q.; Liu, C.C.; Gao, R. Polyaniline: A novel bridge to reduce the fire hazards of epoxy composites. *Compos. Part A Appl. Sci. Manuf.* **2018**, *112*, 432–443. [CrossRef]
53. Shi, Y.Q.; Liu, C.; Duan, Z.P.; Yu, B.; Liu, M.H.; Song, P.A. Interface engineering of MXene towards super-tough and strong polymer nanocomposites with high ductility and excellent fire safety. *Chem. Eng. J.* **2020**, *399*, 125829. [CrossRef]

Article

A Bridge-Linked Phosphorus-Containing Flame Retardant for Endowing Vinyl Ester Resin with Low Fire Hazard

Zihui Xu [1], Jing Zhan [1,*], Zhirong Xu [1], Liangchen Mao [1], Xiaowei Mu [2,3,*] and Ran Tao [1]

1 School of Civil Engineering, Anhui Jianzhu University, Hefei 230601, China
2 State Key Laboratory of Fire Science, University of Science and Technology of China, Hefei 230026, China
3 Anhui Province Key Laboratory of Human Safety, Hefei 230601, China
* Correspondence: zhanjing@ahjzu.edu.cn (J.Z.); xwmu@ustc.edu.cn (X.M.)

Abstract: The high flammability of vinyl ester resin (VE) significantly limits its widespread application in the fields of electronics and aerospace. A new phosphorus-based flame retardant 6,6'-(1-phenylethane-1,2 diyl) bis (dibenzo[c,e][1,2]oxaphosphinine 6-oxide) (PBDOO), was synthesized using 9,10-dihydro-9-oxa-10-phosphaphenanthrene-10-oxide (DOPO) and acetophenone. The synthesized PBDOO was further incorporated with VE to form the VE/PBDOO composites, which displayed an improved flame retardancy with higher thermal stability. The structure of PBDOO was investigated using Fourier transformed infrared spectrometry (FTIR) and nuclear magnetic resonances (NMR). The thermal stability and flame retardancy of VE/PBDOO composites were investigated by thermogravimetric analysis (TGA), vertical burn test (UL-94), limiting oxygen index (LOI), and cone calorimetry. The impacts of PBDOO weight percentage (wt%) on the flame-retardant properties of the formed VE/PBDOO composites were also examined. When applying 15 wt% PBDOO, the formed VE composites can meet the UL-94 V-0 rating with a high LOI value of 31.5%. The peak heat release rate (PHRR) and the total heat release (THR) of VE loaded 15 wt% of PBDOO decreased by 76.71% and 40.63%, respectively, compared with that of untreated VE. In addition, the flame-retardant mechanism of PBDOO was proposed by analyzing pyrolysis behavior and residual carbon of VE/PBDOO composites. This work is expected to provide an efficient method to enhance the fire safety of VE.

Keywords: vinyl ester resin; flame retardant; thermal stability; mechanism analysis

1. Introduction

Vinyl ester resin (VE) is prepared by the reaction of bisphenol-A epoxy resin and methacrylic acid [1]. Due to its excellent electrical insulation, adhesion, chemical resistance, and easy processing characteristics, VE plays a vital role in industries such as coatings, adhesives, anti-corrosion paints, and electronics [2–5]. However, the fire hazard potential of VE is higher than that of bisphenol-A epoxy resin due to its easily degradable vinyl [4,5]. The inflammability of VE limits its application in construction, aviation, and other special fields [6–9]. Therefore, it is imperative to improve the flame-retardant properties of VE.

Phosphorus-based flame retardant has been widely used because of its eco-friendly degradation products and high flame-retardant efficiency [10–13]. 9,10-dihydro-9-oxa-10-phospha-phenanthrene-10-oxide (DOPO) and its derivatives are often used as an efficient reactive phosphorus-based flame retardant for epoxy resins [14–18]. DOPO can exert flame retardants effect in the gas phase and condensed phase by releasing free radicals and promoting carbonation [19]. The limiting oxygen index (LOI) and vertical burning rating of epoxy resin can reach 36.2% and V-0 after incorporating 3 wt% ABD, which is a flame retardant synthesized by acrolein and DOPO [15]. Fang et al. synthesized a new flame retardant TDA containing phosphorus, nitrogen, and silicon-based on DOPO, and 25 wt% TDA can endow EP composites with an LOI value of 33.4% [18]. However, the flame retardancy of VE with DOPO and its derivatives is low, and the LOI of VE composites can only reach 31.5% with 30 wt% DOPO-2-hydroxylethyl acrylates [16].

The main factors that affect the flame-retardant efficiency of DOPO are: (a) Phosphorus content of DOPO derivatives. (b) Other functional groups in DOPO derivatives [17,19]. Epoxy resin with 17.5 wt% phenethyl-bridged DOPO derivative (PN-DOPO) achieved vertical burn test (UL-94) V-0 grade, and its average heat release rate (av-HRR) value was reduced by 32.5% [20]. Acetophenone can react with DOPO to construct a bridging structure between DOPO molecules, thus increasing the overall content of phosphorus and aromatic group in the formed DOPO derivatives [19,20]. Therefore, acetophenone was considered as a sensitizer for DOPO to generate a novel DOPO derivative with improved flame-retardant efficiency.

In this work, a novel phosphorus-based flame retardant named PBDOO was synthesized by DOPO and acetophenone. The thermal stability, mechanical properties, and flame-retardant properties of VE composites with PBDOO were studied in depth. Finally, the flame-retardant mechanism of VE composites was further revealed.

2. Result and Discussion

2.1. Characterization of PBDOO

The Fourier transform infrared (FTIR) spectra of PBDOO is shown in Figure 1a, and the groups corresponding to each absorption peak are listed in Table 1.

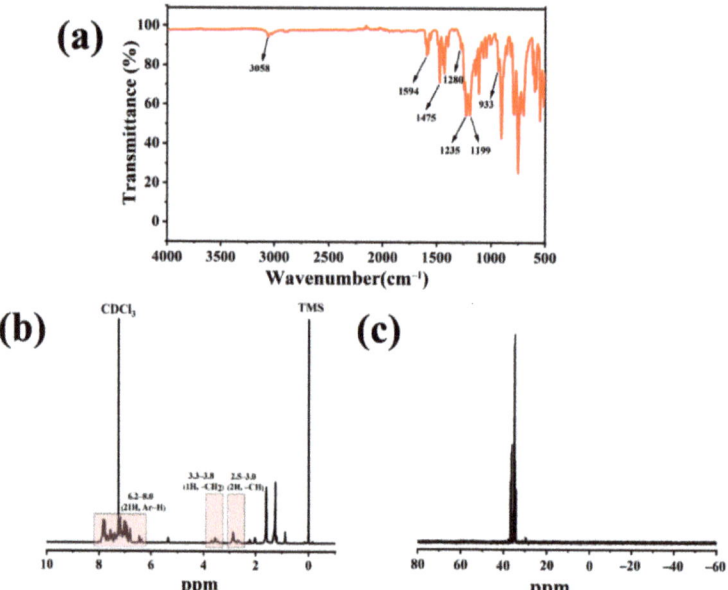

Figure 1. (a) FTIR spectra, (b) ^{1}H NMR spectra and (c) ^{31}P NMR spectra of PBDOO.

Table 1. Wavenumber and corresponding chemical groups in PBDOO.

Wavenumber (cm^{-1})	Corresponding Chemical Groups
3058	C-H stretching vibration peak of the aromatic ring
1594	P-Ar stretching vibration peak
1475	P-Ph absorption peak
1280	O=P-O absorption peak of the benzene ring structure
1235	P=O vibration absorption peak
1199, 933	P-O-Ph vibration absorption peak

It is well-known that the P-H vibration absorption peak at 2400 cm^{-1} is the most obvious characteristic absorption peak of DOPO [21,22]. And acetophenone has a C=O absorption peak of carbonyl around 1727 cm^{-1}. Neither of these absorption peaks is shown in Figure 1a, indicating that the target product is free of DOPO and acetophenone [21]. The new absorption peaks listed in Table 1 indicate that the target product is successfully synthesized [22,23].

^{1}H and ^{31}P nuclear magnetic resonance (NMR) spectra were used to verify the structure of PBDOO. As shown in Figure 1b,c, the characteristic peak at 2.85 ppm in the ^{1}H NMR spectrum of PBDOO corresponds to the hydrogen atom in the -CH$_2$- structure [21,22,24]. In addition, the typical peak at 6.8–8.0 ppm corresponds to the hydrogen atom of the benzene ring [21]. ^{31}P NMR is multimodal at 34.0–37.0 ppm. The presence of these peaks indicates that PBDOO is successfully synthesized.

Thermogravimetric analysis (TGA) and Thermogravimetric differentiation analysis (DTG) curves of PBDOO under nitrogen and air atmospheres are shown in Figure 2. The corresponding data are displayed in Table 2. Among them, $T_{5\%}$ is defined as the temperature at 5% weight loss, T_{max} is defined as the temperature at which the maximum weight loss rate is, and the temperature selected for the char residues is 800 °C.

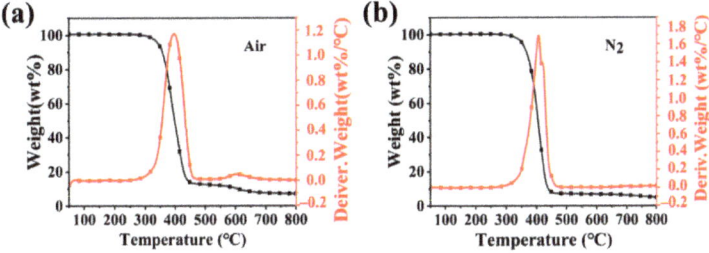

Figure 2. TGA and DTG curves of PBDOO in (**a**) Air and (**b**) N$_2$.

Table 2. TGA and DTG data of PBDOO and the VE composites.

Sample	N$_2$			Air			
	$T_{5\%}$ (°C)	T_{max} (°C)	Residue (wt%)	$T_{5\%}$ (°C)	T_{max1} (°C)	T_{max2} (°C)	Residue (wt%)
PBDOO	352	405	4.9	344	388	–	7.2
VE-0	316	420	6.8	356	422	546	0.0
VE-5	338	414	13.8	358	414	561	0.6
VE-10	319	414	9.1	334	415	562	0.7
VE-15	304	414	9.7	317	413	561	1.3

As shown in Figure 2a, under air atmosphere, the $T_{5\%}$ of PBDOO is 344 °C, the T_{max} of PBDOO is 388 °C, and its char residues at 800 °C are 7.2%. Figure 2b shows the results under a nitrogen atmosphere. The $T_{5\%}$ and T_{max} are 352 °C and 405 °C, respectively. It can be seen that PBDOO has high thermal stability in both air and nitrogen atmosphere, and the initial decomposition temperatures are above 340 °C. The processing temperature of VE is about 150 °C, so PBDOO can meet the processing conditions of VE. Furthermore, the amount of char residue under the air atmosphere is significantly higher than that under a nitrogen atmosphere, which may be because the presence of oxygen promotes the formation of stable carbon layers [24,25].

2.2. Morphologies and Mechanical Properties of the VE Composites

In order to study the effect of PBDOO on the mechanical properties of VE materials, the tensile strength and elongation at break of VE composites with different proportions of flame retardant were tested. The test results are shown in Figure 3. The tensile strength and elongation at the break of the VE without PBDOO are 33 MPa and 25%, respectively. As the content of PBDOO increases, the strength of the cured product first increases from 33 MPa to 36 MPa, and then dropped to 14 MPa for VE-15. The elongation at break changed slightly from 25% of VE-0 to 21% of VE-15. The result indicates that with a small amount of PBDOO involved in the curing process, the VE composites can form a stable cross-linked structure, which gives an increase in tensile strength [26]. When the content of PBDOO further increases, the steric hindrance produced by the rigid DOPO group limited the cross-linking between VE and the curing agent. So the tensile strength and elongation at the break decrease [26,27].

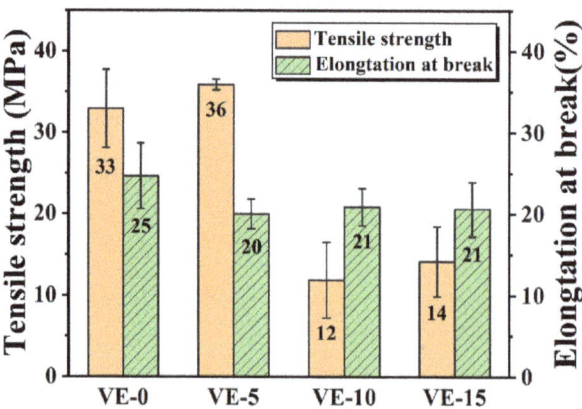

Figure 3. Strength and elongation at break of VE composites.

Figure 4 shows the scanning electron microscope (SEM) image of the cross-section of VE-0 and VE-15. It could be seen that the surface of VE-15 is slightly rougher than that of VE-0, which indicates a minor effect of PBDOO on VE. In addition, there are no noticeable large particles on the fracture surfaces of the two samples, which also indicates that PBDOO could be well dispersed in the VE material. This result shows that PBDOO has good compatibility with VE and can be well dispersed in VE composites.

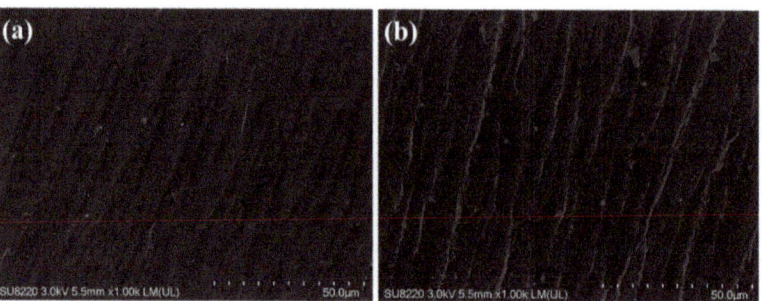

Figure 4. SEM images of the fractured surface of VE composites: (**a**) VE-0, (**b**) VE-15.

2.3. Thermal Stability

TGA and DTG curves of the VE composites under nitrogen and air atmospheres are shown in Figure 5 and the corresponding data are summarized in Table 2. As shown in Figure 5a,c, with the addition of PBDOO, the $T_{5\%}$ temperature of the VE composites moves forward slightly, indicating that the addition of the flame retardant reduces the initial decomposition temperature of VE.

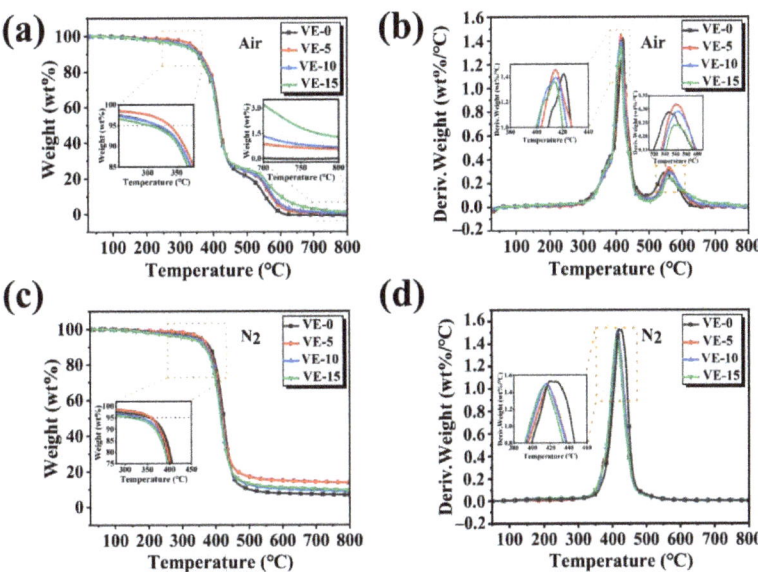

Figure 5. TGA and DTG curves of VE composites in air (**a,b**) and N$_2$ (**c,d**).

It could be observed from Figure 5b that all the samples have a two-stage decomposition process in the air atmosphere. The first decomposition occurs at around 330 °C and reaches T$_{max}$ at 410 °C, which may be due to the decomposition of aromatic rings and alkyl chains [23]. The second degradation of VE composites occurs at about 500 °C and reaches T$_{max}$ at about 550 °C, which may be due to the further thermal-oxidative decomposition of the unstable char layer at high temperatures. The results in Table 2 show that as the amount of PBDOO increases, so does the residual carbon of the VE composites.

2.4. Flame-Retardant Properties of the VE Composites

LOI and UL-94 were used to judge the flame retardancy of the samples, and the data are shown in Table 3. As the content of PBDOO increases, the LOI value of VE composites also increases significantly. The LOI value of VE-15 reached 31.5%, which is much higher than that of VE-0. And the vertical burning rating of VE got V-0 after incorporating 15 wt% PBDOO. Therefore, it could be concluded that PBDOO can effectively improve the flame-retardant properties of VE composites.

Table 3. LOI and UL-94 results of the VE composites.

Sample	VE (g)	BPO (g)	PBDOO (g)	LOI (%)	UL-94
VE-0	50.00	1.50	0.00	22.00	V-2
VE-5	47.50	1.43	2.50	24.50	V-1
VE-10	45.00	1.35	5.00	25.50	V-1
VE-15	42.50	1.28	7.50	31.50	V-0

The combustion behavior of VE was further studied by a cone calorimeter. The total heat release (THR) and heat release rate (HRR) curves of VE composites are shown in Figure 6. Some of the essential parameters are summarized in Table 4. From Figure 6a,b, it can be seen that VE-0 burns quickly and releases a large amount of heat after being ignited. The peak of HRR (PHRR) and THR of VE-15 are 339 kW/m^2 and 47.9 MJ/m^2, respectively, which are reduced by 76.71% and 40.36% compared with those of VE-0.

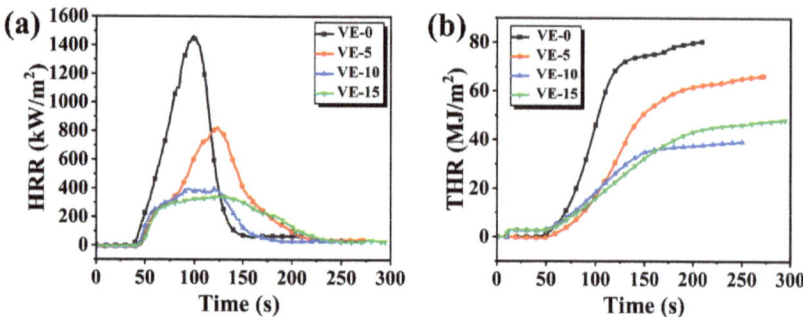

Figure 6. (a) Heat release rate of VE composites, (b) total heat release of VE composites.

Table 4. Cone calorimetry data of VE composites.

Sample	TTI (s)	PHRR (Kw/m^2)	THR (MJ/m^2)	CO$_2$ PR (g/s)	COPR (g/s)
VE-0	59	1455	80.4	1.12	0.041
VE-5	66	822	66.0	0.62	0.038
VE-10	64	385	39.1	0.29	0.052
VE-15	67	339	47.9	0.22	0.044

Note: CO$_2$ PR value in Table 4 is the peak of the CO$_2$ production rate; COPR value in Table 4 is the peak of the CO production rate.

Figure 7 shows the changes in CO production rate (COPR) and CO$_2$ production rate (CO$_2$ PR) of the VE material during the cone tests. The peak value of CO$_2$ PR for VE-0 is 1.12 g/s. With the increase of PBDOO content in VE composites, the peak values of CO$_2$ PR decreased by 44.64%, 74.11%, and 80.35%, respectively. But the values of COPR have no obvious change trend as observed in Figure 7b. The changes in CO$_2$ PR may be result by the thermal decomposition of PBDOO. The decomposition products can quench the active pyrolysis fragments from VE and inhibit the combustion, which reduces the generation of CO$_2$ [25,28,29].

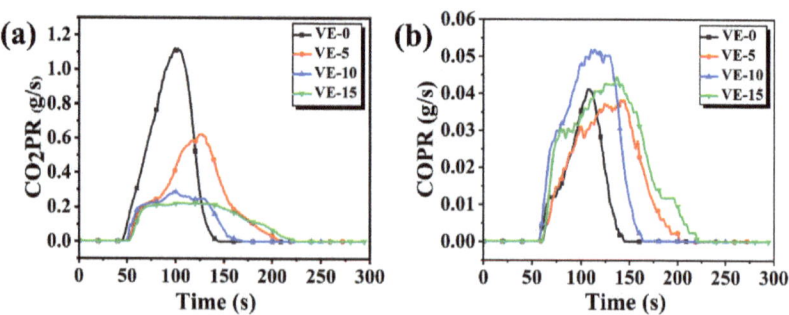

Figure 7. (a) CO$_2$ production rate of VE composites, (b) CO production rate of VE composites.

As seen in Table 3, the addition of PBDOO increases the time to ignition (TTI) of VE composites. The TTI of VE-0 and VE-15 were 59 s and 67 s, respectively, which indicated that the incorporation of highly stable flame retardants plays a good role in physical protection.

2.5. Flame-Retardant Mechanism

2.5.1. Condensed Phase Analysis

Figure 8a–c show the photographs of the VE composites before and after heating at 650 °C for 5 min in a muffle furnace. It could be seen from Figure 8a that there is almost no change in the appearance of VE composites after the addition of PBDOO, which is yellow-brown and transparent. The result in Figure 8b shows that VE-0 has no residue after calcination in the muffle furnace, while VE-15 has a noticeable carbon residue, which is consistent with the thermogravimetric test results. This is probably because the presence of aromatic rings and phosphorus in PBDOO promotes the formation of carbon layers, and therefore more residual carbon is obtained [24,25]. It could be seen from the magnification in Figure 8c that the carbon layer with an expanded three-dimensional structure is loose, porous, and fragile. In order to better characterize the microstructure of the carbonized layer, the carbon residue is observed by SEM, as shown in Figure 8d. It could be seen that the carbon residue is porous, and most of the tiny pores on the surface are cracked. This may be due to a large amount of gas rushing out of the carbon layer during decomposition, which is consistent with the phenomenon observed in the UL-94 test.

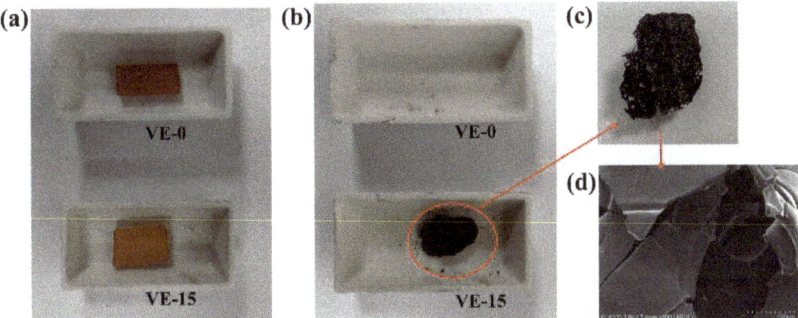

Figure 8. Digital photographs of VE-0 and VE-15 composites (**a**) before and (**b**) after being heated at 650 °C in air; (**c**) Photograph of char carbon of VE-15; (**d**) SEM of the char carbon of VE-15.

2.5.2. Pyrolysis Behaviors of the VE Composites

To better study the degradation process and flame-retardant mechanism of PBDOO in VE, the thermogravimetric analysis-infrared spectroscopy (TG-IR) of VE-0 and VE-15 under nitrogen gas conditions was investigated. Figure 9a shows the FTIR spectra of the pyrolysis gaseous products of VE and its composites at the maximum decomposition rate. The FTIR curve of the VE composites with 15 wt% PBDOO is similar to that of VE-0, and some representative pyrolysis gaseous products are observed. The peaks near 1400–1600 cm^{-1} and 650–900 cm^{-1} correspond to aromatic compounds. The peaks at 1000–1300 cm^{-1} are mainly absorption peaks of C-O compounds. The absorption peaks centered at 1730 cm^{-1} and 2306 cm^{-1} are mainly absorption peaks of carbonyl groups containing aldehydes and CO_2 [23,25–27]. It is worth noting that the peak strength of the VE composites is significantly reduced after the addition of PBDOO.

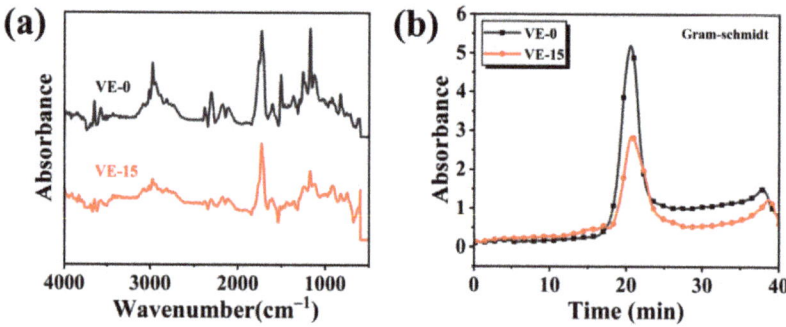

Figure 9. (a) FTIR spectra of pyrolysis gaseous products originated from VE-0 and VE-15 at the maximum evolution rate; (b) Absorbance intensities of VE-0 and VE-15.

The intensity curves of total pyrolysis products are plotted in Figure 9b. It can be seen that the absorption strength of the VE composites with flame retardant is significantly lower than that of the neat samples, which means that fewer gaseous products are detected for VE-15.

To better understand the differences between the gaseous products of VE and VE-15, the absorption intensities of some typical pyrolysis products are shown in Figure 10. The intensities of three absorption peaks of VE-15 at 829 cm^{-1}, 1509 cm^{-1}, and 1610 cm^{-1} are lower than those of VE-0. This result demonstrates that the introduction of PBDOO significantly reduces the production of aromatic compounds, which can participate in the combustion process and release more heat. Therefore, adding PBDOO can effectively increase the fire safety performance of VE composites.

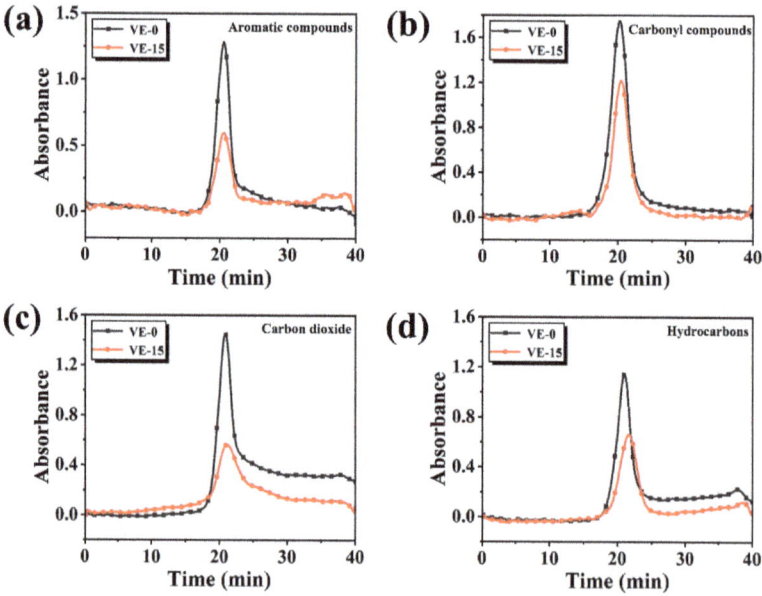

Figure 10. Absorbance intensities of typical pyrolysis products for VE-0 and VE-15: (a) Aromatic compounds, (b) Carbonyl compounds, (c) Carbon dioxide, (d) Hydrocarbons.

The appearance of the C=O absorption peak at 1732 cm^{-1} in Figure 10b may be produced by the decomposition of the ethyl chain segment. The decrease in its intensity also indicates that the decomposition of VE composites is suppressed and the thermal stability is improved after the addition of flame retardant. The hydrocarbon absorption peak curve in Figure 10d also indicates the inhibitory effect of PBDOO on the decomposition of VE composites. Figure 10c shows the absorption intensities of CO_2 at around 2306 cm^{-1}, and it can be seen that the curve of VE-15 is much lower than that of VE-0. This may be due to the decrease in the number of HO· radicals, which is caused by the quenching effect of free radicals are generated from the degradation of PBDOO [30].

As a phosphorus-containing flame retardant, PBDOO mainly functions in gaseous and condensed phases. According to TG data, PBDOO can reduce the initial decomposition temperature of VE composites, which is conducive to the earlier formation of a stable carbon layer to isolate external air and heat. And the decomposition of chemical bonds can release DOPO groups, which are further cleaved to form free radicals such as PO· and PO_2·. These phosphorus-containing radicals can capture active radicals in the flame area, inhibit the chain reaction, and interrupt or slow the combustion of VE composites [30,31]. The DOPO groups in the matrix that have not been cleaved to form radicals and the quenched productions of phosphorus-containing radicals can further aggregate at high temperatures to form phosphorus-containing residues [27]. This residue layer can somewhat prevent the release of combustible gases and heat [32]. The cone calorimeter results show that the HRR and THR values of VE-0 are significantly higher than these of VE-15, which also verifies that PBDOO can suppress the heat release of VE composites during the combustion process. Therefore, in conclusion, the inhibition effect of PBDOO on combustible gas and the formation of condensed phase carbon layer together play a good flame-retardant effect on VE composites. It can be simplified in Figure 11.

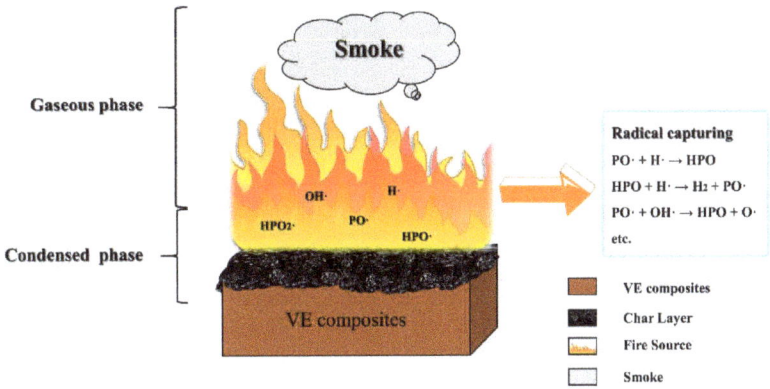

Figure 11. Flame retardant mechanism of PBDOO.

3. Experimental

3.1. Materials

9,10-dihydro-9-oxa-10-phosphaphenanthrene-10-oxide (DOPO), acetophenone, xylene, phosphorus oxychloride, isopropanol, acetone and dibenzoyl peroxide (BPO) were purchased from Aladdin Co., Ltd. (Shanghai, China). Deionized water was self-produced in the laboratory. All chemicals were used as received without any special treatment.

3.2. Synthesis of PBDOO

As shown in Figure 12, a novel phosphorus-containing flame retardant named PBDOO was synthesized by the addition reaction between DOPO and acetophenone. A certain amount of DOPO, acetophenone, and xylene was added to the three-necked flask, and phosphorus oxychloride was slowly added dropwise after being heated to 154 °C under a nitrogen atmosphere. Fractions were collected in a trap at 155 °C. After the dropwise addition of phosphorus oxychloride, stirring was continued for half an hour to make the reaction sufficient. Isopropanol was added to the cooled mixture. The precipitated product was filtered with suction and washed with isopropanol and deionized water to obtain a solid white powder, which was then thoroughly dried at 110 °C.

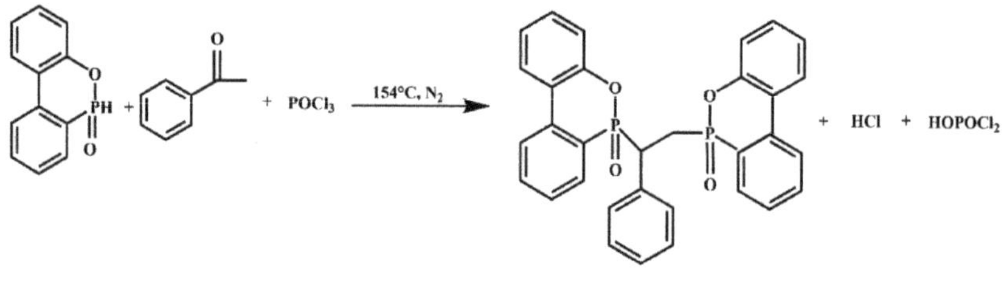

Figure 12. Synthesis route of PBDOO.

3.3. Preparation of the VE Composites

A series of VE composites with different PBDOO content were prepared. The specific formulations are listed in Table 3. BPO is the curing agent.

A certain amount of VE was added into a mixed system of PBDOO and acetone. The mixture was stirred continuously at 60 °C until acetone was volatilized entirely. PBDOO would dissolve into VE during the stirring process, forming a dark brown transparent viscous mixture. Subsequently, the mixture was stirred rapidly for 15 min at 100 °C. After that, a certain amount of initiator was added to the mix and stirred vigorously for 15 s. The resulting mixture was quickly poured into a mold prepared in advance, transferred to the oven, and cured for 2 h at 100 °C, 130 °C, and 150 °C. After cooling to room temperature, the VE composites were taken out of the mold to obtain samples with different PBDOO content.

3.4. Characterization

Instruments Nicolet 6700 Fourier Transform Infrared Spectrometer (Waltham, Massachusetts, USA) was used to record the FTIR spectrum of the sample, and the test wavenumber range was 500–4000 cm^{-1}. The ^{1}H and ^{31}P NMR spectra of the samples were performed on an AVANCE-400 NMR (Bruker company in Zurich, Switzerland) spectrometer, using CDCl$_3$ as the solvent and running in Fourier transform mode. The tensile test was done on the YF-900 computerized tensile testing machine (Jiangsu, China) according to ASTM D3039-08. The sample size was 100 mm × 10 mm × 3 mm, and at least 5 parallel samples were tested for each ratio. Scanning electron microscope (SEM), using JEOL JSM-6700F scanning electron microscope (Tokyo, Japan), the voltage used in the microscope is 10 kV. Before the SEM measurement, the sample was sputtered with a thin layer of gold. Thermogravimetric analysis (TGA) test was carried out on the Q5000 thermal analyzer (TA company in Newcastle, Delaware, USA), measured from 25 °C to 800 °C at a heating rate of 20 °C/min under N$_2$ and Air atmosphere. The cone calorimeter test was tested on the FTT cone calorimeter (UK) using the ISO 5660-1 standard, the sample (size 100 mm × 100 mm × 3 mm) was placed in the aluminum foil, and the heat radiation flux was 35 kW/m^2. Three samples for each ratio were tested. LOI was based on the GB/T 2406.2-2009 standard to evaluate the limiting oxygen index (LOI) value on the HC-2 oxygen index analyzer (Jianning, China). The sample size was 100 mm × 10 mm × 3 mm, and 15 samples were taken from each group. According to the GB/T 2408-2008 standard, the vertical combustion test of the sample was carried out on the CFZ-2 instrument (Jiangning, China). The sample size was 130 mm × 13 mm × 3 mm. Thermogravimetric-infrared spectroscopy (TG-FTIR), measured by PerkinElmer (Waltham, Massachusetts, USA) STA8000 thermogravimetric analyzer connected to PerkinElmer Frontier FT-IR spectrometer, the heating rate was 20 °C/min, the atmosphere was N$_2$, the flow rate was 45 mL/min. The chamber and transfer tube were kept at 300 °C and 280 °C, respectively.

4. Conclusions

In this work, a phosphorous flame retardant based on DOPO was successfully synthesized, characterized, and applied to VE composites. The research indicated that the introduction of PBDOO effectively improved the flame retardancy of VE and reduced its heat release rate of VE during its combustion process. With the addition of 15 wt% PBDOO, the VE composite passed the V-0 level of the UL-94 test and the LOI value reached 31.5%. Compared with neat VE, the PHRR and THR of VE-15 decreased by 76.71% and 40.36%, respectively. The results of TG-IR demonstrated that the added PBDOO significantly reduced the production of aromatic compounds and inhibited the release of combustible gases. In addition, the adoption of PBDOO decreased the tensile strength of VE composites, while the elongation at the break did not change much. Mechanistic analysis shows that adding PBDOO is conducive to forming stabilized carbon layer, which plays an essential role in the early isolation of air and heat. Furthermore, the phosphorus-containing radicals generated by the cleavage of DOPO groups can quickly capture the free radicals in the flame region, delaying or interrupting the combustion. The flame retardant designed by this work is expected to further broaden the application of VE composites in construction, aerospace, and other fields.

Author Contributions: Data analyses, writing original draft preparation, Z.X. (Zihui Xu); conception of the study and revision of draft and funding acquisition, J.Z.; validation, R.T.; testing of the samples, Z.X. (Zhirong Xu); contributing to the tables and figures of the manuscript, L.M.; supervising experiments and final revision of articles, X.M. All authors have read and agreed to the published version of the manuscript.

Funding: This work was funded by Natural Science Research Project of Anhui Province grant number KJ2020A0450; and Natural Science Research Foundation for Introduction of Talent and Doctor grant number 2019QDZ59.

Institutional Review Board Statement: Not applicable.

Informed Consent Statement: Not applicable.

Data Availability Statement: Not applicable.

Conflicts of Interest: The authors declare that they have no known competing financial interests or personal relationships that could have appeared to influence the work reported in this paper.

Sample Availability: The samples of PBDOO and VE composites are available from the authors.

References

1. Pham, S.; Burchill, P.J. Toughening of vinyl ester resins with modified polybutadienes. *Polymer* **1995**, *36*, 3279–3285. [CrossRef]
2. Agarwal, N.; Singh, A.; Varma, I.K.; Choudhary, V. Effect of structure on mechanical properties of vinyl ester resins and their glass fiber-reinforced composites. *J. Appl. Polym. Sci.* **2008**, *108*, 1942–1948. [CrossRef]
3. Da Silva, A.L.N.; Teixeira, S.C.S.; Widal, A.C.C.; Coutinho, F. Mechanical properties of polymer composites based on commercial epoxy vinyl ester resin and glass fiber. *Polym. Test.* **2001**, *20*, 895–899. [CrossRef]
4. Manring, L.E. Thermal degradation of poly (methyl methacrylate). 4. Random side-group scission. *Macromolecules* **1991**, *24*, 3304–3309. [CrossRef]
5. Wu, C.S.; Liu, Y.L.; Chiu, Y.C.; Chiu, Y.S. Thermal stability of epoxy resins containing flame retardant components: An evaluation with thermogravimetric analysis. *Polym. Degrad. Stab.* **2002**, *78*, 41–48. [CrossRef]
6. Shen, D.; Xu, Y.J.; Long, J.W.; Shi, X.H.; Chen, L.; Wang, Y. Epoxy resin flame-retarded via a novel melamine-organophosphinic acid salt: Thermal stability, flame retardance and pyrolysis behavior. *J. Anal. Appl. Pyrolysis* **2017**, *128*, 54–63. [CrossRef]
7. Prabhakar, M.N.; Song, J. Influence of chitosan-centered additives on flammable properties of vinyl ester matrix composites. *Cellulose* **2020**, *27*, 8087–8103. [CrossRef]
8. XU, C. Flame Retard Modification of Vinyl Ester Resin with Phosphorous Polymer Type Flame Retardant. *Chin. J. Mater. Res.* **2021**, *35*, 843–849.
9. Han, X.; Zhang, X.H.; Zhang, S.L.; Zhou, H. Study on flame retardancy and thermal properties of phosphorus-containing POSS modified vinyl resin. *Chin. Chem. Bull.* **2021**, *84*, 1066–1073.
10. Li, H.; Feng, S.Y.; Jin, Z.M. Study on impact composites of glass fiber reinforced vinyl ester Resin. *Eng. Plast. Appl.* **2006**, *34*, 17–20.

11. Huo, S.; Wang, J.; Yang, S.; Cai, H.; Zhang, B.; Chen, X.; Wu, Q.; Yang, L. Synthesis of a novel reactive flame retardant containing phosphaphenanthrene and triazine-trione groups and its application in unsaturated polyester resin. *Mater. Res. Express* **2018**, *5*, 035306. [CrossRef]
12. Zhang, C.; Huang, J.Y.; Liu, S.M.; Zhao, J.Q. The synthesis and properties of a reactive flame-retardant unsaturated polyester resin from a phosphorus-containing diacid. *Polym. Adv. Technol.* **2011**, *22*, 1768–1777. [CrossRef]
13. Bai, Z.; Song, L.; Hu, Y.; Gong, X.; Yuen, R.K. Investigation on flame retardancy, combustion and pyrolysis behavior of flame retarded unsaturated polyester resin with a star-shaped phosphorus-containing compound. *J. Anal. Appl. Pyrolysis* **2014**, *105*, 317–326. [CrossRef]
14. Lin, Y.; Yu, B.; Jin, X.; Song, L.; Hu, Y. Study on thermal degradation and combustion behavior of flame retardant unsaturated polyester resin modified with a reactive phosphorus containing monomer. *RSC Adv.* **2016**, *6*, 49633–49642. [CrossRef]
15. Jin, S.; Qian, L.; Qiu, Y.; Chen, Y.; Xin, F. High-efficiency flame retardant behavior of bi-DOPO compound with hydroxyl group on epoxy resin. *Polym. Degrad. Stab.* **2019**, *166*, 344–352. [CrossRef]
16. Duan, H.; Ji, S.; Yin, T.; Tao, X.; Chen, Y.; Ma, H. Phosphorus–nitrogen-type fire-retardant vinyl ester resin with good comprehensive properties. *J. Appl. Polym. Sci.* **2019**, *136*, 47997. [CrossRef]
17. Yang, S.; Hu, Y.; Zhang, Q. Synthesis of a phosphorus–nitrogen-containing flame retardant and its application in epoxy resin. *High Perform. Polym.* **2019**, *31*, 186–196. [CrossRef]
18. Tao, X.X.; Duan, H.J.; Dong, W.J.; Wang, X. Synthesis of TGICA-DOPO, novel phosphorus and nitrogen flame retardant and its effect on flame retardant properties of vinyl ester resin. *AMCS* **2018**, *35*, 1731–1737.
19. Salmeia, K.A.; Gaan, S. An overview of some recent advances in DOPO-derivatives: Chemistry and flame retardant applications. *Polym. Degrad. Stab.* **2015**, *113*, 119–134. [CrossRef]
20. Huang, W.; He, W.; Long, L.; Yan, W.; He, M.; Qin, S.; Yu, J. Thermal degradation kinetics of flame-retardant glass-fiber-reinforced polyamide 6T composites based on bridged DOPO derivatives. *Polym. Bull.* **2019**, *76*, 2061–2080. [CrossRef]
21. Fang, Y.; Zhou, X.; Xing, Z.; Wu, Y. Flame retardant performance of a carbon source containing DOPO derivative in PET and epoxy. *J. Appl. Polym. Sci.* **2017**, *133*, 44639. [CrossRef]
22. Wu, K.; Kandola, B.K.; Kandare, E.; Hu, Y. Flame retardant effect of polyhedral oligomeric silsesquioxane and triglycidyl isocyanurate on glass fibre-reinforced epoxy composites. *Polym. Compos.* **2011**, *32*, 378–389. [CrossRef]
23. Mu, X.; Zhou, X.; Wang, W.; Xiao, Y.; Liao, C.; Longfei, H.; Kan, Y.; Song, L. Design of compressible flame retardant grafted porous organic polymer based separator with high fire safety and good electrochemical properties. *Chem. Eng. J.* **2021**, *405*, 126946. [CrossRef]
24. Chang, Q.F.; Long, L.J.; He, W.T.; Xiang, Y.S.; Li, J.; Qing, S.H.; Yu, J. Synthesis of a bridge-chain DOPO derivative and flame retardant properties to poly(lactic acid). *Polym. Mater. Sci. Eng.* **2017**, *33*, 30–35.
25. Chen, X.; Zhuo, J.; Jiao, C. Thermal degradation characteristics of flame retardant polylactide using TG-IR. *Polym. Degrad. Stab.* **2012**, *97*, 2143–2147. [CrossRef]
26. Wang, P.; Yang, F.; Li, L.; Cai, Z. Flame retardancy and mechanical properties of epoxy thermosets modified with a novel DOPO-based oligomer. *Polym. Degrad. Stab.* **2016**, *129*, 156–167. [CrossRef]
27. Xiao, Y.; Jiang, G.; Ma, C.; Zhou, X.; Wang, C.; Xu, Z.; Mu, X.; Song, L.; Hu, Y. Construction of multifunctional linear polyphosphazene and molybdenum diselenide hybrids for efficient fire retardant and toughening epoxy resins. *Chem. Eng. J.* **2021**, *426*, 131839. [CrossRef]
28. Zou, B.; Qiu, S.; Ren, X.; Zhou, Y.; Zhou, F.; Xu, Z.; Zhao, Z.; Song, L.; Hu, Y.; Gong, X. Combination of black phosphorus nanosheets and MCNTs via phosphoruscarbon bonds for reducing the flammability of air stable epoxy resin nanocomposites. *J. Hazard. Mater.* **2020**, *383*, 121069. [CrossRef]
29. Zhang, T.; Yan, H.; Shen, L.; Fang, Z.; Zhang, X.; Wang, J.; Zhang, B. A phosphorus-, nitrogen-and carbon-containing polyelectrolyte complex: Preparation, characterization and its flame retardant performance on polypropylene. *RSC Adv.* **2014**, *4*, 48285–48292. [CrossRef]
30. Mu, X.; Jin, Z.; Chu, F.; Cai, W.; Zhu, Y.; Yu, B.; Song, L.; Hu, Y. High-performance flame-retardant polycarbonate composites: Mechanisms investigation and fire-safety evaluation systems establishment. *Compos. Part B: Eng.* **2022**, *238*, 109873. [CrossRef]
31. Qian, L.; Qiu, Y.; Sun, N.; Xu, M.; Xu, G.; Xin, F.; Chen, Y. Pyrolysis route of a novel flame retardant constructed by phosphaphenanthrene and triazine-trione groups and its flame-retardant effect on epoxy resin. *Polym. Degrad. Stab.* **2014**, *107*, 98–105. [CrossRef]
32. Mu, X.; Li, X.; Liao, C.; Yu, H.; Jin, Y.; Yu, B.; Han, L.; Chen, L.; Kan, Y.; Song, L.; et al. Phosphorus-Fixed Stable Interfacial Nonflammable Gel Polymer Electrolyte for Safe Flexible Lithium-Ion Batteries. *Adv. Funct. Mater.* **2022**, *32*, 2203006. [CrossRef]

MDPI
St. Alban-Anlage 66
4052 Basel
Switzerland
www.mdpi.com

Molecules Editorial Office
E-mail: molecules@mdpi.com
www.mdpi.com/journal/molecules

Disclaimer/Publisher's Note: The statements, opinions and data contained in all publications are solely those of the individual author(s) and contributor(s) and not of MDPI and/or the editor(s). MDPI and/or the editor(s) disclaim responsibility for any injury to people or property resulting from any ideas, methods, instructions or products referred to in the content.

www.ingramcontent.com/pod-product-compliance
Lightning Source LLC
LaVergne TN
LVHW070710100526
838202LV00013B/1063